Handbook of Formulas and Software for Plant Geneticists and Breeders

Handbook of Formulas and Software for Plant Geneticists and Breeders

Manjit S. Kang
Editor

Routledge
Taylor & Francis Group

NEW YORK AND LONDON

First published 2003 by Food Products Press®

Published 2021 by Routledge
605 Third Avenue, New York, NY 10017
2 Park Square, Milton Park, Abingdon, Oxon OX14 4RN

Routledge is an imprint of the Taylor & Francis Group, an informa business

Cover design by Marylouise E. Doyle.

Library of Congress Cataloging-in-Publication Data

Handbook of formulas and software for plant geneticists and breeders / Manjit S. Kang, editor.
    p. cm.
  Includes bibliographical references (p.) and index.
  ISBN 1-56022-948-9 (hard : alk. paper) — ISBN 1-56022-949-7 (soft)
    1. Plant genetics—Statistical methods—Computer programs. 2. Plant breeding—Statistical methods—Computer programs. I. Kang, Manjit S.

QK981.5 .H36 2003
581.35'0285—dc21

                                                        2002072056

ISBN 13: 978-1-56022-949-0 (pbk)
ISBN 13: 978-1-56022-948-3 (hbk)

CONTENTS

About the Editor   ix

Contributors   xi

Preface   xiii

Chapter 1. DIALLEL-SAS: A Program for Griffing's Diallel
   Methods   1
       Yudong Zhang
       Manjit S. Kang

Chapter 2. Diallel Analysis for a Seed and Endosperm Model
   with Genotype-by-Environment Interaction Effects   21
       Jun Zhu

Chapter 3. Diallel Analysis for an Additive-Dominance Model
   with Genotype-by-Environment Interaction Effects   39
       Jun Zhu

Chapter 4. Diallel Analysis for an Additive-Dominance-Epistasis
   Model with Genotype-by-Environment Interaction Effects   51
       Jun Zhu

Chapter 5. Diallel Analysis for an Animal Model with Sex-Linked
   and Maternal Effects Along with Genotype-by-Environment
   Interaction Effects   67
       Jun Zhu

Chapter 6. Generation Means Analysis   77
       Michael M. Kenty
       David S. Wofford

Chapter 7. PATHSAS: Path Coefficient Analysis of Quantitative
   Traits   89
       Christopher S. Cramer
       Todd C. Wehner
       Sandra B. Donaghy

Chapter 8. Restricted Maximum Likelihood Procedure to Estimate
Additive and Dominance Genetic Variance Components   97
Agron Collaku

Chapter 9. Calculating Additive Genetic Correlation Using
ANOVA and the Sum Method of Estimating Covariance   109
Blair L. Waldron

Chapter 10. Developmental Analysis for Quantitative Traits   115
Jun Zhu

Chapter 11. Ecovalence and Stability Variance   123
Manjit S. Kang

Chapter 12. Genotype-by-Environment Interaction Variance   129
Robert Magari
Manjit S. Kang

Chapter 13. Code for Simulating Degrees of Freedom
for the Items in a Principal Components Analysis of Variance   137
Walter T. Federer
Russell D. Wolfinger

Chapter 14. Principal Components (PC) and Additive Main Effects
and Multiplicative Interaction (AMMI) Trend Analyses
for Incomplete Block and Lattice Rectangle-Designed
Experiments   145
Walter T. Federer
Russell D. Wolfinger
José Crossa

Chapter 15. A Method for Classifying Observations Using
Categorical and Continuous Variables   153
Jorge Franco
José Crossa

Chapter 16. Mixed Linear Model Approaches for Quantitative
Genetic Models   171
Jixiang Wu
Jun Zhu
Johnie N. Jenkins

Chapter 17. Best Linear Unbiased Prediction (BLUP)
    for Genotype Performance   181
        Mónica Balzarini
        Scott Milligan

Chapter 18. Graphing GE and GGE Biplots   193
        Juan Burgueño
        José Crossa
        Mateo Vargas

Chapter 19. Analysis for Regional Trials with Unbalanced Data   205
        Jun Zhu

Chapter 20. Conditional Mapping of QTL with Epistatic Effects
    and QTL-by-Environment Interaction Effects
    for Developmental Traits   213
        Jun Zhu

Chapter 21. Mapping QTL with Epistatic Effects
    and QTL-by-Environment Interaction Effects   219
        Jun Zhu

Chapter 22. Gene Segregation and Linkage Analysis   231
        Jinsheng Liu
        Todd C. Wehner
        Sandra B. Donaghy

Chapter 23. Mapping Functions   255
        M. Humberto Reyes-Valdés

Chapter 24. Bootstrap and Jackknife for Genetic Diversity
    Parameter Estimates   261
        Julio Di Rienzo
        Mónica Balzarini

Chapter 25. Software on Genetic Linkage and Mapping
    Available Through the Internet   265
        Manjit S. Kang

Chapter 26. Gregor   279
        Todd Krone

Chapter 27. Analysis for an Experiment Designed As Augmented
    Lattice Square Design   283
       Walter T. Federer

Chapter 28. Augmented Row-Column Design and Trend Analyses   291
       Walter T. Federer
       Russell D. Wolfinger

Chapter 29. PROC GLM and PROC MIXED Codes for Trend
    Analyses for Row-Column-Designed Experiments   297
       Walter T. Federer
       Russell D. Wolfinger

Chapter 30. SAS/GLM and SAS/MIXED for Trend Analyses
    Using Fourier and Polynomial Regression for Centered
    and Noncentered Variates   307
       Walter T. Federer
       Murari Singh
       Russell D. Wolfinger

Chapter 31. PROC GLM and PROC MIXED for Trend Analysis
    of Incomplete Block- and Lattice Rectangle-Designed
    Experiments   315
       Walter T. Federer
       Russell D. Wolfinger

Chapter 32. Partitioning Crop Yield into Genetic Components   321
       Vasilia A. Fasoula
       Dionysia A. Fasoula

Short Note 1. Inbreeding Coefficient in Mass Selection
    in Maize   329
       Fidel Márquez-Sánchez

Short Note 2. Regression of Forage Yield Against a Growth Index
    As a Tool for Interpretation of Multiple Harvest Data   331
       Jeffery F. Pedersen

Short Note 3. Tolerance Index   335
       Lajos Bona

Short Note 4. Computer Program to Calculate Population Size   337
       Leví M. Mansur

Index   339

ABOUT THE EDITOR

Manjit S. Kang, PhD, is Professor of Quantitative Genetics in the Department of Agronomy at Louisiana State University. He earned his BSc in agriculture and animal husbandry with honors from the Punjab Agricultural University in India, an MS in biological sciences, majoring in plant genetics from Southern Illinois University at Edwardsville, an MA in botany from Southern Illinois University at Carbondale, and a PhD in crop science (genetics and plant breeding) from the University of Missouri at Columbia.

Dr. Kang is the editor, author, or co-author of hundreds of articles, books, and book chapters. He enjoys an international reputation in genetics and plant breeding. He serves on the editorial boards of *Crop Science, Agronomy Journal, Journal of New Seeds,* and the *Indian Journal of Genetics & Plant Breeding,* as well as The Haworth Food Products Press.

Dr. Kang is a member of Gamma Sigma Delta and Sigma Xi. He was elected a Fellow of the American Society of Agronomy and of the Crop Science Society of America. In 1999 he served as a Fulbright Senior Scholar in Malaysia.

Dr. Kang edited *Genotype-By-Environment Interaction and Plant Breeding* (1990), which resulted from an international symposium that he organized at Louisiana State University in February 1990. He is the author/publisher of *Applied Quantitative Genetics* (1994), which resulted from teaching a graduate-level course on Quantitative Genetics in Plant Improvement. Another book, *Genotype-By-Environment Interaction,* edited by Dr. Kang and Hugh Gauch Jr., was published by CRC Press in 1996. He edited *Crop Improvement for the 21st Century* in 1997 (Research Signpost, India). He recently co-authored *GGE Biplot Analysis: A Geographical Tool for Breeders, Geneticists, and Agronomists* (2002, CRC Press), and he edited *Crop Improvement: Challenges in the Twenty-First Century* (2002, The Haworth Press) and *Quantitative Genetics, Genomics, and Plant Breeding* (2002, CABI Publishing, U.K.).

Dr. Kang's research interests are: genetics of resistance to aflatoxin, weevils, and herbicides in maize; genetics of grain dry-down rate and stalk quality in maize; genotype-by-environment interaction and crop adaptation; interorganismal genetics, and conservation and utilization of plant genetic resources.

Dr. Kang taught Plant Breeding and Plant Genetics courses at Southern Illinois University–Carbondale (1972-1974). He has been teaching a graduate-level Applied Quantitative Genetics course at Louisiana State University since 1986. He developed and taught an intermediary plant genetics course in 1996 and team-taught an Advanced Plant Genetics course (1993-1995). He also taught an Advanced Plant Breeding course at LSU in 2000. He has directed six MS theses and six PhD dissertations. He has been a Full Professor in the Department of Agronomy at LSU since 1990. He has received many invitations to speak at international symposium relative to genetics and plant breeding.

Dr. Kang was recognized for his significant contributions to plant breeding and genetics by Punjab Agricultural University at Ludhiana at its 36th Foundation Day in 1997. He served as President (2000-2001) of the LSU Chapter of Sigma Xi—The Scientific Research Society. He was elected President of the Association of Agricultural Scientists of Indian Origin in 2001 for a two-year term. In addition, he serves as the Chairman of the American Society of Agronomy's Member Services and Retention Committee (2001-2004). Dr. Kang's biographical sketches have been included in *Marquis Who's Who in the South and Southwest, Who's Who in America, Who's Who in the World, Who's Who in Science and Engineering,* and *Who's Who in Medicine and Healthcare.*

CONTRIBUTORS

Mónica Balzarini, PhD, Facultad Ciencias Agropecuarias, Estadística y Biometría, Universidad Nacional de Córdoba, C.C. 509-5000 Córdoba, Argentina.

Lajos Bona, PhD, Cereal Research Institute, Szeged, Hungary.

Juan Burgueño, MS, Biometrics and Statistics Unit, International Maize and Wheat Improvement Center/Centro Internacional para Mejoramiento de Maiz y Trigo (CIMMYT), Mexico City, Mexico.

Agron Collaku, PhD, Department of Pharmacogenomics, Johnson & Johnson Research and Development, Raritan, New Jersey.

Christopher S. Cramer, PhD, Department of Agronomy and Horticulture, New Mexico State University, Las Cruces, New Mexico.

José Crossa, PhD, Biometrics and Statistics Unit, International Maize and Wheat Improvement Center/Centro Internacional para Mejoramiento de Maiz y Trigo (CIMMYT), Mexico City, Mexico.

Sandra B. Donaghy, MS, Department of Statistics, North Carolina State University, Raleigh, North Carolina.

Dionysia A. Fasoula, PhD, Agricultural Research Institute, Nicosia, Cyprus.

Vasilia A. Fasoula, PhD, Center for Applied Genetic Technologies, University of Georgia, Athens, Georgia.

Walter T. Federer, PhD, Department of Biometrics, Cornell University, Ithaca, New York.

Jorge Franco, MS, Universidad de la República, Facultad de Agronomía, Montevideo, Uruguay.

Johnie N. Jenkins, PhD, Crop Science Research Laboratory, U.S. Department of Agriculture, Agricultural Research Station (USDA-ARS), Mississippi State, Mississippi.

Michael M. Kenty, PhD, Helena Chemical Company, Collierville, Tennessee.

Todd Krone, PhD, Pioneer Hi-Bred International, Inc., Johnston, Iowa.

Jinsheng Liu, MS, Jiangsu State Farms Agribusiness Corporation, Agricultural Section, Nanjing, China.

Robert Magari, PhD, Beckman Coulter, Inc., Miami, Florida.

Leví M. Mansur, PhD, Facultad de Agronomia, Universidad Catolica de Valpraiso, Chile.

Fidel Márquez-Sánchez, PhD, Manuel M. Dieguez 113, Sector Hidalgo, Guadalajara, Jalisco, Mexico.

Scott Milligan, PhD, U.S. Sugar Corporation, Clewiston, Florida.

Jeffery F. Pedersen, PhD, U.S. Department of Agriculture, Agricultural Research Station (USDA-ARS), University of Nebraska, Lincoln, Nebraska.

M. Humberto Reyes-Valdés, PhD, Universidad Autonóma Agraria Antonio Narro, Departamento de Fitomejoramiento, Saltillo, Mexico.

Julio Di Rienzo, MS, Facultad Ciencias Agropecuarias, Estadística y Biometría, Universidad Nacional de Córdoba, C.C. 509-5000 Córdoba, Argentina.

Murari Singh, PhD, ICARDA, Aleppo, Syria.

Mateo Vargas, PhD, Biometrics and Statistics Unit, International Maize and Wheat Improvement Center/Centro Internacional para Mejoramiento de Maiz y Trigo (CIMMYT), Mexico City, Mexico.

Blair L. Waldron, PhD, U.S. Department of Agriculture, Agricultural Research Station (USDA-ARS), Forage and Range Research Laboratory, Utah State University, Logan, Utah.

Todd C. Wehner, PhD, Department of Horticultural Science, North Carolina State University, Raleigh, North Carolina.

David S. Wofford, PhD, Department of Agronomy, University of Florida, Gainesville, Florida.

Russell D. Wolfinger, PhD, SAS Institute, Inc., Cary, North Carolina.

Jixiang Wu, MS, Crop Science Research Laboratory, U.S. Department of Agriculture, Agricultural Research Station (USDA-ARS), Mississippi State, Mississippi.

Yudong Zhang, PhD, IBM Printing Systems Division, Boulder, Colorado.

Jun Zhu, PhD, Department of Agronomy, Zhejiang University, Hangzhou, China.

Preface

Statistical techniques and formulas have been part and parcel of genetics and breeding programs. Some techniques and formulas are used routinely while others may be used only occasionally. The ones used frequently have been incorporated into popular statistical packages, such as the Statistical Analysis System (SAS), and are readily available. To meet their needs, researchers began developing specific software not found in statistical packages. For example, in the early 1980s, I began using in my research relative to genotype-by-environment interaction a formula for stability variance that Shukla (1972) developed and a formula for ecovalence that Wricke (1962) developed. Hand calculations for these formulas were tedious and time-consuming. Thus, in my necessitous circumstances I wrote a computer program in the matrix programming language of SAS (Kang, 1985). *The Journal of Heredity* published a description of the program, and I received several hundred requests for the program code from researchers around the world who were working with plants and animals, and I also received requests from those working with humans, e.g., psychiatrists (Kang, 1986). Having published several other programs since, I have realized that there is a need among researchers for special software programs for use in research. For example, geneticists working with many different crops requested the DIALLEL-SAS program (Zhang and Kang, 1997).

Software programs, or descriptions thereof, are occasionally published in international journals such as *The Journal of Heredity* and *Agronomy Journal*. Full-fledged codes for statistical and genetics-related software programs are rarely published in journals because they are generally not the main domain of these journals. I believe it would better serve the scientific community to have published or unpublished programs made more easily accessible in a handbook.

With the intent of making available to researchers and teachers of genetics and breeding a compendium devoted to such specialized programs as DIALLEL-SAS and those which others have created around the world, I turned to Food Products Press, an imprint of The Haworth Press, Inc. (www.haworthpressinc.com), to undertake the publication of *Handbook of Formulas and Software for Plant Geneticists and Breeders*. A questionnaire was sent to some 2,300 members of the C1 Division (geneticists and breeders) of the Crop Science Society of America (CSSA) and many contributors were identified. The response was excellent as evidenced by the various contributions in this book.

This first edition of the handbook is an excellent start to meet the needs of the scientific community. I am sure as researchers, teachers, and students begin to realize the usefulness of this effort, additional contributions will follow, which can, hopefully, be included in a subsequent edition. In this handbook, most contributions of a specific software include program codes and practical examples on how to use the software or formula in question; others direct the reader where to get specific software. Due to the enormous effort devoted to linkage and mapping using molecular markers, I have also included a chapter that lists software on genetic linkage and mapping that can be accessed via the Internet (Chapter 26). Obviously, program codes are not printed for the software listed in that group. I trust the handbook will serve as an up-to-date, ready reference on genetic formulas and software for practicing geneticists/breeders, as well as for students. Please send any comments on this handbook or new contributions to this editor and author, via e-mail, at <mKang@agctr.lsu.edu> or <kang_majit@hotmail.com>.

I thank Dr. Amarjit S. Basra, Editor-in-Chief with Food Products Press, for his encouragement. This project could not have been accomplished without the participation and cooperation of the various authors and the publisher.

REFERENCES

Kang, M.S. (1985). SAS program for calculating stability variance parameters. *Journal of Heredity* 76:142-143.

Kang, M.S. (1986). Update on the stability variance program. *Journal of Heredity* 77:480.

Shukla, G.K. (1972). Some statistical aspects of partitioning genotype-environmental components of variability. *Heredity* 29:237-245.

Wricke, G. (1962). Über eine Methode zur Erfassung der ökologischen Streubreite. *Zeitschrift für Pflanzenzüchtung* 47:92-96.

Zhang, Y. and Kang, M.S. (1997). DIALLEL-SAS: A SAS program for Griffing's diallel analyses. *Agronomy Journal* 89:176-182.

Chapter 1

DIALLEL-SAS: A Program for Griffing's Diallel Methods

Yudong Zhang
Manjit S. Kang

Importance

Diallel mating designs are frequently used in plant breeding research to obtain genetic information, such as general combining ability (GCA) and specific combining ability (SCA), and possibly narrow-sense heritability. Griffing (1956) developed four methods to compute GCA and SCA. These methods have provided valuable information on various important traits in crops (Borges, 1987; Moffatt et al., 1990; Pixley and Bjarnason, 1993; Kang et al., 1995). To obtain more reliable genetic information, multi-environment data are generally needed. The following statistical models illustrate Griffing's methods for analyzing multienvironment data:

The general linear model for Methods 1 and 3 (reciprocal crosses) is:

$$Y_{ijklc} = \mu + \acute{a}_l + b_{kl} + v_{ij} + (\acute{a}v)_{ijl} + e_{ijklc,}$$
where $v_{ij} = g_i + g_j + s_{ij} + r_{ij},$ $(\acute{a}v)_{ijl} = (\acute{a}g)_{il} + (\acute{a}g)_{jl} + (\acute{a}s)_{ijl} + (\acute{a}r)_{ijl,}$
$r_{ij} = m_i + m_j + n_{ij,}$ and $(\acute{a}r)_{ijl} = (\acute{a}m)_{il} + (\acute{a}m)_{jl} + (\acute{a}n)_{ijl}.$

The general linear model for Methods 2 and 4 is:

$$Y_{ijklc} = \mu + \acute{a}_l + b_{kl} + v_{ij} + (\acute{a}v)_{ijl} + e_{ijklc,}$$
where $v_{ij} = g_i + g_j + s_{ij}$ and $(\acute{a}v)_{ijl} = (\acute{a}g)_{il} + (\acute{a}g)_{jl} + (\acute{a}s)_{ijl}.$

In these models, Y_{ijklc} = observed value of each experimental unit, μ = population mean, $\acute{a}_l$ = environment effect, b_{kl} = block or replication effect in each environment, v_{ij} = F_1 hybrid effect, $(\alpha v)_{ijl}$ = interaction between environments and F_1 hybrids, e_{ijklc} = residual effect, g_i = GCA effect for ith parent, g_j = GCA effect for jth parent, s_{ij} = SCA for ijth F_1 hybrid, r_{ij} = reciprocal effect for ijth or jith F_1 hybrid, $(\alpha g)_{il}$ = interaction between GCA effect for ith parent and environments, $(\alpha g)_{jl}$ = interaction between GCA effect for jth parent and environments, $(\alpha s)_{ijl}$ = interaction between SCA effect for ijth F_1 hybrid and environments, $(\alpha r)_{ijl}$ = interaction between reciprocal effect for ijth or jith F_1 hybrid and environments, m_i = maternal effect of parental line i, m_j = maternal effect of parental line j, n_{ij} = nonmaternal effect of ijth or jith F_1 hybrid, $(\alpha m)_{il}$ = interaction between environments and maternal effect of parental line i, $(\alpha m)_{jl}$ = interaction between environments and maternal effect of parental inbred j, and $(\alpha n)_{ijl}$ = interaction between environments and nonmaternal effect of ijth or jith F_1 hybrid.

Definitions

Diallel: A mating design in which all possible two-way combinations are produced among a set of genetically different lines. It is used to estimate GCA (average performance of the progeny) of each parental line in crosses with a set of lines and to estimate SCA (progeny performance of a specific cross).

Program Description

DIALLEL-SAS was developed using SAS (SAS Institute, 1995). It provides a partition of F_1 hybrid (or cross) × environment (E) interaction into GCA × E, SCA × E, and reciprocal × E components for Griffing's Methods 1 and 3, and into GCA × E and SCA × E components for Griffing's Methods 2 and 4. DIALLEL-SAS can be run on any microcomputer with SAS as well as on mainframe computers installed with SAS, via UNIX or TSO. The DIALLEL-SAS output for Methods 1 and 3 includes (1) mean squares for environments, F_1 hybrids, F_1 hybrids × E, GCA, SCA, reciprocal, maternal, nonmaternal, GCA × E, SCA × E, reciprocal × E, maternal × E, nonmaternal × E; (2) estimates of GCA and maternal effects for each parental line; and (3) estimates of SCA, nonmaternal, and reciprocal effects for each F_1 hybrid.

Originator

Griffing, B. (1956). Concept of general and specific combining ability in relation to diallel crossing systems. *Australian Journal of Biological Science* 9:463-493.

Software Availability

Zhang, Y. and Kang, M.S. (1997). DIALLEL-SAS: A SAS program for Griffing's diallel analyses. *Agronomy Journal* 89:176-182.

Key References Where Software Has Been Cited

Goffman, F.D. and Becker, H.C. (2001). Diallel analysis for tocopherol contents in seeds of rapeseed. *Crop Science* 41:1072-1079.

Kang, M.S., Din, A.K., Zhang, Y., and Magari, R. (1999). Combining ability for rind puncture resistance in maize. *Crop Science* 39:368-371.

Le Gouis, J., Beghin, D., Heumez, E., and Pluchard, P. (2002). Diallel analysis of winter wheat at two nitrogen levels. *Crop Science* 42:1129-1134.

Contact

Dr. Manjit S. Kang, 105 Sturgis Hall, Louisiana State University, Baton Rouge, LA 70803-2110, USA. <mKang@agcenter.LSU.edu>.

EXAMPLE

A fictitious diallel data set for Griffing's Method 1 with five parental lines, two environments, and two replications per environment is analyzed using DIALLEL-SAS.

DIALLEL-SAS Method 1 Program Listing with Output

```
OPTIONS PS=56 LS=78;
TITLE 'METHOD 1';
DATA METHOD1;
INPUT I J REP HYBRID YIELD ENV;
DROP N NI NJ P;
P=5;*NUMBER OF PARENTAL LINES;
ARRAY GCA(N) G1 G2 G3 G4;
DO N=1 TO (P-1);
GCA=((I=N)-(I=P))+((J=N)-(J=P));
END;
ARRAY SCA(N) S11 S12 S13 S14 S22 S23 S24 S33 S34 S44;
```

```
N=0;
DO NI=1 TO (P-1);
DO NJ=NI TO (P-1);
N+1;
IF NI=NJ THEN DO;
SCA=(I=NI)*((J=NJ)-(J=P))+(I=P)*((J=P)-(J=NI));END;
ELSE DO;
SCA=(I=NI)*(J=NJ)-(J=P)*((I=NI)+(I=NJ)-(I=P)*2)+(I=NJ)*(J=NI)
-(I=P)*((J=NI)+(J=NJ));
END;END;END;
ARRAY REC(N) R12 R13 R14 R15 R23 R24 R25 R34 R35 R45;
N=0;
DO NI=1 TO (P-1);
DO NJ=(NI+1) TO P;
N+1;
REC=(I=NI)*(J=NJ)-(J=NI)*(I=NJ);
END;END;
ARRAY MAT (N) M1 M2 M3 M4;
DO N=1 TO (P-1);
MAT=(I=N)+(J=P)-(J=N)-(I=P);
END;
ARRAY NONM (N) N12 N13 N14 N23 N24 N34;
N=0;
DO NI=1 TO (P-2);
DO NJ=(NI+1) TO (P-1);
N+1;
NONM=((I=NI)*(J=NJ))-(I=NJ)*(J=NI)-((I=NI)*(J=P))+(I=NJ)*(J=P)
+((I=P)*((J=NI)-(J=NJ)));
END;END;
CARDS;
1 1 1 1   10.5 1
1 1 2 1   10.7 1
1 2 1 2   11.9 1
1 2 2 2   12.0 1
1 3 1 3   14.5 1
1 3 2 3   14.2 1
1 4 1 4    9.0 1
1 4 2 4    8.5 1
1 5 1 5   13.5 1
1 5 2 5   14.2 1
2 1 1 6   12.8 1
2 1 2 6   13.2 1
2 2 1 7   16.2 1
2 2 2 7   17.0 1
2 3 1 8   20.5 1
2 3 2 8   18.2 1
2 4 1 9    9.8 1
2 4 2 9   10.3 1
2 5 1 10  16.5 1
2 5 2 10  18.4 1
3 1 1 11   8.9 1
3 1 2 11  10.2 1
3 2 1 12  15.4 1
3 2 2 12  16.0 1
3 3 1 13  17.8 1
3 3 2 13  18.9 1
3 4 1 14  22.1 1
```

```
3 4 2 14 23.0 1
3 5 1 15 18.4 1
3 5 2 15 20.6 1
4 1 1 16 12.0 1
4 1 2 16 12.2 1
4 2 1 17 13.4 1
4 2 2 17 14.2 1
4 3 1 18 13.5 1
4 3 2 18 13.8 1
4 4 1 19 21.2 1
4 4 2 19 20.7 1
4 5 1 20 17.8 1
4 5 2 20 19.4 1
5 1 1 21 20.1 1
5 1 2 21 21.3 1
5 2 1 22 20.9 1
5 2 2 22 21.2 1
5 3 1 23 19.2 1
5 3 2 23 20.2 1
5 4 1 24 22.2 1
5 4 2 24 21.6 1
5 5 1 25 20.8 1
5 5 2 25 21.3 1
1 1 1 1 11.2 2
1 1 2 1 11.7 2
1 2 1 2 11.2 2
1 2 2 2 12.7 2
1 3 1 3 14.3 2
1 3 2 3 15.2 2
1 4 1 4 9.2 2
1 4 2 4 9.5 2
1 5 1 5 13.0 2
1 5 2 5 14.9 2
2 1 1 6 12.2 2
2 1 2 6 12.8 2
2 2 1 7 16.0 2
2 2 2 7 17.8 2
2 3 1 8 19.5 2
2 3 2 8 18.8 2
2 4 1 9 10.4 2
2 4 2 9 11.3 2
2 5 1 10 16.9 2
2 5 2 10 18.0 2
3 1 1 11 10.8 2
3 1 2 11 11.2 2
3 2 1 12 15.0 2
3 2 2 12 16.6 2
3 3 1 13 17.2 2
3 3 2 13 17.9 2
3 4 1 14 21.1 2
3 4 2 14 22.6 2
3 5 1 15 19.2 2
3 5 2 15 21.6 2
4 1 1 16 12.6 2
4 1 2 16 13.2 2
4 2 1 17 12.4 2
4 2 2 17 13.2 2
```

```
4 3 1 18 14.5 2
4 3 2 18 15.8 2
4 4 1 19 20.2 2
4 4 2 19 20.1 2
4 5 1 20 17.0 2
4 5 2 20 18.4 2
5 1 1 21 22.1 2
5 1 2 21 21.0 2
5 2 1 22 21.9 2
5 2 2 22 20.2 2
5 3 1 23 20.2 2
5 3 2 23 20.8 2
5 4 1 24 21.2 2
5 4 2 24 20.6 2
5 5 1 25 20.2 2
5 5 2 25 21.0 2
;
PROC SORT;BY REP ENV I J;
PROC GLM;CLASS REP ENV HYBRID;MODEL YIELD=ENV REP(ENV)
HYBRID HYBRID*ENV;TEST H=HYBRID E=HYBRID*ENV;
LSMEANS HYBRID;
RUN;
TITLE 'DIALLEL-SAS 1';PROC GLM;CLASS REP ENV HYBRID;
MODEL YIELD=ENV REP(ENV) G1 G2 G3 G4 S11 S12 S13 S14 S22 S23 S24 S33
S34 S44 R12 R13 R14 R15 R23 R24 R25 R34 R35 R45 G1*ENV G2*ENV G3*ENV
G4*ENV S11*ENV S12*ENV S13*ENV S14*ENV S22*ENV S23*ENV S24*ENV S33*ENV
S34*ENV S44*ENV R12*ENV R13*ENV R14*ENV R15*ENV R23*ENV R24*ENV R25*ENV
    R34*ENV R35*ENV R45*ENV;
%MACRO GCASCA;
CONTRAST 'GCA' G1 1,G2 1,G3 1,G4 1;
CONTRAST 'SCA' S11 1,S12 1,S13 1,S14 1,S22 1,S23 1,S24 1,S33 1,S34
    1,S44 1;
ESTIMATE 'G1' G1 1;ESTIMATE 'G2' G2 1;ESTIMATE 'G3' G3 1;
ESTIMATE 'G4' G4 1;
ESTIMATE 'G5' G1 -1 G2 -1 G3 -1 G4 -1;
ESTIMATE 'S11' S11 1; ESTIMATE 'S12' S12 1; ESTIMATE 'S13' S13 1;
ESTIMATE 'S14' S14 1; ESTIMATE 'S22' S22 1; ESTIMATE 'S23' S23 1;
ESTIMATE 'S24' S24 1; ESTIMATE 'S33' S33 1; ESTIMATE 'S34' S34 1;
ESTIMATE 'S44' S44 1;
ESTIMATE 'S15' S11 -1 S12 -1 S13 -1 S14 -1;
ESTIMATE 'S25' S12 -1 S22 -1 S23 -1 S24 -1;
ESTIMATE 'S35' S13 -1 S23 -1 S33 -1 S34 -1;
ESTIMATE 'S45' S14 -1 S24 -1 S34 -1 S44 -1;
ESTIMATE 'S55' S11 1 S12 2 S13 2 S14 2 S22 1 S23 2 S24 2 S33 1 S34 2
    S44 1;
%MEND GCASCA;
%GCASCA
%MACRO INTERACT;
CONTRAST 'GCA*ENV' G1*ENV 1 -1,G2*ENV 1 -1,G3*ENV 1 -1,G4*ENV 1 -1;
CONTRAST 'SCA*ENV' S11*ENV 1 -1,S12*ENV 1 -1,S13*ENV 1 -1,S14*ENV 1 -
    1,S22*ENV 1 -1,S23*ENV 1 -1,S24*ENV 1 -1,S33*ENV 1 -1,S34*ENV 1 -
    1,S44*ENV 1 -1;
%MEND INTERACT;
%INTERACT
CONTRAST 'REC' R12 1, R13 1, R14 1, R15 1, R23 1, R24 1, R25 1, R34 1,
    R35 1, R45 1;
ESTIMATE 'R12' R12 1; ESTIMATE 'R13' R13 1; ESTIMATE 'R14' R14 1;
```

```
ESTIMATE 'R15' R15 1; ESTIMATE 'R23' R23 1; ESTIMATE 'R24' R24 1;
ESTIMATE 'R25' R25 1; ESTIMATE 'R34' R34 1; ESTIMATE 'R35' R35 1;
ESTIMATE 'R45' R45 1;
CONTRAST 'REC*ENV' R12*ENV 1 -1,R13*ENV 1 -1,R14*ENV 1 -1,R15*ENV 1 -
     1,R23*ENV 1 -1,R24*ENV 1 -1,R25*ENV 1 -1,R34*ENV 1 -1,R35*ENV 1 -
     1,R45*ENV 1 -1;
CONTRAST 'MAT SS' R12 1 R13 1 R14 1 R15 1,R12 -1 R23 1 R24 1 R25 1,R13
     -1 R23 -1 R34 1 R35 1,R14 -1 R24 -1 R34 -1 R45 1;
ESTIMATE 'MAT1' R12 1 R13 1 R14 1 R15 1/DIVISOR=4;
ESTIMATE 'MAT2' R12 -1 R23 1 R24 1 R25 1/DIVISOR=4;
ESTIMATE 'MAT3' R13 -1 R23 -1 R34 1 R35 1/DIVISOR=4;
ESTIMATE 'MAT4' R14 -1 R24 -1 R34 -1 R45 1/DIVISOR=4;
ESTIMATE 'MAT5' R15 -1 R25 -1 R35 -1 R45 -1/DIVISOR=4;
RUN;

TITLE 'DIALLEL-SAS 2';PROC GLM;CLASS REP ENV HYBRID;
MODEL YIELD=ENV REP(ENV) G1 G2 G3 G4 S11 S12 S13 S14 S22 S23 S24 S33
S34 S44 M1 M2 M3 M4 N12 N13 N14 N23 N24 N34 G1*ENV G2*ENV G3*ENV
G4*ENV S11*ENV S12*ENV S13*ENV S14*ENV S22*ENV S23*ENV S24*ENV S33*ENV
S34*ENV S44*ENV M1*ENV M2*ENV M3*ENV M4*ENV N12*ENV N13*ENV N14*ENV
N23*ENV N24*ENV N34*ENV;
%GCASCA
%INTERACT
CONTRAST 'MAT SS' M1 1,M2 1,M3 1,M4 1;
CONTRAST 'NONM SS' N12 1,N13 1,N14 1,N23 1,N24 1,N34 1;
CONTRAST 'MAT*ENV' M1*ENV 1 -1,M2*ENV 1 -1,M3*ENV 1 -1,M4*ENV 1 -1;
CONTRAST 'NONM*ENV' N12*ENV 1 -1,N13*ENV 1 -1,N14*ENV 1 -1,N23*ENV 1 -
     1,N24*ENV 1 -1,N34*ENV 1 -1;
ESTIMATE 'M1' M1 1; ESTIMATE 'M2' M2 1; ESTIMATE 'M3' M3 1;
ESTIMATE 'M4' M4 1; ESTIMATE 'M5' M1 -1 M2 -1 M3 -1 M4 -1;
ESTIMATE 'N12' N12 1; ESTIMATE 'N13' N13 1; ESTIMATE 'N14' N14 1;
ESTIMATE 'N23' N23 1; ESTIMATE 'N24' N24 1; ESTIMATE 'N34' N34 1;
ESTIMATE 'N15' N12 -1 N13 -1 N14 -1;
ESTIMATE 'N25' N12 1 N23 -1 N24 -1;
ESTIMATE 'N35' N13 1 N23 1 N34 -1;
ESTIMATE 'N45' N14 1 N24 1 N34 1;
RUN;
```

Output

```
METHOD 1                                                               1
                                          21:17 Sunday, September 2, 2001
The GLM Procedure
Class Level Information
Class   Levels  Values
REP        2    1 2
ENV        2    1 2
HYBRID    25    1 2 3 4 5 6 7 8 9 10 11 12 13 14 15 16 17 18 19 20 21
                22 23 24 25
Number of observations     100
METHOD 1                                                               2
                                          21:17 Sunday, September 2, 2001
The GLM Procedure
Dependent Variable: YIELD
                              Sum of
```

```
Source                   DF      Squares    Mean Square   F Value   Pr > F
Model                    51   1655.034800     32.451663     73.03   <.0001
Error                    48     21.329600      0.444367
Corrected Total  99   1676.364400
R-Square           Coeff Var       Root MSE      YIELD Mean
0.987276           4.098170        0.666608       16.26600
Source                   DF    Type I SS    Mean Square   F Value   Pr > F
ENV                       1      0.384400      0.384400      0.87   0.3570
REP(ENV)                  2      9.130400      4.565200     10.27   0.0002
HYBRID                   24   1631.674400     67.986433    153.00   <.0001
ENV*HYBRID               24     13.845600      0.576900      1.30   0.2171

Source                   DF   Type III SS   Mean Square   F Value   Pr > F
ENV                       1      0.384400      0.384400      0.87   0.3570
REP(ENV)                  2      9.130400      4.565200     10.27   0.0002
HYBRID                   24   1631.674400     67.986433    153.00   <.0001
ENV*HYBRID               24     13.845600      0.576900      1.30   0.2171
Tests of Hypotheses Using the Type III MS for ENV*HYBRID as an Error
     Term
Source                   DF   Type III SS   Mean Square   F Value   Pr > F
HYBRID                   24   1631.674400     67.986433    117.85   <.0001
METHOD 1                                                                 3
                                      21:17 Sunday, September 2, 2001
The GLM Procedure
Least Squares Means
HYBRID      YIELD LSMEAN
1             11.0250000
2             11.9500000
3             14.5500000
4              9.0500000
5             13.9000000
6             12.7500000
7             16.7500000
8             19.2500000
9             10.4500000
10            17.4500000
11            10.2750000
12            15.7500000
13            17.9500000
14            22.2000000
15            19.9500000
16            12.5000000
17            13.3000000
18            14.4000000
19            20.5500000
20            18.1500000
21            21.1250000
22            21.0500000
23            20.1000000
24            21.4000000
25            20.8250000
DIALLEL-SAS 1                                                            4
                                      21:17 Sunday, September 2, 2001
The GLM Procedure
Class Level Information
Class Levels  Values
REP      2   1 2
```

```
ENV      2   1 2
HYBRID  25   1 2 3 4 5 6 7 8 9 10 11 12 13 14 15 16 17 18 19 20 21 22
             23 24 25
Number of observations       100
DIALLEL-SAS 1                                                          5
                                21:17 Sunday, September 2, 2001
The GLM Procedure
Dependent Variable: YIELD
                                Sum of
Source              DF         Squares    Mean Square   F Value   Pr > F
Model               51      1655.034800     32.451663     73.03   <.0001
Error               48        21.329600      0.444367
Corrected Total     99      1676.364400
```

R-Square	Coeff Var	Root MSE	YIELD Mean
0.987276	4.098170	0.666608	16.26600

Source	DF	Type I SS	Mean Square	F Value	Pr > F
ENV	1	0.3844000	0.3844000	0.87	0.3570
REP(ENV)	2	9.1304000	4.5652000	10.27	0.0002
G1	1	887.7781250	887.7781250	1997.85	<.0001
G2	1	9.6400417	9.6400417	21.69	<.0001
G3	1	50.0520833	50.0520833	112.64	<.0001
G4	1	0.0060500	0.0060500	0.01	0.9076
S11	1	10.0806250	10.0806250	22.69	<.0001
S12	1	10.6666667	10.6666667	24.00	<.0001
S13	1	12.5563021	12.5563021	28.26	<.0001
S14	1	27.3195312	27.3195312	61.48	<.0001
S22	1	6.2084028	6.2084028	13.97	0.0005
S23	1	9.9487674	9.9487674	22.39	<.0001
S24	1	52.2200104	52.2200104	117.52	<.0001
S33	1	2.1267361	2.1267361	4.79	0.0336
S34	1	61.6333333	61.6333333	138.70	<.0001
S44	1	115.8852250	115.8852250	260.79	<.0001
R12	1	1.2800000	1.2800000	2.88	0.0961
R13	1	36.5512500	36.5512500	82.25	<.0001
R14	1	23.8050000	23.8050000	53.57	<.0001
R15	1	104.4012500	104.4012500	234.94	<.0001
R23	1	24.5000000	24.5000000	55.13	<.0001
R24	1	16.2450000	16.2450000	36.56	<.0001
R25	1	25.9200000	25.9200000	58.33	<.0001
R34	1	121.6800000	121.6800000	273.83	<.0001
R35	1	0.0450000	0.0450000	0.10	0.7517
R45	1	21.1250000	21.1250000	47.54	<.0001
G1*ENV	1	1.5401250	1.5401250	3.47	0.0688
G2*ENV	1	0.5320417	0.5320417	1.20	0.2793
G3*ENV	1	0.0700833	0.0700833	0.16	0.6930
G4*ENV	1	0.9384500	0.9384500	2.11	0.1527
S11*ENV	1	0.0756250	0.0756250	0.17	0.6818
S12*ENV	1	0.5400000	0.5400000	1.22	0.2758

```
DIALLEL-SAS 1                                                          6
                                21:17 Sunday, September 2, 2001
The GLM Procedure
Dependent Variable: YIELD
```

Source	DF	Type I SS	Mean Square	F Value	Pr > F
S13*ENV	1	0.0567187	0.0567187	0.13	0.7225
S14*ENV	1	1.4987813	1.4987813	3.37	0.0725
S22*ENV	1	0.3500694	0.3500694	0.79	0.3792

S23*ENV	1	0.5941840	0.5941840	1.34	0.2533
S24*ENV	1	0.1575938	0.1575938	0.35	0.5543
S33*ENV	1	1.8677778	1.8677778	4.20	0.0458
S34*ENV	1	0.3307500	0.3307500	0.74	0.3926
S44*ENV	1	0.2209000	0.2209000	0.50	0.4842
R12*ENV	1	0.1250000	0.1250000	0.28	0.5983
R13*ENV	1	0.5512500	0.5512500	1.24	0.2709
R14*ENV	1	0.0200000	0.0200000	0.05	0.8329
R15*ENV	1	0.2812500	0.2812500	0.63	0.4302
R23*ENV	1	0.0450000	0.0450000	0.10	0.7517
R24*ENV	1	1.6200000	1.6200000	3.65	0.0622
R25*ENV	1	0.0000000	0.0000000	0.00	1.0000
R34*ENV	1	2.4200000	2.4200000	5.45	0.0238
R35*ENV	1	0.0050000	0.0050000	0.01	0.9160
R45*ENV	1	0.0050000	0.0050000	0.01	0.9160

Source	DF	Type III SS	Mean Square	F Value	Pr > F
ENV	1	0.3844000	0.3844000	0.87	0.3570
REP(ENV)	2	9.1304000	4.5652000	10.27	0.0002
G1	1	595.4700500	595.4700500	1340.04	<.0001
G2	1	25.9920500	25.9920500	58.49	<.0001
G3	1	47.1906125	47.1906125	106.20	<.0001
G4	1	0.0060500	0.0060500	0.01	0.9076
S11	1	17.2432562	17.2432562	38.80	<.0001
S12	1	0.7710118	0.7710118	1.74	0.1940
S13	1	22.2103059	22.2103059	49.98	<.0001
S14	1	48.4334235	48.4334235	108.99	<.0001
S22	1	23.1842250	23.1842250	52.17	<.0001
S23	1	11.3796735	11.3796735	25.61	<.0001
S24	1	157.5091882	157.5091882	354.46	<.0001
S33	1	0.4192563	0.4192563	0.94	0.3363
S34	1	13.5576735	13.5576735	30.51	<.0001
S44	1	115.8852250	115.8852250	260.79	<.0001
R12	1	1.2800000	1.2800000	2.88	0.0961
R13	1	36.5512500	36.5512500	82.25	<.0001
R14	1	23.8050000	23.8050000	53.57	<.0001
R15	1	104.4012500	104.4012500	234.94	<.0001
R23	1	24.5000000	24.5000000	55.13	<.0001
R24	1	16.2450000	16.2450000	36.56	<.0001
R25	1	25.9200000	25.9200000	58.33	<.0001
R34	1	121.6800000	121.6800000	273.83	<.0001
R35	1	0.0450000	0.0450000	0.10	0.7517

DIALLEL-SAS 1   7

The GLM Procedure
Dependent Variable: YIELD

Source	DF	Type III SS	Mean Square	F Value	Pr > F
R45	1	21.1250000	21.1250000	47.54	<.0001
G1*ENV	1	2.1632000	2.1632000	4.87	0.0322
G2*ENV	1	0.2592000	0.2592000	0.58	0.4488
G3*ENV	1	0.2485125	0.2485125	0.56	0.4582
G4*ENV	1	0.9384500	0.9384500	2.11	0.1527
S11*ENV	1	0.0175563	0.0175563	0.04	0.8433
S12*ENV	1	1.2274000	1.2274000	2.76	0.1030
S13*ENV	1	0.1751059	0.1751059	0.39	0.5331
S14*ENV	1	0.5539882	0.5539882	1.25	0.2697

```
S22*ENV          1      0.3364000      0.3364000      0.76    0.3886
S23*ENV          1      0.0860029      0.0860029      0.19    0.6620
S24*ENV          1      0.1106941      0.1106941      0.25    0.6200
S33*ENV          1      2.2725563      2.2725563      5.11    0.0283
S34*ENV          1      0.4920029      0.4920029      1.11    0.2980
S44*ENV          1      0.2209000      0.2209000      0.50    0.4842
R12*ENV          1      0.1250000      0.1250000      0.28    0.5983
R13*ENV          1      0.5512500      0.5512500      1.24    0.2709
R14*ENV          1      0.0200000      0.0200000      0.05    0.8329
R15*ENV          1      0.2812500      0.2812500      0.63    0.4302
R23*ENV          1      0.0450000      0.0450000      0.10    0.7517
R24*ENV          1      1.6200000      1.6200000      3.65    0.0622
R25*ENV          1      0.0000000      0.0000000      0.00    1.0000
R34*ENV          1      2.4200000      2.4200000      5.45    0.0238
R35*ENV          1      0.0050000      0.0050000      0.01    0.9160
R45*ENV          1      0.0050000      0.0050000      0.01    0.9160


Contrast        DF    Contrast SS    Mean Square    F Value     Pr > F
GCA              4    947.4763000    236.8690750    533.05     <.0001
SCA             10    308.6456000     30.8645600     69.46     <.0001
GCA*ENV          4      3.0807000      0.7701750      1.73     0.1581
SCA*ENV         10      5.6924000      0.5692400      1.28     0.2678
REC             10    375.5525000     37.5552500     84.51     <.0001
REC*ENV         10      5.0725000      0.5072500      1.14     0.3527
MAT SS           4    112.5565000     28.1391250     63.32     <.0001
                               Standard
Parameter        Estimate        Error      t Value    Pr > |t|
G1           -3.45100000     0.09427265      -36.61      <.0001
G2           -0.72100000     0.09427265       -7.65      <.0001
G3            0.97150000     0.09427265       10.31      <.0001
G4           -0.01100000     0.09427265       -0.12      0.9076
G5            3.21150000     0.09427265       34.07      <.0001
S11           1.66100000     0.26664333        6.23      <.0001
DIALLEL-SAS 1                                                       8
                         21:17 Sunday, September 2, 2001
```

The GLM Procedure
Dependent Variable: YIELD

```
                               Standard
Parameter        Estimate        Error      t Value    Pr > |t|
S12           0.25600000     0.19434806        1.32      0.1940
S13          -1.37400000     0.19434806       -7.07      <.0001
S14          -2.02900000     0.19434806      -10.44      <.0001
S22           1.92600000     0.26664333        7.22      <.0001
S23           0.98350000     0.19434806        5.06      <.0001
S24          -3.65900000     0.19434806      -18.83      <.0001
S33          -0.25900000     0.26664333       -0.97      0.3363
S34           1.07350000     0.19434806        5.52      <.0001
S44           4.30600000     0.26664333       16.15      <.0001
S15           1.48600000     0.19434806        7.65      <.0001
S25           0.49350000     0.19434806        2.54      0.0144
S35          -0.42400000     0.19434806       -2.18      0.0341
S45           0.30850000     0.19434806        1.59      0.1190
S55          -1.86400000     0.26664333       -6.99      <.0001
R12          -0.40000000     0.23568164       -1.70      0.0961
R13           2.13750000     0.23568164        9.07      <.0001
R14          -1.72500000     0.23568164       -7.32      <.0001
R15          -3.61250000     0.23568164      -15.33      <.0001
```

```
R23                      1.75000000      0.23568164       7.43      <.0001
R24                     -1.42500000      0.23568164      -6.05      <.0001
R25                     -1.80000000      0.23568164      -7.64      <.0001
R34                      3.90000000      0.23568164      16.55      <.0001
R35                     -0.07500000      0.23568164      -0.32      0.7517
R45                     -1.62500000      0.23568164      -6.89      <.0001
MAT1                    -0.90000000      0.11784082      -7.64      <.0001
MAT2                    -0.26875000      0.11784082      -2.28      0.0271
MAT3                    -0.01562500      0.11784082      -0.13      0.8951
MAT4                    -0.59375000      0.11784082      -5.04      <.0001
MAT5                     1.77812500      0.11784082      15.09      <.0001
```

DIALLEL-SAS 2   9
                                        21:17 Sunday, September 2, 2001

The GLM Procedure
Class Level Information
Class   Levels   Values
REP   2   1 2
ENV   2   1 2
HYBRID   25   1 2 3 4 5 6 7 8 9 10 11 12 13 14 15 16 17 18 19 20 21
                 22 23 24 25
Number of observations   100
DIALLEL-SAS 2   10
                                        21:17 Sunday, September 2, 2001

The GLM Procedure
Dependent Variable: YIELD
                              Sum of
Source   DF   Squares   Mean Square   F Value   Pr > F
Model   51   1655.034800   32.451663   73.03   <.0001
Error   48   21.329600   0.444367
Corrected Total   99   1676.364400
R-Square   Coeff Var   Root MSE   YIELD Mean
0.987276   4.098170   0.666608   16.26600
Source   DF   Type I SS   Mean Square   F Value   Pr > F
ENV   1   0.3844000   0.3844000   0.87   0.3570
REP(ENV)   2   9.1304000   4.5652000   10.27   0.0002
G1   1   887.7781250   887.7781250   1997.85   <.0001
G2   1   9.6400417   9.6400417   21.69   <.0001
G3   1   50.0520833   50.0520833   112.64   <.0001
G4   1   0.0060500   0.0060500   0.01   0.9076
S11   1   10.0806250   10.0806250   22.69   <.0001
S12   1   10.6666667   10.6666667   24.00   <.0001
S13   1   12.5563021   12.5563021   28.26   <.0001
S14   1   27.3195312   27.3195312   61.48   <.0001
S22   1   6.2084028   6.2084028   13.97   0.0005
S23   1   9.9487674   9.9487674   22.39   <.0001
S24   1   52.2200104   52.2200104   117.52   <.0001
S33   1   2.1267361   2.1267361   4.79   0.0336
S34   1   61.6333333   61.6333333   138.70   <.0001
S44   1   115.8852250   115.8852250   260.79   <.0001
M1   1   91.8061250   91.8061250   206.60   <.0001
M2   1   8.5503750   8.5503750   19.24   <.0001
M3   1   0.9187500   0.9187500   2.07   0.1570
M4   1   11.2812500   11.2812500   25.39   <.0001
N12   1   5.3204167   5.3204167   11.97   0.0011
N13   1   81.2500521   81.2500521   182.84   <.0001
N14   1   7.2902812   7.2902812   16.41   0.0002
N23   1   7.2226563   7.2226563   16.25   0.0002
```
```

```
N24            1      4.3605104       4.3605104      9.81    0.0030
N34            1    157.5520833     157.5520833    354.55    <.0001
G1*ENV         1      1.5401250       1.5401250      3.47    0.0688
G2*ENV         1      0.5320417       0.5320417      1.20    0.2793
G3*ENV         1      0.0700833       0.0700833      0.16    0.6930
G4*ENV         1      0.9384500       0.9384500      2.11    0.1527
S11*ENV        1      0.0756250       0.0756250      0.17    0.6818
S12*ENV        1      0.5400000       0.5400000      1.22    0.2758
```
DIALLEL-SAS 2   11
                            21:17 Sunday, September 2, 2001
The GLM Procedure
Dependent Variable: YIELD

Source	DF	Type I SS	Mean Square	F Value	Pr > F
S13*ENV	1	0.0567187	0.0567187	0.13	0.7225
S14*ENV	1	1.4987813	1.4987813	3.37	0.0725
S22*ENV	1	0.3500694	0.3500694	0.79	0.3792
S23*ENV	1	0.5941840	0.5941840	1.34	0.2533
S24*ENV	1	0.1575938	0.1575938	0.35	0.5543
S33*ENV	1	1.8677778	1.8677778	4.20	0.0458
S34*ENV	1	0.3307500	0.3307500	0.74	0.3926
S44*ENV	1	0.2209000	0.2209000	0.50	0.4842
M1*ENV	1	0.2101250	0.2101250	0.47	0.4950
M2*ENV	1	0.1450417	0.1450417	0.33	0.5705
M3*ENV	1	0.0440833	0.0440833	0.10	0.7542
M4*ENV	1	0.0612500	0.0612500	0.14	0.7121
N12*ENV	1	0.2604167	0.2604167	0.59	0.4477
N13*ENV	1	0.0713021	0.0713021	0.16	0.6905
N14*ENV	1	0.0300313	0.0300313	0.07	0.7960
N23*ENV	1	0.0689062	0.0689062	0.16	0.6955
N24*ENV	1	1.1412604	1.1412604	2.57	0.1156
N34*ENV	1	3.0400833	3.0400833	6.84	0.0119

Source	DF	Type III SS	Mean Square	F Value	Pr > F
ENV	1	0.3844000	0.3844000	0.87	0.3570
REP(ENV)	2	9.1304000	4.5652000	10.27	0.0002
G1	1	595.4700500	595.4700500	1340.04	<.0001
G2	1	25.9920500	25.9920500	58.49	<.0001
G3	1	47.1906125	47.1906125	106.20	<.0001
G4	1	0.0060500	0.0060500	0.01	0.9076
S11	1	17.2432562	17.2432562	38.80	<.0001
S12	1	0.7710118	0.7710118	1.74	0.1940
S13	1	22.2103059	22.2103059	49.98	<.0001
S14	1	48.4334235	48.4334235	108.99	<.0001
S22	1	23.1842250	23.1842250	52.17	<.0001
S23	1	11.3796735	11.3796735	25.61	<.0001
S24	1	157.5091882	157.5091882	354.46	<.0001
S33	1	0.4192563	0.4192563	0.94	0.3363
S34	1	13.5576735	13.5576735	30.51	<.0001
S44	1	115.8852250	115.8852250	260.79	<.0001
M1	1	25.9200000	25.9200000	58.33	<.0001
M2	1	2.3112500	2.3112500	5.20	0.0271
M3	1	0.0078125	0.0078125	0.02	0.8951
M4	1	11.2812500	11.2812500	25.39	<.0001
N12	1	0.1470000	0.1470000	0.33	0.5679
N13	1	107.9203333	107.9203333	242.86	<.0001
N14	1	29.2053333	29.2053333	65.72	<.0001
N23	1	50.8300833	50.8300833	114.39	<.0001

N24	1	37.8563333	37.8563333	85.19	<.0001

DIALLEL-SAS 2   12
21:17 Sunday, September 2, 2001

The GLM Procedure
Dependent Variable: YIELD

Source	DF	Type III SS	Mean Square	F Value	Pr > F
N34	1	157.5520833	157.5520833	354.55	<.0001
G1*ENV	1	2.1632000	2.1632000	4.87	0.0322
G2*ENV	1	0.2592000	0.2592000	0.58	0.4488
G3*ENV	1	0.2485125	0.2485125	0.56	0.4582
G4*ENV	1	0.9384500	0.9384500	2.11	0.1527
S11*ENV	1	0.0175563	0.0175563	0.04	0.8433
S12*ENV	1	1.2274000	1.2274000	2.76	0.1030
S13*ENV	1	0.1751059	0.1751059	0.39	0.5331
S14*ENV	1	0.5539882	0.5539882	1.25	0.2697
S22*ENV	1	0.3364000	0.3364000	0.76	0.3886
S23*ENV	1	0.0860029	0.0860029	0.19	0.6620
S24*ENV	1	0.1106941	0.1106941	0.25	0.6200
S33*ENV	1	2.2725563	2.2725563	5.11	0.0283
S34*ENV	1	0.4920029	0.4920029	1.11	0.2980
S44*ENV	1	0.2209000	0.2209000	0.50	0.4842
M1*ENV	1	0.2812500	0.2812500	0.63	0.4302
M2*ENV	1	0.1250000	0.1250000	0.28	0.5983
M3*ENV	1	0.0703125	0.0703125	0.16	0.6926
M4*ENV	1	0.0612500	0.0612500	0.14	0.7121
N12*ENV	1	0.8333333	0.8333333	1.88	0.1772
N13*ENV	1	0.6750000	0.6750000	1.52	0.2238
N14*ENV	1	0.0480000	0.0480000	0.11	0.7438
N23*ENV	1	0.3520833	0.3520833	0.79	0.3778
N24*ENV	1	2.5230000	2.5230000	5.68	0.0212
N34*ENV	1	3.0400833	3.0400833	6.84	0.0119

Contrast	DF	Contrast SS	Mean Square	F Value	Pr > F
GCA	4	947.4763000	236.8690750	533.05	<.0001
SCA	10	308.6456000	30.8645600	69.46	<.0001
GCA*ENV	4	3.0807000	0.7701750	1.73	0.1581
SCA*ENV	10	5.6924000	0.5692400	1.28	0.2678
MAT SS	4	112.5565000	28.1391250	63.32	<.0001
NONM SS	6	262.9960000	43.8326667	98.64	<.0001
MAT*ENV	4	0.4605000	0.1151250	0.26	0.9027
NONM*ENV	6	4.6120000	0.7686667	1.73	0.1344

Parameter	Estimate	Standard Error	t Value	Pr > |t|
G1	-3.45100000	0.09427265	-36.61	<.0001
G2	-0.72100000	0.09427265	-7.65	<.0001
G3	0.97150000	0.09427265	10.31	<.0001
G4	-0.01100000	0.09427265	-0.12	0.9076
G5	3.21150000	0.09427265	34.07	<.0001

DIALLEL-SAS 2   13
21:17 Sunday, September 2, 2001

The GLM Procedure
Dependent Variable: YIELD

Parameter	Estimate	Standard Error	t Value	Pr > |t|
S11	1.66100000	0.26664333	6.23	<.0001
S12	0.25600000	0.19434806	1.32	0.1940

S13	-1.37400000	0.19434806	-7.07	<.0001
S14	-2.02900000	0.19434806	-10.44	<.0001
S22	1.92600000	0.26664333	7.22	<.0001
S23	0.98350000	0.19434806	5.06	<.0001
S24	-3.65900000	0.19434806	-18.83	<.0001
S33	-0.25900000	0.26664333	-0.97	0.3363
S34	1.07350000	0.19434806	5.52	<.0001
S44	4.30600000	0.26664333	16.15	<.0001
S15	1.48600000	0.19434806	7.65	<.0001
S25	0.49350000	0.19434806	2.54	0.0144
S35	-0.42400000	0.19434806	-2.18	0.0341
S45	0.30850000	0.19434806	1.59	0.1190
S55	-1.86400000	0.26664333	-6.99	<.0001
M1	-0.72000000	0.09427265	-7.64	<.0001
M2	-0.21500000	0.09427265	-2.28	0.0271
M3	-0.01250000	0.09427265	-0.13	0.8951
M4	-0.47500000	0.09427265	-5.04	<.0001
M5	1.42250000	0.09427265	15.09	<.0001
N12	0.10500000	0.18255821	0.58	0.5679
N13	2.84500000	0.18255821	15.58	<.0001
N14	-1.48000000	0.18255821	-8.11	<.0001
N23	1.95250000	0.18255821	10.70	<.0001
N24	-1.68500000	0.18255821	-9.23	<.0001
N34	3.43750000	0.18255821	18.83	<.0001
N15	-1.47000000	0.18255821	-8.05	<.0001
N25	-0.16250000	0.18255821	-0.89	0.3778
N35	1.36000000	0.18255821	7.45	<.0001
N45	0.27250000	0.18255821	1.49	0.1421

DIALLEL-SAS Method 2 Program Listing

```
DATA METHOD2;TITLE 'METHOD 2';
INPUT I J REP HYBRID YIELD ENV;
DROP N NI NJ P;
P=5;*NUMBER OF PARENTAL LINES;
ARRAY GCA(N) G1 G2 G3 G4;
DO N=1 TO (P-1);
GCA=((I=N)-(I=P))+((J=N)-(J=P));
END;
ARRAY SCA(N) S11 S12 S13 S14 S22 S23 S24 S33 S34 S44;
N=0;
DO NI=1 TO (P-1);
DO NJ=NI TO (P-1);
N+1;
IF NI=NJ THEN DO;
SCA=(I=NI)*((J=NJ)-(J=P)*2)+(I=P)*(J=P);END;
ELSE DO;
SCA=(I=NI)*(J=NJ)-(J=P)*((I=NI)+(I=NJ)-(I=P));
END;END;END;
CARDS;
```

[Your data here]

```
;
```

```
PROC SORT;BY REP ENV I J;
PROC GLM;CLASS REP ENV HYBRID;MODEL YIELD=ENV REP(ENV)
HYBRID HYBRID*ENV;TEST H=HYBRID E=HYBRID*ENV;
LSMEANS HYBRID;
PROC GLM;CLASS REP ENV HYBRID;
MODEL YIELD=ENV REP(ENV) G1 G2 G3 G4 S11 S12 S13 S14 S22 S23 S24 S33
S34 S44 G1*ENV G2*ENV G3*ENV G4*ENV S11*ENV S12*ENV S13*ENV S14*ENV
S22*ENV S23*ENV S24*ENV S33*ENV S34*ENV S44*ENV;
CONTRAST 'GCA' G1 1,G2 1,G3 1,G4 1;
CONTRAST 'SCA' S11 1,S12 1,S13 1,S14 1,S22 1,S23 1,S24 1,S33 1,S34 1,
    S44 1;
ESTIMATE 'G1' G1 1;ESTIMATE 'G2' G2 1;ESTIMATE 'G3' G3 1;
ESTIMATE 'G4' G4 1;
ESTIMATE 'G5' G1 -1 G2 -1 G3 -1 G4 -1;
ESTIMATE 'S11' S11 1; ESTIMATE 'S12' S12 1; ESTIMATE 'S13' S13 1;
ESTIMATE 'S14' S14 1; ESTIMATE 'S22' S22 1; ESTIMATE 'S23' S23 1;
ESTIMATE 'S24' S24 1; ESTIMATE 'S33' S33 1; ESTIMATE 'S34' S34 1;
ESTIMATE 'S44' S44 1;
ESTIMATE 'S15' S11 -1 S12 -1 S13 -1 S14 -1;
ESTIMATE 'S25' S12 -1 S22 -1 S23 -1 S24 -1;
ESTIMATE 'S35' S13 -1 S23 -1 S33 -1 S34 -1;
ESTIMATE 'S45' S14 -1 S24 -1 S34 -1 S44 -1;
ESTIMATE 'S55' S11 1 S12 2 S13 2 S14 2 S22 1 S23 2 S24 2 S33 1 S34 2
    S44 1;
CONTRAST 'GCA*ENV' G1*ENV 1 -1,G2*ENV 1 -1,G3*ENV 1 -1,G4*ENV 1 -1;
CONTRAST 'SCA*ENV' S11*ENV 1 -1,S12*ENV 1 -1,S13*ENV 1 -1,S14*ENV 1 -
    1, S22*ENV 1 -1,S23*ENV 1 -1,S24*ENV 1 -1,S33*ENV 1 -1, S34*ENV 1
    -1,S44*ENV 1 -1;
RUN;
```

DIALLEL Method 3 Program Listing

```
TITLE 'METHOD 3';
DATA METHOD3;
INPUT I J REP HYBRID YIELD ENV;
DROP N NI NJ P;
P=5;*NUMBER OF PARENTAL LINES;
ARRAY GCA(N) G1 G2 G3 G4;
DO N=1 TO (P-1);
GCA=((I=N)-(I=P))+((J=N)-(J=P));
END;
ARRAY SCA(N) S12 S13 S14 S23 S24;
N=0;
DO NI=1 TO (P-3);
DO NJ=NI+1 TO (P-1);
N+1;
SCA=(I=NI)*(J=NJ)-((I=NI)+(I=NJ))*(J=P)+(J=NI)*(I=NJ)
-(I=P)*((J=NI)+(J=NJ));
IF ((I>=(P-2))&(J>=(P-1)))|((I>=(P-1))&(J>=(P-2))) THEN DO;
SCA=-(I=(P-2))*(J=(P-1))+(I>=(P-2))*(J=P)*(I NE NJ)-(J=(P-2))*(I=(P-
    1))+(J>=(P-2))*(I=P)*(J NE NJ);
END;END;END;
ARRAY REC(N) R12 R13 R14 R15 R23 R24 R25 R34 R35 R45;
N=0;
DO NI=1 TO (P-1);
```

```
DO NJ=(NI+1) TO P;
N+1;
REC=(I=NI)*(J=NJ)-(J=NI)*(I=NJ);
END;END;
ARRAY MAT(N) M1 M2 M3 M4;
DO N=1 TO (P-1);
MAT=(I=N)+(J=P)-(J=N)-(I=P);
END;
ARRAY NONM(N) N12 N13 N14 N23 N24 N34;
N=0;
DO NI=1 TO (P-2);
DO NJ=(NI+1) TO (P-1);
N+1;
NONM=((I=NI)*(J=NJ))-(I=NJ)*(J=NI)+(((I=NJ)-(I=NI))*(J=P))
+((I=P)*((J=NI)-(J=NJ)));
END;END;
CARDS;
```

[Your data here]

```
;
PROC SORT;BY ENV REP I J;
PROC GLM;CLASS REP ENV HYBRID;MODEL YIELD=ENV REP(ENV) HYBRID
    ENV*HYBRID;TEST H=HYBRID E=ENV*HYBRID;TEST H=ENV E=REP(ENV);
    LSMEANS HYBRID;RUN;
TITLE 'DIALLEL-SAS 1';
PROC GLM;CLASS REP HYBRID ENV;
MODEL YIELD=ENV REP(ENV) G1 G2 G3 G4 S12 S13 S14 S23 S24 R12
R13 R14 R15 R23 R24 R25 R34 R35 R45 G1*ENV G2*ENV G3*ENV G4*ENV
S12*ENV S13*ENV S14*ENV S23*ENV S24*ENV R12*ENV R13*ENV R14*ENV
R15*ENV R23*ENV R24*ENV R25*ENV R34*ENV R35*ENV R45*ENV;
%MACRO GCASCA;
CONTRAST 'GCA' G1 1,G2 1,G3 1,G4 1;
CONTRAST 'SCA' S12 1,S13 1,S14 1,S23 1,S24 1;
ESTIMATE 'G1' G1 1;ESTIMATE 'G2' G2 1;ESTIMATE 'G3' G3 1;
ESTIMATE 'G4' G4 1;
ESTIMATE 'G5' G1 -1 G2 -1 G3 -1 G4 -1;
ESTIMATE 'S12' S12 1; ESTIMATE 'S13' S13 1; ESTIMATE 'S14' S14 1;
ESTIMATE 'S23' S23 1; ESTIMATE 'S24' S24 1;
ESTIMATE 'S15' S12 -1 S13 -1 S14 -1;ESTIMATE 'S25' S12 -1 S23 -1 S24 -
    1;
ESTIMATE 'S34' S12 -1 S13 -1 S14 -1 S23 -1 S24 -1;
ESTIMATE 'S35' S12 1 S14 1 S24 1;ESTIMATE 'S45' S12 1 S13 1 S23 1;
%MEND GCASCA;
%GCASCA
CONTRAST 'REC' R12 1,R13 1,R14 1,R15 1,R23 1,R24 1,R25 1,R34 1,R35 1,
    R45 1;
ESTIMATE 'R12' R12 1; ESTIMATE 'R13' R13 1; ESTIMATE 'R14' R14 1;
ESTIMATE 'R15' R15 1; ESTIMATE 'R23' R23 1; ESTIMATE 'R24' R24 1;
ESTIMATE 'R25' R25 1; ESTIMATE 'R34' R34 1; ESTIMATE 'R35' R35 1;
ESTIMATE 'R45' R45 1;
CONTRAST 'MAT SS' R12 1 R13 1 R14 1 R15 1,R12 -1 R23 1 R24 1 R25 1,
    R13 -1 R23 -1 R34 1 R35 1,R14 -1 R24 -1 R34 -1 R45 1;
ESTIMATE 'MAT1' R12 1 R13 1 R14 1 R15 1/DIVISOR=4;
ESTIMATE 'MAT2' R12 -1 R23 1 R24 1 R25 1/DIVISOR=4;
ESTIMATE 'MAT3' R13 -1 R23 -1 R34 1 R35 1/DIVISOR=4;
```

```
ESTIMATE 'MAT4' R14 -1 R24 -1 R34 -1 R45 1/DIVISOR=4;
ESTIMATE 'MAT5' R15 -1 R25 -1 R35 -1 R45 -1/DIVISOR=4;
%MACRO INTERACT;
CONTRAST 'GCA*ENV' G1*ENV 1 -1,G2*ENV 1 -1,G3*ENV 1 -1,G4*ENV 1 -1;
CONTRAST 'SCA*ENV' S12*ENV 1 -1,S13*ENV 1 -1,S14*ENV 1 -1, S23*ENV 1 -
    1,S24*ENV 1 -1;
%MEND INTERACT;
%INTERACT
CONTRAST 'REC*ENV'  R12*ENV 1 -1,R13*ENV 1 -1,R14*ENV 1 -1, R15*ENV 1
    -1,R23*ENV 1 -1,R24*ENV 1 -1, R25*ENV 1 -1,R34*ENV 1 -1,R35*ENV 1
    -1, R35*ENV 1 -1,R45*ENV 1 -1;
RUN;
TITLE 'DIALLEL-SAS 2';
PROC GLM;CLASS REP HYBRID ENV;
MODEL YIELD=ENV REP(ENV) G1 G2 G3 G4 S12 S13 S14 S23 S24 M1 M2 M3
M4 N12 N13 N14 N23 N24 N34 G1*ENV G2*ENV G3*ENV G4*ENV
S12*ENV S13*ENV S14*ENV S23*ENV S24*ENV M1*ENV M2*ENV M3*ENV
M4*ENV N12*ENV N13*ENV N14*ENV N23*ENV N24*ENV N34*ENV;
%GCASCA
%INTERACT
CONTRAST 'MAT' M1 1,M2 1,M3 1,M4 1;
CONTRAST 'NONM' N12 1,N13 1,N14 1,N23 1,N24 1,N34 1;
ESTIMATE 'M1' M1 1;   ESTIMATE 'M2' M2 1;   ESTIMATE 'M3' M3 1;
ESTIMATE 'M4' M4 1;   ESTIMATE 'M5' M1 -1 M2 -1 M3 -1 M4 -1;
ESTIMATE 'N12' N12 1;ESTIMATE 'N13' N13 1;ESTIMATE 'N14' N14 1;
ESTIMATE 'N23' N23 1;ESTIMATE 'N24' N24 1;ESTIMATE 'N34' N34 1;
ESTIMATE 'N15' N12 -1 N13 -1 N14 -1;
ESTIMATE 'N25' N12 1 N23 -1 N24 -1;
ESTIMATE 'N35' N13 1 N23 1 N34 -1;
ESTIMATE 'N45' N14 1 N24 1 N34 1;
CONTRAST 'MAT*ENV' M1*ENV 1 -1,M2*ENV 1 -1,M3*ENV 1 -1, M4*ENV 1 -1;
CONTRAST 'NONM*ENV' N12*ENV 1 -1,N13*ENV 1 -1,N14*ENV 1 -1, N23*ENV 1
    -1,N24*ENV 1 -1,N34*ENV 1 -1;
RUN;
```

DIALLEL Method 4 Program Listing

```
DATA METHOD4;TITLE 'METHOD 4';
INPUT I J REP HYBRID YIELD ENV;
DROP N NI NJ P;
P=5;*NUMBER OF PARENTAL LINES;
ARRAY GCA(N) G1 G2 G3 G4;
DO N=1 TO (P-1);
GCA=((I=N)-(I=P))+((J=N)-(J=P));
END;
ARRAY SCA(N) S12 S13 S14 S23 S24;
N=0;
DO NI=1 TO (P-3);
DO NJ=NI+1 TO (P-1);
N+1;
SCA=(I=NI)*(J=NJ)-((I=NI)+(I=NJ))*(J=P);
IF  ((I>=(P-2))&(J>=(P-1)))|((I>=(P-1))&(J>=(P-2))) THEN DO;
SCA=-(I=(P-2))*(J=(P-1))+(I>=(P-2))*(J=P)*(I NE NJ);
END;END;END;
CARDS;
```

[Your data here]

```
;
PROC SORT;BY REP ENV I J;
PROC GLM;CLASS REP ENV HYBRID;MODEL YIELD=ENV REP(ENV) HYBRID
     HYBRID*ENV;TEST H=HYBRID E=HYBRID*ENV; LSMEANS HYBRID;
PROC GLM;CLASS REP ENV HYBRID;
MODEL YIELD=ENV REP(ENV) G1 G2 G3 G4 S12 S13 S14 S23 S24
G1*ENV G2*ENV G3*ENV G4*ENV S12*ENV S13*ENV S14*ENV S23*ENV S24*ENV;

CONTRAST 'GCA' G1 1,G2 1,G3 1,G4 1;
CONTRAST 'SCA' S12 1,S13 1,S14 1,S23 1,S24 1;
CONTRAST 'GCA*ENV' G1*ENV 1 -1,G2*ENV 1 -1,G3*ENV 1 -1,G4*ENV 1 -1;
CONTRAST 'SCA*ENV' S12*ENV 1 -1,S13*ENV 1 -1,S14*ENV 1 -1,S23*ENV 1 -
     1,S24*ENV 1 -1;
ESTIMATE 'G1' G1 1;ESTIMATE 'G2' G2 1;ESTIMATE 'G3' G3 1;
ESTIMATE 'G4' G4 1;
ESTIMATE 'G5' G1 -1 G2 -1 G3 -1 G4 -1;
ESTIMATE 'S12' S12 1; ESTIMATE 'S13' S13 1; ESTIMATE 'S14' S14 1;
ESTIMATE 'S23' S23 1; ESTIMATE 'S24' S24 1;
ESTIMATE 'S15' S12 -1 S13 -1 S14 -1;
ESTIMATE 'S25' S12 -1 S23 -1 S24 -1;
ESTIMATE 'S34' S12 -1 S13 -1 S14 -1 S23 -1 S24 -1;
ESTIMATE 'S35' S12 1 S14 1 S24 1;
ESTIMATE 'S45' S12 1 S13 1 S23 1;
RUN;
```

REFERENCES

Borges, O.L.F. (1987). Diallel analysis of maize resistance to sorghum downy mildew. *Crop Science* 27:178-180.

Griffing, B. (1956). Concept of general and specific combining ability in relation to diallel crossing systems. *Australian Journal of Biological Science* 9:463-493.

Kang, M.S., Zhang, Y., and Magari, R. (1995). Combining ability for maize weevil preference of maize grain. *Crop Science* 35:1556-1559.

Moffatt, J.M., Sears, R.G., Cox, T.S., and Paulsen, G.M. (1990). Wheat high temperature tolerance during reproductive growth. II. Genetic analysis of chlorophyll fluorescence. *Crop Science* 30:886-889.

Pixley, K.V. and Bjarnason, M.S. (1993). Combining ability for yield and protein quality among modified-endosperm *opaque-2* tropical maize inbreds. *Crop Science* 33:1229-1234.

SAS Institute Inc. (1995). *SAS Language and Procedure: Usage,* Version 6, First Edition. SAS Institute, Cary, NC.

Chapter 2

Diallel Analysis for a Seed and Endosperm Model with Genotype-by-Environment Interaction Effects

Jun Zhu

Purpose

To analyze balanced or unbalanced data of diploid seed and triploid endosperm models for estimating components of variance, covariance, heritability, and selection response.

Definitions

Mating Design

A set of inbred lines are sampled from a reference population. These parents are used to produce F_1 and F_2 seeds. Experiments with parents, F_1s, and F_2s are conducted in multiple environments using a randomized complete-block design.

Genetic Model

The genetic model for genetic entry of the kth type of generation derived from parents i and j in the lth block within the hth environment is

$$y_{hijkl} = \mu + E_h + G_{ijk} + GE_{hijk} + B_{hl} + e_{hijkl}$$

where μ = population mean, E_h = environment effect, G_{ijk} = total genotypic effect, GE_{hijk} = genotype × environment interaction effect, B_{hl} = block effect, and e_{hijkl} = residual effect.

Genetic partitioning for the diploid seed model (Zhu and Weir, 1994a; Zhu, 1996):

For parent (P_i, $k = 0$):

$$G_{ii0} + GE_{hii0} = 2A_i + D_{ii} + C_i + 2Am_i + Dm_{ii} + 2AE_{hi} + DE_{hiii} + CE_{hi} \\ + 2AmE_{hi} + DmE_{hii}$$

For F_1 ($P_i \times P_j$, $k = 1$):

$$G_{ij}1 + GE_{hij2} = A_i + A_j + D_{ij} + C_i + 2Am_i + Dm_{ii} + AE_{hi} + AE_{hj} + DE_{hij} \\ + CE_{hi} + 2AmE_{hi} + DmB_{hii}$$

For F_2 ($F_1\otimes$, $k = 2$):

$$G_{ij2} + GE_{hij2} = A_i + A_j + \tfrac{1}{4}D_{ii} + \tfrac{1}{4}D_{jj} + \tfrac{1}{2}D_{ij} + C_i + Am_i + Am_j + Dm_{ij} \\ + AE_{hi} + AE_{hj} + \tfrac{1}{4}DE_{hii} + \tfrac{1}{4}DE_{hjj} + \tfrac{1}{2}DE_{hij} + CE_{hi} \\ + AmE_{hi} + AmE_{hj} + DmE_{hij}$$

Genetic partitioning for triploid endosperm model (Zhu and Weir, 1994b; Zhu, 1996):

For parent (P_i, $k = 0$):

$$G_{ii0} + GE_{hii0} = 3A_i + 3D_{ii} + C_i + 2Am_i + Dm_{ii} + 3AE_{hi} + 3DE_{hii} + CE_{hi} \\ + 2AmE_{hi} + DmE_{hii}$$

For F_1 ($P_i \times P_j$, $k = 1$):

$$G_{ij1} + GE_{hij1} = 2A_i + A_j + D_{ii} + 2D_{ij} + C_i + 2Am_i + Dm_{ii} + 2AE_{hi} \\ + AE_{hj} + DE_{hii} + 2DE_{hij} + CE_{hi} + 2AmE_{hi} + DmE_{hii}$$

For F_2 ($F_1 \otimes$, $k = 2$):

$$G_{ij2} + GE_{hij2} = 1\tfrac{1}{2}A_i + 1\tfrac{1}{2}A_j + D_{ii} + D_{jj} + D_{ij} + C_i + Am_i + Am_j + Dm_{ij}$$
$$+ 1\tfrac{1}{2}AE_{hi} + 1\tfrac{1}{2}AE_{hj} + DE_{hii} + DE_{hjj} + DE_{hij} + CE_{hi}$$
$$+ AmE_{hi} + AmE_{hj} + DmE_{hij}$$

where A = direct additive effect, D = direct dominance effect, C = cytoplasm effect, Am = maternal additive effect, Dm = maternal dominance effect, AE = direct additive by environment interaction effect, DE = direct dominance by environment interaction effect, CE = cytoplasm by environment interaction effect, AmE = maternal additive by environment interaction effect, DmE = maternal dominance by environment interaction effect.

Other generations, such as BC_1s and BC_2s and their reciprocals (RBC_1s and RBC_2s) can also be used for analyzing seed traits (Zhu and Weir, 1994a; Zhu, 1996).

Analysis Methodology

Mixed Linear Model

The phenotypic mean of the seed genetic model can be expressed by a mixed linear model as

$$y = Xb + U_A e_A + U_D e_D + U_C e_C + U_{Am} e_{Am} + U_{Dm} e_{Dm} + U_{AE} e_{AE} + U_{DE} e_{DE}$$
$$+ U_{CE} e_{CE} + U_{AmE} e_{AmE} + U_{DmE} e_{DmE} + U_B e_B + e_e$$
$$= Xb + \sum_{u}^{12} U_u e_u$$

with variance-covariance matrix

$$\mathrm{var}(y) = \sigma_A^2 V_1 + \sigma_D^2 V_2 + \sigma_C^2 V_3 + \sigma_{Am}^2 V_4 + \sigma_{Dm}^2 V_5 + \sigma_{AE}^2 V_6 + \sigma_{DE}^2 V_7$$
$$+ \sigma_{CE}^2 V_8 + \sigma_{AmE}^2 V_9 + \sigma_{DmE}^2 V_{10} + \sigma_B^2 V_{11} + \sigma_{A.Am} V_{12} + \sigma_{D.Dm} V_{13}$$
$$+ \sigma_{AE.AmE} V_{14} + \sigma_{DE.DmE} V_{15} + \sigma_e^2 V_{16}$$
$$= \sum_{u=1}^{16} \theta_u V_u$$

where $V_u = U_u U_u^T (u = 1,2,...,11)$, $V_{12} = \left(U_1 U_4^T + U_4 U_1^T \right)$, $V_{13} = \left(U_2 U_5^T + U_5 U_2^T \right)$, $V_{14} = \left(U_6 U_9^T + U_9 U_6^T \right)$, $V_{15} = \left(U_7 U_{10}^T + U_{10} U_7^T \right)$, $V_{16} = I$.

Variance Components

Unbiased estimation of variances and covariances of the same trait can be obtained by the following MINQUE(0/1) equations (Zhu, 1992; Zhu and Weir, 1994a):

$$\left[tr\left(Q_{(0/1)} V_u Q_{(0/1)} V_v \right) \right]\left[\hat{\theta}_u \right] = \left[y^T Q_{(0/1)} V_u Q_{(0/1)} y \right]$$

where

$$Q_{(0/1)} = V_{(0/1)}^{-1} - V_{(0/1)}^{-1} X \left(X^T V_{(0/1)}^{-1} X \right)^+ X^T V_{(0/1)}^{-1}$$

$$V_{(0/1)} = \sum_{u=1}^{11} U_u U_u^T + I$$

For diploid seed of F_2, genetic variance and covariance components can be obtained by $V_A = 2\sigma_A^2$, $V_D = \frac{3}{8}\sigma_D^2$, $V_C = \sigma_C^2$, $V_{Am} = 2\sigma_{Am}^2$, $V_{Dm} = \sigma_D^2$, $V_{AE} = 2\sigma_{AE}^2$, $V_{DE} = \frac{3}{8}\sigma_{DE}^2$, $V_{CE} = \sigma_{CE}^2$, $V_{AmE} = 2\sigma_{AmE}^2$, $V_{DmE} = \sigma_{DE}^2$, $V_e = \sigma_e^2$, $C_{A.Am} = 2\sigma_{A.Am}$, $C_{D.Dm} = \frac{1}{2}\sigma_{D.Dm}$, $C_{AE.AmE} = 2\sigma_{AE.AmE}$, $C_{DE.DmE} = \frac{1}{2}\sigma_{DE.DmE}$.

For triploid endosperm of F_2, genetic variance and covariance components can be obtained by $V_A = 4\frac{1}{2}\sigma_A^2$, $V_D = 3\sigma_D^2$, $V_C = \sigma_C^2$, $V_{AM} = 2\sigma_{Am}^2$, $V_{Dm} = \sigma_D^2$, $V_{AE} = 4\frac{1}{2}\sigma_{AE}^2$, $V_{DE} = 3\sigma_{DE}^2$, $V_{CE} = \sigma_{CE}^2$, $V_{AmE} = 2\sigma_{AmE}^2$, $V_{DmE} = \sigma_{DE}^2$, $V_e = \sigma_e^2$, $C_{A.Am} = 3\sigma_{A.Am}$, $C_{D.Dm} = \sigma_{D.Dm}$, $C_{AE.AmE} = 3\sigma_{AE.AmE}$, $C_{DE.DmE} = \sigma_{DE.DmE}$.

The total phenotypic variance is $V_P = V_A + V_D + V_C + V_{Am} + V_{Dm} + V_{AE} + V_{DE} + V_{CE} + V_{AmE} + V_{DmE} + 2C_{A.Am} + 2C_{D.Dm} + 2C_{AE.AmE} + 2C_{DE.DmE} + V_e$, where $C_{A.Am}$ and $C_{D.Dm}$ are the covariances between direct effects (A and D) and maternal effects (Am and Dm) of the same trait, $C_{AE.AmE}$ and $C_{DE.DmE}$ are the covariances between direct by environment interaction effect (AE and DE) and maternal by environment interaction effect (AmE and DmE) of the same trait.

Covariance Components and Correlation

Unbiased estimation of covariances between two traits (y_1 and y_2) can be obtained by MINQUE(0/1) approaches (Zhu, 1992; Zhu and Weir, 1994a).

$$\left[tr\left(Q_{(0/1)} V_u Q_{(0/1)} V_v \right) \right]\left[\hat{\theta}_{u/u} \right] = \left[y_1^T Q_{(0/1)} V_u Q_{(0/1)} y_2 \right]$$

For diploid seed of F_2, genetic covariance components can be obtained by $C_A = 2\sigma_{A/A}$, $C_D = \frac{3}{8}\sigma_{D/D}$, $C_C = \sigma_{C/C}$, $C_{Am} = 2\sigma_{Am/Am}$, $C_{Dm} = \sigma_{D/D}$, $C_{AE} = 2\sigma_{AE/AE}$, $C_{DE} = \frac{3}{8}\sigma_{DE/DE}$, $C_{CE} = \sigma_{CE/CE}$, $C_{AmE} = 2\sigma_{AmE/AmE}$, $C_{DmE} = \sigma_{DE/DE}$, $C_e = \sigma_{e/e}$, $C_{A/Am} = 2\sigma_{A/Am}$, $C_{D/Dm} = \frac{1}{2}\sigma_{D/Dm}$, $C_{AE/AmE} = 2\sigma_{AE/AmE}$, $C_{DE/DmE} = \frac{1}{2}\sigma_{DE/DmE}$.

For triploid endosperm of F_2, genetic covariance components can be obtained by $C_A = 4\frac{1}{2}\sigma_{A/A}$, $C_D = 3\sigma_{D/D}$, $C_C = \sigma_{C/C}$, $C_{Am} = 2\sigma_{Am/Am}$, $C_{Dm} = \sigma_{D/D}$, $C_{AE} = 4\frac{1}{2}\sigma_{AE/AE}$, $C_{DE} = 3\sigma_{DE/DE}$, $C_{CE} = \sigma_{CE/CE}$, $C_{AmE} = 2\sigma_{AmE/AmE}$, $C_{DmE} = \sigma_{DE/DE}$, $C_e = \sigma_{e/e}$, $C_{A/Am} = 3\sigma_{A/Am}$, $C_{D/Dm} = \sigma_{D/Dm}$, $C_{AE/AmE} = 3\sigma_{AE/AmE}$, $C_{DE/DmE} = \sigma_{DE/DmE}$.

The total phenotypic covariance is $C_P = C_A + C_D + C_C + C_{Am} + C_{Dm} + C_{AE} + C_{DE} + C_{CE} + C_{AmE} + C_{DmE} + 2C_{A/Am} + 2C_{D/DM} + 2C_{AE/AmE} + 2C_{DE/DmE} + C_e$. For trait 1 and trait 2, correlation coefficients of genetic components can be estimated by $r_A = C_A / \sqrt{V_{A(1)} V_{A(2)}}$, $r_D = C_D / \sqrt{V_{D(1)} V_{D(2)}}$, $r_C = C_C / \sqrt{V_{C(1)} V_{C(2)}}$, $r_{Am} = C_{Am} / \sqrt{V_{Am(1)} V_{Am(2)}}$, $r_{Dm} = C_{Dm} / \sqrt{V_{Dm(1)} V_{Dm(2)}}$, $r_{AE} = C_{AE} / \sqrt{V_{AE(1)} V_{AE(2)}}$, $r_{DE} = C_{DE} / \sqrt{V_{DE(1)} V_{DE(2)}}$, $r_{CE} = C_{CE} / \sqrt{V_{CE(1)} V_{CE(2)}}$, $r_{AmE} = C_{AmE} / \sqrt{V_{AmE(1)} V_{AmE(2)}}$, $r_{DmE} = C_{DmE} / \sqrt{V_{DmE(1)} V_{DmE(2)}}$, $r_e = C_e / \sqrt{V_{e(1)} V_{e(2)}}$.

Heritability Components

The total heritability (h^2) can be partitioned into general heritability (h_G^2) and interaction heritability (h_{GE}^2) with their components (Zhu, 1997),

$$h^2 = h_G^2 + h_{GE}^2$$
$$= h_O^2 + h_C^2 + h_M^2 + h_{OE}^2 + h_{CE}^2 + h_{ME}^2$$

where $h_O^2 = (V_A + C_{A.Am})/V_P$ is direct general heritability, $h_C^2 = V_C/V_P$ is cytoplasm general heritability, and $h_M^2 = (V_{Am} + C_{A.Am})/V_P$ is maternal general heritability; $h_{OE}^2 = (V_{AE} + C_{AE.AmE})/V_P$ is direct interaction heritability, $h_{CE}^2 = V_{CE}/V_P$ is cytoplasm interaction heritability, and $h_{ME}^2 = (V_{AmE} + C_{AE.AmE})/V_P$ is maternal interaction heritability.

Selection Response

The total selection response $(R = ih^2\sqrt{V_P})$ can be partitioned into several components (Zhu, 1997):

$$R = R_G + R_{GE}$$
$$= (R_O + R_C + R_M) + (R_{OE} + R_{CE} + R_{ME})$$

where $R_G = ih_G^2\sqrt{V_P}$ is general response, which consists of direct general response $(R_O = ih_O^2\sqrt{V_P})$, cytoplasm general response $(R_C = ih_C^2\sqrt{V_P})$, and maternal general response $(R_M = ih_M^2\sqrt{V_P})$; $R_{GE} = ih_{GE}^2\sqrt{V_P}$ is interaction response, which consists of direct interaction response $(R_{OE} = ih_{OE}^2\sqrt{V_P})$, cytoplasm interaction response $(R_{CE} = ih_{CE}^2\sqrt{V_P})$, and maternal interaction response $(R_{ME} = ih_{ME}^2\sqrt{V_P})$.

Heterosis Components

Prediction of genetic merits can be obtained using the linear unbiased prediction (LUP) method (Zhu, 1992; Zhu and Weir, 1996) or the adjusted unbiased prediction (AUP) method (Zhu, 1993a; Zhu and Weir, 1996). Predicted genotypic effects and *GE* interaction effects can be further used in analyzing heterosis of different generations (Zhu, 1997). Heterosis in specific environments consists of two components. General heterosis is due to genotypic effects and can be expected in overall environments, and interaction heterosis is a deviant of *GE* interaction relative to specific environments. The two components of heterosis relative to midparent or female parent can be calculated as ($x = 1$ for diploid seed and $x = 2$ for triploid endosperm):

General heterosis of F_n relative to midparent:

$$H_M(F_n) = H_{MO} + H_{MC} + H_{MM}$$
$$= \left(\tfrac{1}{2}\right)^{n-x} \Delta_O + \tfrac{1}{2}\varpi_C + \left(\tfrac{1}{2}\right)^{n-2} \Delta_M$$

Interaction heterosis of F_n relative to midparent:

$$H_{ME}(F_n) = H_{MOE} + H_{MCE} + H_{MME}$$
$$+ \left(\tfrac{1}{2}\right)^{n-x} \Delta_{OE} + \tfrac{1}{2}\varpi_{CE} + \left(\tfrac{1}{2}\right)^{n-2} \Delta_{ME}$$

General heterosis of F_n relative to female parent (P_i):

$$H_F(F_n) = H_{FO} + H_{FM}$$
$$= \left[\left(\tfrac{1}{2}\right)^{n-x} \Delta_O - \tfrac{1}{2}\varpi_O \right] + \left[\left(\tfrac{1}{2}\right)^{n-2} \Delta_M - \tfrac{1}{2}\varpi_M \right]$$

Interaction heterosis of F_n relative to female parent (P_i):

$$H_{FE}(F_n) = H_{FOE} + H_{FME}$$
$$+ \left[\left(\tfrac{1}{2}\right)^{n=x} \Delta_{OE} - \tfrac{1}{2}\varpi_{OE} \right] + \left[\left(\tfrac{1}{2}\right)^{n-2} \Delta_{ME} - \tfrac{1}{2}\varpi_{ME} \right]$$

where $\Delta_O = D_{ij} - \tfrac{1}{2}\left(D_{ii} + D_{jj}\right)$, $\Delta_M = Dm_{ij} - \tfrac{1}{2}\left(Dm_{ii} + DM_{jj}\right)$, $\Delta_{OE} = DE_{hij}$ $- \tfrac{1}{2}\left(DE_{hii} + DE_{hjj}\right)$, $\Delta_{ME} = DmE_{hij} - \tfrac{1}{2}\left(DmE_{hii} + DmE_{hjj}\right)$, $\varpi_O = 2\left(A_i - A_j\right)$ $+ \left(D_{ii} - D_{jj}\right)$ for diploid and $\varpi_O = 3\left(A_i - A_j\right) + 3\left(D_{ii} - D_{jj}\right)$ for triploid endosperm, $\varpi_C = C_i - C_j$, $\varpi_M = 2\left(Am_i - Am_j\right) + \left(Dm_{ii} - Dm_{jj}\right)$.

Heterosis based on population mean ($H_{PM} = \tfrac{1}{\mu}H_M$, $H_{PME} = \tfrac{1}{\mu}H_{ME}$, $H_{PF} = \tfrac{1}{\mu}H_F$, or $H_{PFE} = \tfrac{1}{\mu}H_{FE}$) can be used to compare proportion of heterosis among different traits.

Covariances Between Seed Quality Trait and Plant Agronomic Trait

In plant breeding, breeders usually want to improve seed quality traits while keeping the genetic merit of yield traits. Therefore, understanding the genetic relationship between seed quality traits and plant yield traits is of importance. Seed models and plant models have unequal design matrices.

Zhu (1993b) developed a new method for estimating genetic covariance components between seed traits ($\mathbf{y_s}$) **and** plant traits ($\mathbf{y_p}$). For seed model:

$$y_s = Xb_{(S)} + U_A e_{A(S)} + U_D e_{D(S)} + U_C e_{C(S)} + U_{Am} e_{Am(S)} + U_{Dm} e_{Dm(S)}$$
$$+ U_{AE} e_{AE(S)} + U_{DE} e_{DE(S)} + U_{CE} e_{CE(S)} + U_{AmE} e_{AmE(S)} + U_{DmE} e_{DmE(S)}$$
$$+ U_B e_{B(S)} + e_{e(S)}$$
$$= Xb_{(S)} + \sum_u^{12} U_u e_{u(S)}$$

The corresponding plants bearing the seeds will have the following mixed linear model:

$$y_p = Xb_{(P)} + U_C e_{C(P)} + U_{Am} e_{Am(P)} + U_{Dm} e_{Dm(P)}$$
$$+ U_{CE} e_{CE(P)} + U_{AmE} e_{AmE(P)} + U_{DmE} e_{DmE(P)}$$
$$+ U_B e_{B(P)} + e_{e(P)}$$
$$= Xb_{(P)} + \sum_u^{8} U_u e_{u(P)}$$

There are covariances between random factors of seed traits and those of plant traits: $\sigma_{A/Am}$ = covariance between seed direct additive effects and plant additive effects, $\sigma_{D/Dm}$ = covariance between seed direct dominance effects and plant dominance effects, $\sigma_{C/C}$ = covariance between seed cytoplasm effects and plant cytoplasm effects, $\sigma_{Am/Am}$ = covariance between seed maternal additive effects and plant additive effects, $\sigma_{Dm/Dm}$ = covariance between seed maternal dominance effects and plant dominance effects, $\sigma_{AE/AmE}$ = covariance between seed AE effects and plant AmE effects, $\sigma_{DE/DmE}$ = covariance between seed DE effects and plant DmE effects, $\sigma_{CE/CE}$ = covariance between seed CE effects and plant CE effects, $\sigma_{AmE/AmE}$ = covariance between seed AmE effects and plant AmE effects, $\sigma_{DmE/DmE}$ = covariance between seed DmE effects and plant DmE effects, $\sigma_{B/B}$ = covariance between seed block effects and plant block effects, $\sigma_{e/e}$ = covariance between seed residual effects and plant residual effects.

If we define $F_1 = (U_A U_{Am}^T + U_{Am} U_A^T)$, $F_s = (U_D U_{Dm}^T + U_{Dm} U_D^T)$, $F_3 = (2U_C U_C^T)$, $F_4 = (2U_{Am} U_{Am}^T)$, $F_5 = (2U_{Dm} U_{Dm}^T)$, $F_6 = (U_{AE} U_{AmE}^T + U_{AmE} U_{AE}^T)$, $F_7 = (U_{DE} U_{DmE}^T + U_{DmE} U_{DE}^T)$, $F_8 = (2U_{CE} U_{CE}^T)$, $F_9 = (2U_{AmE} U_{AmE}^T)$, $F_{10} = (2U_{DmE} U_{DmE}^T)$, $F_{11} = (2U_B U_B^T)$, and $F_{12} = 2I$, covariance components between a seed trait and a plant trait can then be estimated by the following equations:

$$\left[tr(Q_{(0/1)}F_u Q_{(0/1)}F_v)\right]\left[\hat{\sigma}_{u/u}\right]=\left[2y_s^T Q_{(0/1)}F_u Q_{(0/1)}y_p\right]$$

where

$$Q_{(0/1)}=V_{(0/1)}^{-1}-V_{(0/1)}^{-1}X(X^T V_{(0/1)}^{-1}X)^+ X^T V_{(0/1)}^{-1}$$

$$V_{(0/1)}=2[U_C U_C^T +U_{Am}U_{Am}^T +U_{Dm}U_{Dm}^T +U_{CE}U_{CE}^T +U_{AmE}U_{AmE}^T +U_{DmE}U_{DmE}^T$$
$$+U_B U_B^T +I]$$

Originators

Zhu, J. (1992). Mixed model approaches for estimating genetic variances and covariances. *Journal of Biomathematics* 7(1):1-11.

Zhu, J. (1993a). Methods of predicting genotype value and heterosis for offspring of hybrids (Chinese). *Journal of Biomathematics* 8(1):32-44.

Zhu, J. (1993b). Mixed model approaches for estimating covariances between two traits with unequal design matrices (Chinese). *Journal of Biomathematics* 8(3):24-30.

Zhu, J. (1996). Analysis methods for seed models with genotype × environment interactions (Chinese). *Acta Genetica Sinica* 23(1):56-68.

Zhu, J. (1997). *Analysis Methods for Genetic Models*. Agricultural Publication House of China, Beijing.

Zhu, J. and Weir, B.S. (1994a). Analysis of cytoplasmic and maternal effects. I. A genetic model for diploid plant seeds and animals. *Theoretical and Applied Genetics* 89:153-159.

Zhu, J. and Weir, B.S. (1994b). Analysis of cytoplasmic and maternal effects. II. Genetic models for triploid endosperm. *Theoretical and Applied Genetics* 89:160-166.

Zhu, J. and Weir, B.S. (1996). Diallel analysis for sex-linked and maternal effects. *Theoretical and Applied Genetics* 92(1):1-9.

Software Available

Zhu, J. (1997). GENDIPLD.EXE for constructing seed model, GENVAR0.EXE for estimating components of variance and heritability, GENCOV0.EXE for estimating components of covariance and correlation, GENHET0.EXE for predicting genetic effects and components of heterosis. *Analysis Methods for Genetic Models* (pp. 256-278), Agricultural Publication House of China, Beijing (program free of charge). Contact Dr. Jun Zhu, Department of Agronomy, Zhejiang University, Hangzhou, China. E-mail: <jzhu@zju.edu.cn>.

EXAMPLE

Unbalanced data (COTSEEDM.TXT) to be analyzed (Parent = 5, Year = 2, Generation = P, F_1, F_2, Blk = 1):

Year	Fema	Male	Gene	Blk	Pro%	Oil%
1	1	1	0	1	37.6	37.4
1	1	3	1	1	37.5	36.5
1	1	3	2	1	38.3	36.1
1	1	4	1	1	38.4	34.6
1	1	4	2	1	37.8	35.3
1	2	2	0	1	42.9	32.5
1	2	3	1	1	39.8	33.1
1	2	3	2	1	39.2	35.5
1	3	1	2	1	37.2	35.9
1	3	2	2	1	37.1	37.1
1	3	3	0	1	38	34.8
1	3	5	1	1	40.6	35.4
1	3	5	2	1	39.8	36.1
1	4	1	1	1	37.2	36.8
1	4	1	2	1	37	36.1
1	4	2	1	1	39.2	38.1
1	4	2	2	1	38.1	35.3
1	4	4	0	1	38.9	35.5
1	4	5	1	1	41	38.1
1	4	5	2	1	40.1	35.6
1	5	5	0	1	45.8	34.5
2	1	1	0	1	37.7	36.5
2	1	3	1	1	37.2	36.5
2	1	3	2	1	37.2	35.6
2	1	4	1	1	36	36.5
2	1	4	2	1	35.9	36.2
2	2	2	0	1	40.5	34.8
2	2	3	1	1	37.4	36.9
2	2	3	2	1	37	36.8
2	2	4	1	1	38.3	36.3
2	2	4	2	1	37.2	36.9
2	3	3	0	1	38.6	35.4
2	3	5	1	1	38.3	35.7
2	3	5	2	1	37.8	35.8
2	4	4	0	1	39.7	35.1
2	4	5	1	1	38.9	35.6
2	4	5	2	1	38.6	34.6
2	5	5	0	1	44	31.2

1. Use one of the following two programs for generating a mating design matrix and data:

GENDIPLD.EXE for traits of diploid seeds or animals.

GENTRIPL.EXE for traits of triploid endosperm.

Before running these programs, create a data file (COTSEEDM. TXT) for your analysis with five design columns followed by trait columns. The five design columns should be labeled (1) environment, (2) maternal, (3) paternal, (4) generation, and (5) replication. There is a limitation (<100 traits) for the number of trait columns.

2. Run programs for variance and covariance analyses. Standard errors of estimates are calculated by jackknifing over cell means.

3. You should always run GENVAR0C.EXE for estimating variance components and predicting genetic effects before estimating covariance and correlation. This program will allow you to choose the prediction methods (LUP or AUP). You also need to input coefficients (1, 0, or −1) for conducting linear contrasts for genetic effects.

4. After finishing variance analysis, run GENCOV0C.EXE for estimating covariance components and coefficients of correlation among all the traits analyzed.

5. If you want to predict heterosis and genotypic value for F_2 seed, you can run GENHET0C.EXE.

6. All results will be automatically stored in text files for later use or printing. Examples of result files are provided with the names COTSEEDM.VAR for analysis of variance and genetic effects, COTSEEDM.PRE for predicting genotype values and heterosis, and COTSEEDM.COR for analysis of covariances and correlation.

Output 1 for Single Trait Test

```
Traits =, 2
Variance components = , 15
Degree of freedom = , 37
File name is cotseedm.VAR
Date and Time for Analysis: Fri Jun 23 21:06:32 2000

Variance Components Estimated by MINQUE(0/1) with GENHET0C.EXE.
Predicting Genetic Effects by Adjusted Unbiased Prediction (AUP)
     Method.
Jackknifing Over Block Conducted for Estimating S.E.

NS = Not significant; S+ = Significant at 0.10 level.
S* = Significant at 0.05 level; S** = Significant at 0.01 level.

Linear Contrasting Test:
  + <1> + <2> + <3> − <4> − <5>

Genetic Analysis of 1 Trait, Pro%,  Public Users.
Var Comp            Estimate      S. E.      P-value
Direct Additive     4.22578       0.971094   5.11522e-005    S**
Direct Dominance    0.661729      0.187698   0.000572911     S**
Cytoplasm           0.979784      0.316941   0.00188757      S**
Maternal Additive   0             0          0.5             NS
```

```
Maternal Dominance   2.14103     0.491783    5.08401e-005   S**
D Add. × Env.        4.14108     1.08163     0.00024065     S**
D Dom. × Env.        0.226718    0.0427548   2.75864e-006   S**
Cyto × Env.          3.08077     0.725817    7.05334e-005   S**
M Add. × Env.        2.1         0.520779    0.000132504    S**
M Dom. × Env.        0           0           0.5            NS
A.Am                 0           0.271806    1              NS
D.Dm                 0.684771    0.400178    0.0954231      S+
AE.AmE              -2.45518     1.3901      1.91438        NS
DE.DmE               0           0           1              NS
Residual             1.56207     0.341266    2.58189e-005   S**
Var(Phenotype)       15.5781     2.49795     1.50314e-007   S**
```

```
Heritability                          Estimate    S. E.      P-value
General Heritability N(A)             0.271263    0.0372287  5.88087e-009  S**
General Heritability B(A+D)           0.357699    0.0380067  -1.37905e-011 S**
General Heritability N(C)             0.0628948   0.0226378  0.00426618    S**
General Heritability N(Am)            0           0          0.5           NS
General Heritability B(Am+Dm)         0.181395    0.0240944  2.80833e-009  S**
Interaction Heritability N(AE)        0.108222    0.0325895  0.00101347    S**
Interaction Heritability B(AE+DE)     0.28038     0.0788095  0.000523146   S**
Interaction Heritability N(CE)        0.197762    0.0388026  5.23056e-006  S**
Interaction Heritability N(AmE)      -0.0227996   0.0252744  0.186424      NS
Interaction Heritability B(AmE+DmE)   0.134804    0.0530296  0.00766977    S**
```

```
Genetic Predictor,  S. E. , P-value of Two Tail t-test
<1>: Random Effect is Direct Additive
A1               -1.024558   0.602650   0.0975     S+
A2                0.849192   0.492454   0.093      S+
A3               -0.959033   0.579315   0.106      NS
A4                0.023544   0.447671   0.958      NS
A5                1.110611   0.833896   0.191      NS
Linear Contrast  -1.74484    1.15265    0.138582   NS

<2>: Random Effect is Direct Dominance
D1*1              0.430174   0.474310   0.37       NS
D2*2              2.060002   1.236368   0.104      NS
D3*3              0.896631   0.591905   0.138      NS
D4*4              1.637057   0.873815   0.0689     S+
D5*5              1.830764   0.882766   0.0451     S*
D1*3             -0.478229   0.456267   0.301      NS
D1*4             -2.177030   1.321475   0.108      NS
D2*3             -2.215217   1.325812   0.103      NS
D2*4             -0.755464   0.577872   0.199      NS
D3*5             -0.781604   0.580198   0.186      NS
D4*5             -0.447118   0.831929   0.594      NS
Heterosis <Delta> -2.30767   1.86916    0.225      NS

<3>: Random Effect is Cytoplasm
C1               -0.278965   0.268528   0.306      NS
C2                0.125445   0.666036   0.852      NS
C3               -0.434191   1.217060   0.723      NS
C4                0.022447   0.667609   0.973      NS
C5                0.565182   1.479941   0.705      NS
Linear Contrast  -1.32754    4.23754    0.755827   NS

<4>: Random Effect is Maternal Additive
```

No Significant Effects.


```
<5>: Random Effect is Maternal Dominance
Dm1*1                  0.328107    0.347168  0.351       NS
Dm2*2                  1.334076    0.415525  0.00274     S**
Dm3*3                  1.422192    0.519389  0.00944     S**
Dm4*4                  1.798117    0.562728  0.00285     S**
Dm5*5                  1.842966    0.707409  0.0131      S*
Dm1*3                  0.355737    0.403627  0.384       NS
Dm1*4                 -0.581557    0.660676  0.384       NS
Dm2*3                 -1.434893    0.704201  0.0488      S*
Dm2*4                 -1.937210    0.780548  0.0177      S*
Dm3*5                 -1.322985    0.768254  0.0934      S+
Dm4*5                 -1.804685    0.762519  0.0233      S*
Heterosis <Delta>  -2.05554        0.153624 -5.09e-011   S**

<6>: Random Effect is D Add. × Env.
AE1 in E1             -2.254065    0.889523  0.0156      S*
AE2 in E1              0.550176    0.647403  0.401       NS
AE3 in E1             -1.761268    0.848635  0.045       S*
AE4 in E1              0.449786    0.592606  0.453       NS
AE5 in E1              3.015095    1.252800  0.0212      S*
AE1 in E2              0.632488    0.575894  0.279       NS
AE2 in E2              0.874021    0.831909  0.3         NS
AE3 in E2              0.005609    0.629156  0.993       NS
AE4 in E2             -0.596792    0.806521  0.464       NS
AE5 in E2             -0.915384    0.884746  0.308       NS
Linear Contrast      -5.3845       1.32097   0.000233    S**

<7>: Random Effect is D Dom. × Env.
DE1*1 in E1          -0.159850     0.800087  0.843       NS
DE2*2 in E1           0.407181     1.084693  0.71        NS
DE3*3 in E1          -0.132742     0.696827  0.85        NS
DE4*4 in E1           0.001560     0.707096  0.998       NS
DE5*5 in E1           0.110871     0.599268  0.854       NS
DE1*3 in E1          -0.214554     0.706944  0.763       NS
DE1*4 in E1          -0.558690     0.675476  0.413       NS
DE2*3 in E1          -0.403386     1.054141  0.704       NS
DE2*4 in E1           0.075915     0.482125  0.876       NS
DE3*5 in E1           0.105371     1.002477  0.917       NS
DE4*5 in E1           0.768304     1.301119  0.558       NS
DE1*1 in E2           0.508888     1.015302  0.619       NS
DE2*2 in E2           0.623124     1.464434  0.673       NS
DE3*3 in E2           0.853831     1.248241  0.498       NS
DE4*4 in E2           0.310951     1.774209  0.862       NS
DE5*5 in E2           0.149658     1.260388  0.906       NS
DE1*3 in E2          -0.242427     0.585483  0.681       NS
DE1*4 in E2          -0.388505     2.125310  0.856       NS
DE2*3 in E2          -0.489308     1.570559  0.757       NS
DE2*4 in E2          -0.455681     0.679028  0.506       NS
DE3*5 in E2          -0.711946     1.394222  0.613       NS
DE4*5 in E2          -0.158580     1.360425  0.908       NS
Heterosis <Delta>    1.00932       2.70601   0.711       NS

<8>: Random Effect is Cyto × Env.
```

```
CE1 in E1          0.776034   0.458734  0.0991    S+
CE2 in E1          2.165781   1.241707  0.0894    S+
CE3 in E1         -1.300904   1.205518  0.288     NS
CE4 in E1         -1.491405   0.967790  0.132     NS
CE5 in E1         -0.149753   1.093888  0.892     NS
CE1 in E2         -1.444760   0.686140  0.0421    S*
CE2 in E2         -1.573186   0.812138  0.0604    S+
CE3 in E2         -0.538525   0.951240  0.575     NS
CE4 in E2          0.589799   1.194278  0.624     NS
CE5 in E2          2.966626   1.385019  0.0388    S*
Linear Contrast    2.09064    1.35549   0.131499  NS


<9>   : Random Effect is M Add. × Env.
AmE1 in E1         0.164594   0.421542  0.698     NS
AmE2 in E1        -0.132997   0.439265  0.764     NS
AmE3 in E1         1.211076   0.516050  0.0244    S*
AmE4 in E1        -2.478739   1.117579  0.0328    S*
AmE5 in E1         1.235679   0.732488  0.1       NS
AmE1 in E2        -1.227880   0.750977  0.111     NS
AmE2 in E2        -1.586811   0.633135  0.0167    S*
AmE3 in E2        -0.309359   0.516714  0.553     NS
AmE4 in E2        -0.003356   0.573114  0.995     NS
AmE5 in E2         3.127315   1.394990  0.0311    S*
Linear Contrast    2.71224    3.94209   0.495730  NS


<10>: Random Effect is M Dom. × Env.
No Significant Effects.

Results of Oil% are not presented.

Time Used (Hour) = 0.001389
```

Output 2 for Covariance Analysis

```
Traits =, 2
Variance components = , 15
Degree of freedom = , 37
File name is cotseedm.COV
Date and Time for Analysis: Fri Jun 23 21:06:49 2000

Variance Components Estimated by MINQUE(0/1) with GENHETOC.EXE.
Jackknifing Over Block Conducted for Estimating S.E.
For statistical methods, see the following references:

NS = Not significant; S+ = Significant at 0.10 level.
S* = Significant at 0.05 level; S** = Significant at 0.01 level.
```

Covariances and Correlations Between, Pro% &, Oil% for, Public Users.:

Covariances	Estimates	S.E.	P-value	
Direct Additive Cov	-0.12332	0.785577	0.876	NS
Direct Dominance Cov	-0.18046	0.223768	0.425	NS
Cytoplasm Cov	-0.0560615	0.946456	0.953	NS
Maternal Additive Cov	-0.247106	0.497206	0.622	NS
Maternal Dominance Cov	-0.76196	0.363677	0.0431	S*
D Add. × Env. Cov	0.238753	1.25031	0.85	NS

```
D Dom. × Env. Cov       0.0696955     0.201047      0.731     NS
Cyto × Env. Cov        -0.748348      0.966985      0.444     NS
M Add. × Env. Cov      -0.567987      0.712327      0.43      NS
M Dom. × Env. Cov       0.139297      0.400729      0.73      NS
A.Am Cov                0.0209709     0.592716      0.972     NS
D.Dm Cov               -0.317895      0.181318      0.0878    S+
AE.AmE Cov              0.538553      0.864524      0.537     NS
DE.DmE Cov             -0.104535      0.150625      0.492     NS
Residual Cov           -0.125117      0.335582      0.711     NS


Cov <1=Genotypic>
Cov <2=Phenotypic>      Estimates      S.E.         P-value
Cov 2                  -2.08843       1.00453       0.0446    S*
Cov 1                  -1.96331       1.0037        0.058     S+

Correlation             Estimates      S.E.         P-value
Direct Additive Cor     0.000000      0             1         NS
Direct Dominance Cor   -0.313393      0.0713042     8.97e-005 S**
Cytoplasm  Cor          0.000000      0             1         NS
Maternal Additive Cor   0.000000      0             1         NS
Maternal Dominance Cor -0.527223      0.0751039     2.66e-008 S**
D Add. × Env. Cor       0.064792      0.0855313     0.454     NS
D Dom. × Env. Cor       0.225062      0.0669711     0.00182   S**
Cyto × Env. Cor        -0.302136      0.0755769     0.000294  S**
M Add. × Env. Cor      -0.271364      0.0710278     0.000493  S**
M Dom. × Env. Cor       0.000000      0             1         NS
Residual Cor           -0.379882      0.0634317     6.5e-007  S**


Cor <1=Genotypic>
<2=Phenotypic>          Estimates      S.E.         P-value
Cor 2                  -0.157513      0.0580681     0.0101    S*
Cor 1                  -0.154315      0.0587436     0.0125    S*

Time Used (Hour) = 0.000556
```

Output 3 for Heterosis Analysis

```
Traits =, 2
Variance components = , 15
Degree of freedom = , 37
File name is cotseedm.PRE
Date and Time for Analysis: Fri Jun 23 21:07:07 2000

Variance Components Estimated by MINQUE(0/1) with GENHET0C.EXE.
Predicting Genetic Effects by Adjusted Unbiased Prediction (AUP)
Method.
Jackknifing Over Block Conducted for Estimating S.E.

NS = Not significant; S+ = Significant at 0.10 level.
S* = Significant at 0.05 level; S** = Significant at 0.01 level.

Genetic Analysis of 1 Trait, Pro%, for Public Users.
```

```
Var Comp                    Estimate      S. E.        P-value
Direct Additive             4.22543       0.971079     5.12e-005    S**
Direct Dominance            0.661737      0.187698     0.000573     S**
Cytoplasm                   0.979678      0.316932     0.00189      S**
Maternal Additive           0             0            0.5          NS
Maternal Dominance          2.141         0.491782     5.08e-005    S**
D Add. × Env.               4.14101       1.08162      0.000241     S**
D Dom. × Env.               0.226713      0.042754     2.76e-006    S**
Cyto × Env.                 3.08084       0.725821     7.05e-005    S**
M Add. × Env.               2.10002       0.520781     0.000132     S**
M Dom. × Env.               0             0            0.5          NS
A.Am                        0             0            1            NS
D.Dm                        0.684787      0.400178     0.0954       S+
AE.AmE                      -2.45513      1.3901       0.0856       S+
DE.DmE                      0             0            1            NS
Residual                    1.5621        0.341269     2.58e-005    S**
Var(Phenotype)              15.5779       2.49795      1.5e-007     S**


Genetic Advance(for 0.05)         Estimate   S. E.       P-value
General Genetic Advance(A)         5.69401    0.844563    3.13324e-008   S**
General Genetic Advance(C)         1.32017    0.382756    0.000709897    S**
General Genetic Advance(Am)        0          0           0.5            NS
Interaction Genetic Advance(AE)    2.27182    0.576013    0.00017183     S**
Interaction Genetic Advance(CE)    4.15161    0.768543    2.0263e-006    S**
Interaction Genetic Advance(AmE)   0.478526   0.466339    0.155745       NS
```

Heterosis Analysis of Trait, Pro%, for F2 Seeds with total mean =,
38.731644

```
No.     Cro      G(T) S.E.  Pv   Sig  G(O)  S.E. Pv   Sig  G(C)  S.E. Pv   Sig  G(M)  S.E. Pv   Sig
Cro<1>  <1*3> -1.81 0.71  0.01 S *  -1.89 0.69 0.01 S ** -0.28 0.19 0.15 NS  0.36  0.34 0.30 NS
Cro<2>  <1*4> -2.43 0.67  0.00 S ** -1.57 0.59 0.01 S *  -0.28 0.19 0.15 NS  -0.58 0.40 0.16 NS
Cro<3>  <2*3> -1.79 0.75  0.02 S *  -0.48 0.37 0.21 NS   0.13  0.40 0.76 NS  -1.43 0.66 0.04 S *
Cro<4>  <2*4> -0.39 0.99  0.69 NS   1.42  0.57 0.02 S *  0.13  0.40 0.76 NS  -1.94 0.76 0.02 S *
Cro<5>  <3*5> -1.31 1.39  0.35 NS   0.44  0.32 0.18 NS  -0.43 0.89 0.63 NS  -1.32 0.76 0.09 S +
Cro<6>  <4*5>  0.00 1.04  1.00 NS   1.78  0.58 0.00 S ** 0.02  0.40 0.96 NS  -1.80 0.74 0.02 S *

No.     Cro     Hm(T) S.E. Pv   Sig  Hm(O) S.E. Pv   Sig  Hm(C) S.E. Pv   Sig  Hm(M) S.E. Pv   Sig
Cro<1>  <1*3>-0.03 0.02  0.22 NS   -0.01 0.01 0.07 S +  0.00  0.01 0.83 NS  -0.01 0.01 0.33 NS
Cro<2>  <1*4>-0.09 0.02  0.00 S ** -0.04 0.02 0.04 S *  0.00  0.00 0.24 NS  -0.04 0.02 0.02 S *
Cro<3>  <2*3>-0.11 0.04  0.01 S *  -0.05 0.02 0.06 S +  0.01  0.02 0.66 NS  -0.07 0.02 0.00 S **
Cro<4>  <2*4>-0.12 0.04  0.00 S ** -0.03 0.02 0.05 S *  0.00  0.01 0.89 NS  -0.09 0.03 0.00 S **
Cro<5>  <3*5>-0.12 0.05  0.04 S *  -0.03 0.01 0.03 S *  -0.01 0.03 0.62 NS  -0.08 0.03 0.02 S *
Cro<6>  <4*5>-0.13 0.05  0.01 S ** -0.03 0.01 0.00 S ** -0.01 0.02 0.73 NS  -0.09 0.03 0.00 S **

No.     Cro     Hf(T) S.E. Pv   Sig  Hf(O) S.E. Pv   Sig  Hf(M) S.E. Pv   Sig  Gen.  S.E. Pv   Sig
Cro<1>  <1*3>-0.01 0.02  0.75 NS   -0.01 0.01 0.60 NS   0.00  0.01 0.96 NS  0.00  0.00 2.00 NS
Cro<2>  <1*4>-0.02 0.02  0.33 NS    0.00 0.02 0.95 NS  -0.02 0.02 0.15 NS  0.00  0.00 2.00 NS
Cro<3>  <2*3>-0.18 0.06  0.01 S ** -0.11 0.05 0.03 S *  -0.07 0.02 0.00 S ** 0.00  0.00 2.00 NS
Cro<4>  <2*4>-0.14 0.05  0.01 S *  -0.06 0.04 0.11 NS  -0.08 0.03 0.00 S ** 0.00  0.00 2.00 NS
Cro<5>  <3*5>-0.03 0.03  0.33 NS    0.04 0.03 0.15 NS  -0.07 0.03 0.02 S *  0.00  0.00 2.00 NS
Cro<6>  <4*5>-0.09 0.03  0.00 S ** 0.00  0.02 0.90 NS  -0.09 0.03 0.00 S ** 0.00  0.00 2.00 NS
```

Interaction Heterosis Analysis of Trait, Pro%, for F2 Seeds with total
mean =, 38.731644

```
No.     Cro     GE(T) S.E. Pv   Sig  GE(O) S.E. Pv   Sig  GE(C) S.E. Pv   Sig  GE(M) S.E. Pv   Sig
Env.<1> <1*3>-2.04 1.06  0.06 S +  -4.20 1.25 0.00 S ** 0.78  0.32 0.02 S *  1.38  0.50 0.01 S **
```

Env. <1> <1 * 4>-3.66 1.30 0.01 S** -2.12 0.77 0.01 S** 0.78 0.43 0.08 S+ -2.31 1.12 0.05 S*
Env. <1> <2 * 3>1.90 1.69 0.27 NS -1.34 0.87 0.13 NS 2.17 1.82 0.24 NS 1.08 0.98 0.28 NS
Env. <1> <2 * 4>0.70 1.99 0.73 NS 1.14 1.11 0.31 NS 2.17 1.82 0.24 NS -2.61 1.03 0.02 S*
Env. <1> <3 * 5>2.45 1.57 0.13 NS 1.30 0.95 0.18 NS -1.30 1.23 0.30 NS 2.45 0.91 0.01 S*
Env. <1> <4 * 5>1.14 1.06 0.29 NS 3.88 1.45 0.01 S* -1.49 0.69 0.04 S* -1.24 0.99 0.22 NS
Env. <2> <1 * 3>-2.12 1.11 0.06 S+ 0.86 1.17 0.47 NS -1.44 0.65 0.03 S* -1.54 0.85 0.08 S+
Env. <2> <1 * 4>-2.63 1.41 0.07 S+ 0.05 0.83 0.96 NS -1.44 0.65 0.03 S* -1.23 0.61 0.05 S+
Env. <2> <2 * 3>-2.47 1.47 0.10 S+ 1.00 0.70 0.16 NS -1.57 1.03 0.14 NS -1.90 0.84 0.03 S*
Env. <2> <2 * 4>-2.88 1.55 0.07 S+ 0.28 0.73 0.70 NS -1.57 1.03 0.14 NS -1.59 0.84 0.07 S+
Env. <2> <3 * 5>1.27 1.24 0.31 NS -1.01 0.69 0.15 NS -0.54 0.39 0.17 NS 2.82 1.24 0.03 S*
Env. <2> <4 * 5>2.24 1.67 0.19 NS -1.48 0.83 0.08 S+ 0.59 0.80 0.46 NS 3.12 1.48 0.04 S*

No.	Cro HmE(T)	S.E.	Pv	Sig	HmE(O)	S.E.	Pv	Sig	HmE(C)	S.E.	Pv	Sig	HmE (M)	S.E.	Pv	Sig
Env.<1> <1 * 3>0.03	0.01	0.02	S*	0.00	0.00	0.37	NS	0.03	0.01	0.01	S*	0.00	0.00	2.00	NS
Env. <1> <1 * 4>0.02	0.04	0.59	NS	-0.01	0.03	0.81	NS	0.03	0.01	0.02	S*	0.00	0.00	2.00	NS
Env. <1> <2 * 3>0.04	0.05	0.49	NS	-0.01	0.01	0.53	NS	0.04	0.04	0.28	NS	0.00	0.00	2.00	NS
Env. <1> <2 * 4>0.05	0.03	0.18	NS	0.00	0.02	0.94	NS	0.05	0.03	0.13	NS	0.00	0.00	2.00	NS
Env. <1> <3 * 5>-0.01	0.02	0.54	NS	0.00	0.01	0.85	NS	-0.01	0.02	0.38	NS	0.00	0.00	2.00	NS
Env. <1> <4 * 5>-0.01	0.03	0.79	NS	0.01	0.02	0.62	NS	-0.02	0.01	0.11	NS	0.00	0.00	2.00	NS
Env. <2> <1 * 3>-0.02	0.01	0.10	S+	-0.01	0.01	0.06	S+	-0.01	0.01	0.28	NS	0.00	0.00	2.00	NS
Env. <2> <1 * 4>-0.04	0.05	0.48	NS	-0.01	0.04	0.78	NS	-0.03	0.02	0.14	NS	0.00	0.00	2.00	NS
Env. <2> < * 3> -0.03	0.05	0.53	NS	-0.02	0.04	0.71	NS	-0.01	0.02	0.41	NS	0.00	0.00	2.00	NS
Env. <2> <2 * 4>-0.04	0.05	0.47	NS	-0.01	0.03	0.71	NS	-0.03	0.02	0.24	NS	0.00	0.00	2.00	NS
Env. <2> <3 * 5>-0.06	0.06	0.29	NS	-0.02	0.04	0.67	NS	-0.05	0.01	0.00	S**	0.00	0.00	2.00	NS
Env. <2> <4 * 5>-0.04	0.05	0.45	NS	-0.01	0.03	0.84	NS	-0.03	0.01	0.00	S**	0.00	0.00	2.00	NS

No.	Cro HfE(T)	S.E.	Pv	Sig	HfE(O)	S.E.	Pv	Sig	HfE(C)	S.E.	Pv	Sig	HfE (M)	S.E.	Pv	Sig
Env. <1> <1 * 3>0.04	0.02	0.07	S+	0.01	0.02	0.47	NS	0.03	0.01	0.05	S+	0.00	0.00	2.00	NS
Env. <1> <1 * 4>0.00	0.08	0.97	NS	0.07	0.09	0.47	NS	-0.07	0.04	0.09	S+	0.00	0.00	2.00	NS
Env. <1> <2 * 3>-0.04	0.07	0.58	NS	-0.07	0.04	0.06	S+	0.03	0.05	0.50	NS	0.00	0.00	2.00	NS
Env. <1> <2 * 4>-0.07	0.05	0.21	NS	-0.01	0.05	0.79	NS	-0.06	0.05	0.23	NS	0.00	0.00	2.00	NS
Env. <1> <3 * 5>0.13	0.04	0.00	S**	0.13	0.04	0.00	S**	0.00	0.02	0.98	NS	-4.88	0.27	0.00	S**
Env. <1> <4 * 5>0.17	0.04	0.00	S**	0.08	0.03	0.01	S**	0.10	0.05	0.09	S+	-2.54	0.25	0.00	S**
Env. <2> <1 * 3>0.00	0.04	1.00	NS	-0.02	0.03	0.42	NS	0.02	0.02	0.23	NS	0.00	0.00	2.00	NS
Env. <2> <1 * 4>-0.01	0.04	0.77	NS	-0.04	0.05	0.41	NS	0.03	0.03	0.24	NS	0.00	0.00	2.00	NS
Env. <2> <2 * 3>0.00	0.05	0.97	NS	-0.04	0.06	0.55	NS	0.03	0.02	0.14	NS	0.00	0.00	2.00	NS
Env. <2> <2 * 4>-0.01	0.05	0.80	NS	-0.05	0.06	0.36	NS	0.04	0.02	0.11	NS	0.00	0.00	2.00	NS
Env. <2> <3 * 5>0.04	0.05	0.44	NS	-0.05	0.04	0.24	NS	0.09	0.04	0.04	S*	0.00	0.00	2.00	NS
Env. <2> <4 * 5>0.07	0.05	0.19	NS	-0.02	0.02	0.53	NS	0.08	0.04	0.04	NS	0.00	0.00	2.00	NS

Results of Oil% are not presented.

Time Used (Hour) = 0.000278

Chapter 3

Diallel Analysis for an Additive-Dominance Model with Genotype-by-Environment Interaction Effects

Jun Zhu

Purpose

To analyze balanced or unbalanced data of an additive x dominance (AD) genetic model for estimating components of variance, covariance, heritability, and selection response.

Definitions

Mating Design

A set of inbred lines are sampled from a reference population. These parents are used to produce F_1 crosses. If it is difficult to use F_1 crosses for some crops, F_2 crosses can be used as an alternative. Experiments with parents and F_1s (or F_2s) are conducted in multiple environments using a randomized complete block design.

Genetic Model

The genetic model for a genetic entry derived from parents i and j in the kth block within the hth environment is

$$y_{hijk} = \mu + E_h + G_{ij} + GE_{hij} + B_{hk} + e_{hijk}$$

where μ = population mean, E_h = environment effect, G_{ij} = total genotypic effect, GE_{hij} = genotype × environment interaction effect, B_{hk} = block effect, and e_{hijk} = residual effect.

For parent (P_i):

$$G_{ii} + GE_{hii} = 2A_i + D_{ii} + 2AE_{hi} + DE_{hii}$$

For F_1 $(P_i \times P_j)$:

$$G_{ij} + GE_{hij} = A_i + A_j + D_{ij} + AE_{hi} + AE_{hj} + DE_{hij}$$

For F_2 $(F_1 \otimes)$:

$$G_{ij} + GE_{hij} = A_i + A_j + \tfrac{1}{4}D_{ij} + \tfrac{1}{4}D_{ii} + \tfrac{1}{2}D_{jj}$$
$$+ AE_{hi} + AE_{hj} + \tfrac{1}{4}DE_{hii} + \tfrac{1}{4}DE_{hjj} + \tfrac{1}{2}DE_{hij}$$

where A = additive effect, D = dominance effect, AE = additive by environment interaction effect, DE = dominance by environment interaction effect.

Analysis Methodology

Mixed Linear Model

The phenotypic mean of the genetic model can be expressed by a mixed linear model as

$$y = Xb + U_A e_A + U_D e_D + U_{AE} e_{AE} + U_{DE} e_{DE} + U_B e_B + e_e$$
$$= Xb + \sum_{u}^{6} U_u e_u$$

with variance-covariance matrix

$$\mathrm{var}(y) = \sigma_A^2 U_A U_A^T + \sigma_D^2 U_D U_D^T + \sigma_{AE}^2 U_{AE} U_{AE}^T + \sigma_{DE}^2 U_{DE} U_{DE}^T$$
$$+ \sigma_B^2 U_B U_B^T + \sigma_e^2 I$$
$$= \sum_{u=1}^{6} \sigma_u^2 U_u U_u^T = \sum_{u=1}^{6} \sigma_u^2 V_u.$$

Variance Components

Unbiased estimation of variances can be obtained by the following MINQUE(1) equations (Zhu, 1992; Zhu and Weir, 1996):

$$\left[tr(Q_{(1)}V_u Q_{(1)}V_v)\right]\left[\hat{\sigma}_u^2\right]=\left[y^T Q_{(1)}V_u Q_{(1)}y\right]$$

where

$$Q_{(1)} =V_{(1)}^{-1} -V_{(1)}^{-1}X(X^T V_{(1)}^{-1}X)^+ X^T V_{(1)}^{-1}$$
$$V_{(1)} =\sum_u V_u =\sum_u U_u U_u^T$$

When experimental variances (σ_u^2) are estimated, genetic variance components can be obtained by $V_A =2\sigma_A^2$, $V_D =\sigma_D^2$, $V_{AE} =2\sigma_{AE}^2$, $V_{DE} =\sigma_{DE}^2$, and $V_e =\sigma_e^2$. The total phenotypic variance is $V_P =V_A +V_D +V_{AE} +V_{DE} +V_e$.

Covariance Components and Correlation

Unbiased estimation of covariances $\sigma_{u/u}$ between two traits $(y_1$ and $y_2)$ can be obtained by MINQUE(1) approaches (Zhu, 1992; Zhu and Weir, 1996):

$$\left[tr(Q_{(1)}V_u Q_{(1)}V_v)\right]\left[\hat{\sigma}_{u/u}\right]=\left[y_1^T Q_{(1)}V_u Q_{(1)}y_2\right]$$

When experimental covariances $(\sigma_{u/u})$ are estimated, genetic covariance components can be obtained by $C_A =2\sigma_{A/A}$, $C_D =\sigma_{D/D}$, $C_{AE} =2\sigma_{AE/AE}$, $C_{DE} =\sigma_{DE/DE}$, and $C_e =\sigma_{e/e}$. The total phenotypic covariance is $C_P =C_A +C_D +C_{AE} +C_{DE} +C_e$. For trait 1 and trait 2, correlation coefficient of genetic components can be estimated by $r_A =C_A /\sqrt{V_{A(1)}V_{A(2)}}$, $r_D =C_D /\sqrt{V_{D(1)}V_{D(2)}}$, $r_{AE} =C_{AE} /\sqrt{V_{AE(1)}V_{AE(2)}}$, $r_{DE} =C_{DE} /\sqrt{V_{DE(1)}V_{DE(2)}}$, and $r_e =C_e /\sqrt{V_{e(1)}V_{e(2)}}$.

Heritability Components

For the genetic model with *GE* interaction effects, the total heritability (h^2) can be partitioned into two components $(h^2 =h_G^2 +h_{GE}^2)$, where $h_G^2 =V_A /V_P$ is general heritability and $h_{GE}^2 =V_{AE}V_P$ is interaction heritability

(Zhu, 1997). General heritability is applicable to multiple environments whereas interaction heritability is applicable only to specific environments.

Selection Response

The total selection response $(R = ih^2 \sqrt{V_P})$ can be partitioned into two components (Zhu, 1997):

$$R = R_G + R_{GE}$$

where $R_G = ih_G^2 \sqrt{V_P}$ is general response and $R_{GE} = ih_{GE}^2 \sqrt{V_P}$ is interaction response.

Heterosis Components

Prediction of genetic merits can be obtained by using the linear unbiased prediction (LUP) method (Zhu, 1992; Zhu and Weir, 1996) or adjusted unbiased prediction (AUP) method (Zhu, 1993; Zhu and Weir, 1996). Predicted genotypic effects and *GE* interaction effects can be further used in analyzing heterosis of different generations (Zhu, 1997). Heterosis in specific environments consists of two components. General heterosis is due to genotypic effects and can be expected in overall environments, and interaction heterosis is a deviant of *GE* interaction relative to specific environments. The two components of heterosis based on midparent or better parent can be calculated as

General heterosis of F_n relative to midparent: $H_M(F_n) = (\frac{1}{2})^{n-1} \Delta_D$
Interaction heterosis of F_n relative to midparent: $H_{ME}(F_n) = (\frac{1}{2})^{n-1} \Delta_{DE}$
General heterosis of F_n relative to better parent (P_i):
$$H_B(F_n) = (\tfrac{1}{2})^{n-1} \Delta_D - \tfrac{1}{2} \varpi_G$$
Interaction heterosis of F_n relative to better parent (P_i):
$$H_{BE}(F_n) = (\tfrac{1}{2})^{n-1} \Delta_{DE} - \tfrac{1}{2} \varpi_{GE}$$

where $\Delta_D = D_{ij} - \frac{1}{2}(D_{ii} + D_{jj})$ is dominance heterosis, $\Delta_{DE} = DE_{hij} - \frac{1}{2}(DE_{hii} + DE_{hjj})$ is *DE* interaction heterosis, $\varpi_G = |G(P_i) - G(P_j)|$ is parental genotypic difference, and $\varpi_{GE} = |GE(P_i) - GE(P_j)|$ is parental interaction difference.

Heterosis based on population mean ($H_{PM} = \frac{1}{\mu}H_M$, $H_{PME} = \frac{1}{\mu}H_{ME}$, $H_{PB} = \frac{1}{\mu}H_B$, or $H_{PBE} = \frac{1}{\mu}H_{BE}$) can be used to compare proportion of heterosis among different traits.

Originators

Zhu, J. (1992). Mixed model approaches for estimating genetic variances and covariances. *Journal of Biomathematics* 7(1):1-11.

Zhu, J. (1993). Methods of predicting genotype value and heterosis for offspring of hybrids (Chinese). *Journal of Biomathematics* 8(1):32-44.

Zhu, J. (1997). *Analysis Methods for Genetic Models.* Agricultural Publication House of China, Beijing.

Zhu, J. and Weir, B.S. (1996). Diallel analysis for sex-linked and maternal effects. *Theoretical and Applied Genetics* 92(1):1-9.

Software Available

Zhu, J. (1997). GENAD.EXE for constructing AD model, GENVAR1.EXE for estimating components of variance and heritability, GENCOV1.EXE for estimating components of covariance and correlation, GENHET1.EXE for predicting genetic effects and components of heterosis. *Analysis Methods for Genetic Models* (pp. 278-285), Agricultural Publication House of China, Beijing (program free of charge). Contact: Dr. Jun Zhu, Department of Agronomy, Zhejiang University, Hangzhou, China. E-mail: <jzhu @zju. edu.cn>.

EXAMPLE

Unbalanced data (COTDATA.TXT) to be analyzed (Parent = 4, Year = 2, Blk = 2):

Env	Fem	Male	Cross	Blk	Bolls	Fiber Yield
1	1	1	0	1	14.5	54.4
1	1	1	0	2	11.2	29.7
1	1	2	1	1	10.9	54.3
1	1	2	1	2	12.4	55.1
1	1	3	1	1	12.7	43.7
1	1	3	1	2	10.4	51.2
1	1	4	1	1	15.5	58.3
1	1	4	1	2	14.3	38.5
1	2	1	1	1	15.9	62.5
1	2	1	1	2	14.0	56.8
1	2	2	0	1	14.0	34.8
1	2	2	0	2	14.9	35.0
1	2	3	1	1	12.7	34.3

1	2	3	1	2	10.0	24.1
1	2	4	1	1	14.7	34.9
1	2	4	1	2	18.2	34.1
1	3	1	1	1	10.0	26.9
1	3	1	1	2	11.4	28.1
1	3	2	1	1	13.9	23.9
1	3	2	1	2	11.1	33.5
1	3	3	0	1	6.3	12.5
1	3	3	0	2	9.1	22.3
1	3	4	1	1	11.4	19.8
1	3	4	1	2	11.0	21.4
1	4	1	1	1	13.3	43.8
1	4	1	1	2	12.0	42.0
1	4	2	1	1	15.9	31.5
1	4	2	1	2	16.7	40.2
1	4	3	1	1	13.6	39.9
1	4	3	1	2	14.9	19.6
1	4	4	0	1	10.0	28.5
1	4	4	0	2	15.0	28.1
2	1	1	0	1	19.4	55.1
2	1	1	0	2	24.1	56.3
2	1	2	1	1	21.7	69.2
2	1	2	1	2	25.1	79.5
2	1	3	1	1	15.1	76.8
2	1	3	1	2	16.6	42.7
2	1	4	1	1	22.9	72.7
2	1	4	1	2	19.2	62.7
2	2	2	0	1	17.2	60.6
2	2	2	0	2	19.6	71.6
2	2	3	1	1	18.9	36.8
2	2	3	1	2	17.2	47.8
2	2	4	1	1	32.8	61.9
2	2	4	1	2	30.7	78.3
2	3	3	0	1	13.6	27.8
2	3	3	0	2	8.4	19.1
2	3	4	1	1	16.8	37.9
2	3	4	1	2	17.0	34.2
2	4	4	0	1	21.5	49.5
2	4	4	0	2	19.9	57.1

1. Run GENAD.EXE to create mating design matrix files and data files for additive-dominance (AD) model. Before running this program, you should create a file for your analysis with five design columns followed by trait columns. The first five columns are: (1) environment, (2) maternal, (3) paternal, (4) generation, and (5) replication. There is a limitation (<100 traits) for the number of trait columns. An example of a data file is provided with the name COTDATA.TXT.

2. Run programs for variance and covariance analyses. Standard errors of estimates are calculated using jackknife procedures. If you have multiple blocks for your experiments, you can use GENVAR1R.EXE or GENCOV1R.EXE for jackknifing over blocks. Otherwise you can use

GENVAR1C.EXE or GENCOV1C.EXE for jackknifing over cell means.
3. Run GENVAR1R.EXE or GENVAR1C.EXE for estimating variance components and predicting genetic effects before estimating covariance and correlation. These two programs will allow you to choose the parental type (inbred or outbred) and the prediction methods (LUP or AUP). You also need to input coefficients (1, 0, or −1) for conducting linear contrasts for genetic effects of parents.
4. After you finish variance analysis, you can run GENCOV1R.EXE or GENCOV1C.EXE for estimating covariance components and coefficients of correlation among all the traits analyzed.
5. If you want to predict heterosis and genotypic value for each F_1 or F_2 cross by an AD model, you can run GENHET1R.EXE or GENHET1C.EXE.
6. The results from the analyses will be automatically stored in text files for later use or printing. Examples of result files are provided with the names COTDATA.VAR for analysis of variance and genetic effects, COTDATA.PRE for heterosis, and COTDATA.COR for analysis of covariances and correlation.

Output 1 for Variance Analysis

```
Traits =, 2
 Variance components = , 6
 Degree of freedom = , 3
 File name is cotdata.VAR
 Date and Time for Analysis: Thu Jun 22 21:43:19 2000

 Variance Components Estimated by MINQUE(1) with GENVAR1R.EXE.
 Jackknifing Over Block Conducted for Estimating S.E.
 Predicting Genetic Effects by Adjusted Unbiased Prediction (AUP)
     Method.

NS = Not significant; S+ = Significant at 0.10 level.
 S* = Significant at 0.05 level; S** = Significant at 0.01 level.

 Linear Contrasts:
 + <1> + <2> - <3> + <4>

Diallel Analysis of Trait 'Bolls' for Public Users

Var Comp                  Estimate      S. E.         P-value
(1): Additive Var         7.30438       1.4425        0.00743      S**
(2): Dominance Var        3.29038       0.935764      0.0195       S*
(3): Add. * Env. Var      0.866547      0.683906      0.147        NS
(4): Dom. * Env. Var      4.82384       1.71492       0.0336       S*
(6): Residual Var         4.18772       0.555651      0.00242      S**
```

```
(7): Var(Pheno.)            20.4729      2.79941      0.00264    S**

Proportion of Var(G)/Var(T)Estimate    S. E.        P-value
(1): Additive Var/Vp       0.356783     0.0888328    0.0139     S*
(2): Dominance Var/Vp      0.160719     0.0275068    0.005      S**
(3): Add. * Env. Var/Vp    0.0423266    0.0244       0.0906     S+
(4): Dom. * Env. Var/Vp    0.235621     0.0667933    0.0194     S*
(6): Residual Var/Vp       0.20455      0.0209314    0.00114    S**

Heritability               Estimate     S. E.        P-value
(7): Heritability(N)       0.356783     0.0888328    0.0139     S*
(8): Heritability(B)       0.517502     0.0661515    0.00217    S**
(9): Heritability(NE)      0.0423266    0.0244       0.0906     S+
(10): Heritability(BE)     0.277948     0.0575567    0.00846    S**

Genetic Predictor,  S. E. , P-value  for Two-tail t-test
(1): Random Effect is Additive Effect
A1                          0.392019     0.632218     0.579     NS
A2                          1.326158     0.211643     0.0082    S**
A3                         -2.917664     0.353084     0.00371   S**
A4                          1.199422     0.548514     0.117     NS
Linear Contrast             3.05342      0.470211     0.00741   S**

(2): Random Effect is Dominance Effect
D1*1                        0.263665     0.437706     0.589     NS
D2*2                       -2.176718     0.509854     0.0236    S*
D3*3                       -1.244060     0.786988     0.212     NS
D4*4                       -1.836773     0.495630     0.0341    S*
D1*2                        0.388935     0.399988     0.403     NS
D1*3                       -0.349309     0.311425     0.344     NS
D1*4                       -0.273600     0.915662     0.785     NS
D2*3                        0.210065     0.829160     0.816     NS
D2*4                        4.888072     0.707825     0.00622   S**
D3*4                        0.129694     0.666849     0.858     NS
Heterosis <Delta>           1.37653      0.337286     0.0266    S*

(3): Random Effect is Add. * Env. Effect
AE1 in E1                   0.089686     0.496245     0.868     NS
AE2 in E1                   0.242420     0.070750     0.0416    S*
AE3 in E1                  -0.079016     0.182773     0.695     NS
AE4 in E1                  -0.253113     0.407134     0.578     NS
AE1 in E2                  -0.136729     0.705758     0.859     NS
AE2 in E2                   0.342004     0.648600     0.634     NS
AE3 in E2                  -1.257717     1.654640     0.502     NS
AE4 in E2                   1.052433     0.508793     0.13      NS
Linear Contrast            -1.76747e-005 6.70577e-006 0.0779    S+

(4): Random Effect is Dom. * Env. Effect
DE11 in E1                 -0.349740     0.559982     0.577     NS
DE22 in E1                  2.063640     0.515148     0.0279    S*
DE33 in E1                 -0.889553     1.186821     0.508     NS
DE44 in E1                 -0.318622     1.006553     0.772     NS
DE12 in E1                 -1.294414     0.520453     0.0887    S+
DE13 in E1                  0.562069     0.405746     0.26      NS
DE14 in E1                  1.114442     1.050572     0.367     NS
DE23 in E1                 -0.217579     1.660683     0.904     NS
DE24 in E1                 -2.368299     0.651167     0.0358    S*
```

```
DE34 in E1                    1.698030     0.881687    0.15      NS
DE11 in E2                    0.751202     0.780806    0.407     NS
DE22 in E2                   -4.653504     1.121175    0.0254    S*
DE33 in E2                   -0.744776     1.086469    0.542     NS
DE44 in E2                   -1.890872     0.707435    0.0755    S+
DE12 in E2                    1.815819     0.552480    0.0462    S*
DE13 in E2                   -0.926615     0.698261    0.276     NS
DE14 in E2                   -1.611676     1.678116    0.408     NS
DE23 in E2                    0.437429     0.800245    0.623     NS
DE24 in E2                    8.238749     1.841553    0.0208    S*
DE34 in E2                   -1.415766     1.261023    0.343     NS
Heterosis <Delta>            0.971038     0.40929     0.0983    S+
```

```
Fixed Effect , 12.8719
Fixed Effect , 19.885
```

```
Results of Fiber Yield are not presented.
```

```
Time Used (Hour) = 0.001389
```

Output 2 for Covariance Analysis

```
Traits =, 2
Variance components = , 6
Degree of freedom = , 3
File name is cotdata.COV
Date and Time for Analysis: Thu Jun 22 22:00:24 2000
```

```
Variance Components Estimated by MINQUE(1) with GENVAR1R.EXE.
Jackknifing Over Block Conducted for Estimating S.E.
```

```
NS = Not significant; S+ = Significant at 0.10 level.
 S* = Significant at 0.05 level; S** = Significant at 0.01 level.
```

```
Covariances and Correlations Between Bolls & FibYield, for Public
     Users:
```

Covariances	Estimates	S.E.	P-value	
Additive Cov	26.4031	15.0422	0.177	NS
Dominance Cov	3.68996	4.09328	0.434	NS
Add. * Env. Cov	5.2684	5.4782	0.407	NS
Dom. * Env. Cov	0.725936	6.15727	0.914	NS
Residual Cov	5.74197	4.7088	0.31	NS

```
Cov 1=Genotypic
```

Cov2=Phenotypic	Estimates	S.E.	P-value	
Cov 2	41.8294	20.9549	0.14	NS
Cov 1	36.0874	21.3383	0.189	NS

Correlation	Estimates	S.E.	P-value	
Additive Cor	0.905414	0.326942	0.0696	S+
Dominance Cor	0.275228	0.243147	0.34	NS
Add. * Env. Cor	1.000000	0.288675	0.0405	S*
Dom. * Env. Cor	0.000000	0	1	NS

```
Cro 3      <1 * 4>    16.630 1.075 0.308  NS    16.374 0.748  0.251  NS
Cro 4      <2 * 3>    13.931 1.064 0.285  NS    12.971 0.832  0.067  S+
Cro 5      <2 * 4>    22.726 0.625 0.001  S**   19.279 0.832  0.018  S*
Cro 6      <3 * 4>    13.724 0.814 0.146  NS    12.889 1.033  0.101  NS
```

No.	Cro		$H_{pm}(F_1)$	S.E.	P-value	Signif.	$H_{pm}(F_2)$	S.E.	P-value	Signif.
Cro 1 <E1>	<1 * 2>		-0.140	0.034	0.027	S*	-0.070	0.017	0.027	S*
Cro 2 <E1>	<1 * 3>		0.077	0.030	0.079	S+	0.039	0.015	0.079	S+
Cro 3 <E1>	<1 * 4>		0.095	0.087	0.355	NS	0.047	0.043	0.355	NS
Cro 4 <E1>	<2 * 3>		-0.053	0.140	0.733	NS	-0.026	0.070	0.733	NS
Cro 5 <E1>	<2 * 4>		-0.212	0.059	0.038	S*	-0.106	0.030	0.038	S*
Cro 6 <E1>	<3 * 4>		0.150	0.115	0.283	NS	0.075	0.058	0.283	NS
Mean for			-0.014	0.059	0.830	NS	-0.007	0.029	0.830	NS
Env. Cro No. = 6										
Cro 7 <E2>	<1 * 2>		0.246	0.052	0.018	S*	0.123	0.026	0.018	S*
Cro 8 <E2>	<1 * 3>		-0.061	0.056	0.361	NS	-0.030	0.028	0.361	NS
Cro 9 <E2>	<1 * 4>		-0.068	0.153	0.686	NS	-0.034	0.076	0.686	NS
Cro 10 <E2>	<2 * 3>		0.205	0.074	0.071	S+	0.102	0.037	0.071	S+
Cro 11 <E2>	<2 * 4>		0.752	0.200	0.033	S*	0.376	0.100	0.033	S*
Cro 12 <E2>	<3 * 4>		-0.006	0.113	0.958	NS	-0.003	0.057	0.958	NS

Significance of F1 or F2 is over Population Mean 15.312341

No.	Cro		$H_{pm}(F_1)$ (G)	S.E.	P-value	Signif.	$H_{pm}(F_2)$ (G)	S.E.	P-value	nif.
Cro 1	<1 * 2>		0.088	0.031	0.067	S+	0.044	0.016	0.067	S+
Cro 2	<1 * 3>		0.009	0.027	0.754	NS	0.005	0.013	0.754	NS
Cro 3	<1 * 4>		0.033	0.078	0.696	NS	0.017	0.039	0.696	NS
Cro 4	<2 * 3>		0.125	0.072	0.181	NS	0.063	0.036	0.181	NS
Cro 5	<2 * 4>		0.450	0.083	0.012	S*	0.225	0.042	0.012	S*
Cro 6	<3 * 4>		0.109	0.075	0.241	NS	0.055	0.037	0.241	NS

No.	Cro		$H_{pb}(F_1)$	S.E.	P-value	Signif.	$H_{pb}(F_2)$	S.E.	P-value	Signif.
Cro 1 <E1>	<1 * 2>		-0.229	0.003	0.000	S**	-0.159	0.019	0.004	S**
Cro 2 <E1>	<1 * 3>		0.049	0.012	0.025	S*	0.010	0.017	0.595	NS
Cro 3 <E1>	<1 * 4>		0.073	0.107	0.542	NS	0.026	0.069	0.732	NS
Cro 4 <E1>	<2 * 3>		-0.170	0.128	0.277	NS	-0.144	0.068	0.123	NS
Cro 5 <E1>	<2 * 4>		-0.322	0.077	0.025	S*	-0.216	0.054	0.029	S*
Cro 6 <E1>	<3 * 4>		0.143	0.114	0.297	NS	0.068	0.056	0.315	NS
Cro 7 <E2>	<1 * 2>		0.101	0.048	0.128	NS	-0.022	0.034	0.563	NS
Cro 8 <E2>	<1 * 3>		-0.183	0.118	0.219	NS	-0.152	0.095	0.206	NS
Cro 9 <E2>	<1 * 4>		-0.077	0.163	0.670	NS	-0.043	0.087	0.657	NS
Cro 10 <E2>	<2 * 3>		0.182	0.116	0.215	NS	0.079	0.126	0.573	NS
Cro 11 <E2>	<2 * 4>		0.615	0.151	0.027	S*	0.239	0.055	0.023	S*
Cro 12 <E2>	<3 * 4>		-0.120	0.165	0.520	NS	-0.117	0.111	0.369	NS

Significance of F1 or F2 is over Population Mean 15.312341

No.	Cro		$H_{pb}(F_1)$ (G)	S.E.	P-value	Signif.	$H_{pb}(F_2)$ (G)	S.E.	P-value	nif.
Cro 1	<1 * 2>		0.069	0.040	0.183	NS	0.025	0.029	0.445	NS
Cro 2	<1 * 3>		-0.256	0.081	0.050	S+	-0.261	0.091	0.064	S+
Cro 3	<1 * 4>		0.018	0.088	0.854	NS	0.001	0.049	0.987	NS
Cro 4	<2 * 3>		-0.121	0.092	0.278	NS	-0.184	0.071	0.081	S+
Cro 5	<2 * 4>		0.447	0.076	0.010	S**	0.222	0.036	0.008	S**
Cro 6	<3 * 4>		-0.140	0.082	0.184	NS	-0.195	0.050	0.030	S*

Significance of Heterosis is over Population Mean 15.312341

Pre(F_1)	16.4871	0.580926	0.136	NS
Pre(F_2)	15.4467	0.665146	0.853	NS
Hpm(F_1)	0.135267	0.030647	0.0216	S*
Hpm(F_2)	0.067634	0.015324	0.0216	S*
Hpb(F_1)	-0.04533	0.042593	0.365	NS
Hpb(F_2)	-0.11297	0.035282	0.0493	S*

Results of Fiber yield are not presented.

Time Used (Hour) = 0.000278

Chapter 4

Diallel Analysis for an Additive-Dominance-Epistasis Model with Genotype-by-Environment Interaction Effects

Jun Zhu

Purpose

To analyze balanced or unbalanced data of an additive x dominance (AD) + additive x additive (AA) genetic model for estimating components of variance, covariance, heritability, and selection response.

Definitions

Mating Design

A set of inbred lines is sampled from a reference population. Parents are used to produce F_1 crosses and their F_2. Experiments with parents, F_1s, and F_2s are conducted in multiple environments using a randomized complete block design.

Genetic Model

The genetic model for genetic entry of the kth type of generation derived from parents i and j in the lth block within the hth environment is

$$y_{hijkl} = \mu + E_h + G_{ijk} + GE_{hijk} + B_{hl} + e_{hijkl}$$

where μ = population mean, E_h = environment effect, G_{ijk} = total genotypic effect, GE_{hijk} = genotype × environment interaction effect, B_{hl} = block effect, and e_{hijkl} = residual effect.

For parent (P_i, $k = 0$):

$$G_{ii0} + GE_{hii0} = 2A_i + D_{ii} + 4AA_{ii} + 2AE_{hi} + DE_{hii} + 4AAE_{hii}$$

For F_1 ($P_i \times P_j$, $k = 1$):

$$G_{ij1} + GE_{hij1} = A_i + A_j + D_{ij} + AA_{ii} + AA_{jj} + 2AA_{ij} + AE_{hi}$$
$$+ AE_{hj} + DE_{hij} + AAE_{hii} + AAE_{hjj} + 2AAE_{hij}$$

For F_2 ($F_1 \otimes$, $k = 2$):

$$G_{ij2} + GE_{hij2} = A_i + A_j + \tfrac{1}{4}D_{ij} + \tfrac{1}{4}D_{ii} + \tfrac{1}{2}D_{jj} + AA_{ii} + AA_{jj} + 2AA_{ij}$$
$$+ AE_{hi} + AE_{hj} + \tfrac{1}{4}DE_{hii} + \tfrac{1}{4}DE_{hjj} + \tfrac{1}{2}DE_{hij}$$
$$+ AAE_{hii} + AAE_{hjj} + 2AAE_{hij}$$

where A = additive effect, D = dominance effect, AA = additive by additive epistatic effect, AE = additive by environment interaction effect, DE = dominance by environment interaction effect, and AAE = epistasis by environment interaction effect.

Analysis Methodology

Mixed Linear Model

The phenotypic mean of the genetic model can be expressed by a mixed linear model as

$$y = Xb + U_A e_A + U_D e_D + U_{AA} e_{AA} + U_{AE} e_{AE} + U_{DE} e_{DE} + U_{AAE} e_{AAE} + U_B e_B + e_e$$
$$= Xb + \sum_u^8 U_u e_u$$

with variance-covariance matrix

$$\mathrm{var}(y) = \sigma_A^2 U_A U_A^T + \sigma_D^2 U_D U_D^T + \sigma_{AA}^2 U_{AA} U_{AA}^T + \sigma_{AE}^2 U_{AE} U_{AE}^T$$
$$+ \sigma_{DE}^2 U_{DE} U_{DE}^T + \sigma_{AAE}^2 U_{AAE} U_{AAE}^T + \sigma_B^2 U_B U_B^T + \sigma_e^2 I$$
$$= \sum_{u=1}^8 \sigma_u^2 U_u U_u^T.$$

Variance Components

Unbiased estimation of variances can be obtained by restricted maximum likelihood (REML) or MINQUE(1) approaches. When experimental variances (σ_u^2) are estimated, genetic variance components can be obtained by $V_A = 2\sigma_A^2$, $V_D = \sigma_D^2$, $V_{AA} = 4\sigma_{AA}^2$, $V_{AE} = 2\sigma_{AE}^2$, $V_{DE} = \sigma_{DE}^2$, $V_{AAE} = 4\sigma_{AAE}^2$, $V_e = \sigma_e^2$. The total phenotypic variance is $V_P = V_A + V_D + V_{AA} + V_{AE} + V_{DE} + V_{AAE} + V_e$.

Covariance Components and Correlation

Unbiased estimation of covariances can be obtained by MINQUE(1) approaches (Zhu, 1992; Zhu and Weir, 1996). When experimental covariances $(\sigma_{u/u})$ are estimated, genetic covariance components can be obtained by $C_A = 2\sigma_{A/A}$, $C_D = \sigma_{D/D}$, $C_{AA} = 4\sigma_{AA/AA}$, $C_{AE} = 2\sigma_{AE/AE}$, $C_{DE} = \sigma_{DE/DE}$, $C_{AAE} = 4\sigma_{AAE/AAE}$, $C_e = \sigma_{e/e}$. The total phenotypic covariance is $C_P = C_A + C_D + C_{AA} + C_{AE} + C_{DE} + C_{AAE} + C_e$. For trait 1 and trait 2, correlation coefficients of genetic components can be estimated by $r_A = C_A / \sqrt{V_{A(1)} V_{A(2)}}$, $r_D = C_D / \sqrt{V_{D(1)} V_{D(2)}}$, $r_{AA} = C_{AA} / \sqrt{V_{AA(1)} V_{AA(2)}}$, $r_{AE} = C_{AE} / \sqrt{V_{AE(1)} V_{AE(2)}}$, $r_{DE} = C_{DE} / \sqrt{V_{DE(1)} V_{DE(2)}}$, $r_{AAE} = C_{AAE} / \sqrt{V_{AAE(1)} V_{AAE(2)}}$, and $r_e = C_e / \sqrt{V_{e(1)} V_{e(2)}}$.

Heritability Components

The total heritability (h^2) can be partitioned into two components $(h^2 = h_G^2 + h_{GE}^2)$, where $h_G^2 = (V_A + V_{AA})/V_P$ is general heritability and $h_{GE}^2 = (V_{AE} + V_{AAE})/V_P$ is interaction heritability (Zhu, 1997).

Selection Response

The total selection response $(R = ih^2\sqrt{V_P})$ can be partitioned into two components (Zhu, 1997):

$$R = R_G + R_{GE}$$

where $R_G = ih_G^2\sqrt{V_P}$ is general response and $R_{GE} = ih_{GE}^2\sqrt{V_P}$ is interaction response.

Heterosis Components

Prediction of genetic merits can be obtained by use of the linear unbiased prediction (LUP) method (Zhu, 1992; Zhu and Weir, 1996) or the adjusted unbiased prediction (AUP) method (Zhu, 1993; Zhu and Weir, 1996). Predicted genotypic effects and *GE* interaction effects can be further used in analyzing heterosis of different generations (Zhu, 1997). Heterosis in specific environments consists of two components. General heterosis is due to genotypic effects and can be expected in overall environments, and interaction heterosis is a deviant of *GE* interaction relative to specific environments. The two components of heterosis relative to midparent or relative to better parent can be calculated as follows:

General heterosis of F_n relative to midparent:
$$H_M(F_n) = (\tfrac{1}{2})^{n-1} \Delta_D + 2\Delta_{AA}$$
Interaction heterosis of F_n relative to midparent:
$$H_{ME}(F_n) = (\tfrac{1}{2})^{n-1} \Delta_{DE} + 2\Delta_{AAE}$$
General heterosis of F_n relative to better parent (P_i):
$$H_B(F_n) = H_M(F_n) - \tfrac{1}{2}\varpi_G$$
Interaction heterosis of F_n relative to better parent (P_i):
$$H_{BE}(F_n) = H_{ME}(F_n) - \tfrac{1}{2}\varpi_{GE}$$

where $\Delta_D = D_{ij} - \tfrac{1}{2}(D_{ii} + D_{jj})$ is dominance heterosis, $\Delta_{DE} = DE_{hij} - \tfrac{1}{2}(DE_{hii} + DE_{hjj})$ is *DE* interaction heterosis, $\varpi_G = |G(P_i) - G(P_j)|$ is parental genotypic difference, and $\varpi_{GE} = |GE(P_i) - GE(P_j)|$ is parental interaction difference.

Heterosis based on population mean ($H_{PM} = \tfrac{1}{\mu}H_M$, $H_{PME} = \tfrac{1}{\mu}H_{ME}$, $H_{PB} = \tfrac{1}{\mu}H_B$, or $H_{PBE} = \tfrac{1}{\mu}H_{BE}$) can be used to compare proportion of heterosis among different traits.

Originators

Zhu, J. (1992). Mixed model approaches for estimating genetic variances and covariances. *Journal of Biomathematics* 7(1):1-11.

Zhu, J. (1993). Methods of predicting genotype value and heterosis for offspring of hybrids (Chinese). *Journal of Biomathematics* 8(1):32-44.

Zhu, J. (1997). *Analysis Methods for Genetic Models.* Agricultural Publication House of China, Beijing.

Zhu, J. and Weir, B.S. (1996). Diallel analysis for sex-linked and maternal effects. *Theoretical and Applied Genetics* 92(1):1-9.

Software Available

Zhu, J. (1997). GENAD.EXE for constructing AD model, GENVAR1.EXE for estimating components of variance and heritability, GENCOV1.EXE for estimating components of covariance and correlation, GENHET1.EXE for predicting genetic effects and components of heterosis. *Analysis Methods for Genetic Models* (pp. 278-285). Agricultural Publication House of China, Beijing (program free of charge). Contact Dr. Jun Zhu, Department of Agronomy, Zhejiang University, Hangzhou, China. E-mail: <jzhu@zju.edu.cn>.

EXAMPLE

Unbalanced data (COTADAA.TXT) to be analyzed (Parent = 10, Year = 2, Generation = P, F_1, F_2, Blk = 1):

Year	Male	Fem	Gen	Blk	Bolls	Lint%
1	1	1	0	1	10.39	37.16
1	1	6	1	1	16.69	39.29
1	1	6	2	1	15.05	37.68
1	1	7	1	1	18.27	40.92
1	1	7	2	1	14.44	38.35
1	1	9	1	1	13.36	36.43
1	1	9	2	1	12.37	36.1
1	1	10	1	1	14.57	33.45
1	1	10	2	1	11.52	34.81
1	2	2	0	1	18.06	34.95
1	2	6	1	1	16.65	38.28
1	2	6	2	1	15.43	39.5
1	2	7	1	1	17.67	39.27
1	2	7	2	1	18.82	38.43
1	2	8	1	1	19.89	38.22
1	2	8	2	1	12.65	35.44
1	2	9	1	1	18.03	34.57
1	2	9	2	1	15.45	35.51
1	2	10	1	1	17.08	33.69
1	2	10	2	1	16.1	29.89
1	3	3	0	1	11.03	39.53
1	3	7	1	1	17.52	42.46
1	3	7	2	1	13.99	39.38
1	3	9	1	1	14.56	37.04
1	3	9	2	1	12.28	38.27
1	3	10	1	1	13.27	37.83
1	3	10	2	1	16.42	39.14
1	4	4	0	1	16.54	40.8
1	4	6	1	1	17.11	40.34
1	4	6	2	1	14.58	40.77
1	4	8	1	1	16.7	40.92
1	4	8	2	1	14.7	39.72
1	4	9	1	1	17.1	38.7
1	4	9	2	1	17.34	38.41

1	4	10	1	1	14.14	36.54
1	4	10	2	1	13.86	36.99
1	5	5	0	1	13.89	40.49
1	5	7	1	1	18.57	41.6
1	5	7	2	1	14.53	41.53
1	5	8	1	1	17.27	40.33
1	5	8	2	1	16.1	39.9
1	5	9	1	1	16.31	39.16
1	5	9	2	1	14.82	39.92
1	5	10	1	1	16.98	37.65
1	5	10	2	1	12.22	37.3
1	6	6	0	1	16.66	39.1
1	7	7	0	1	18.35	42.04
1	8	8	0	1	13.49	38.81
1	9	9	0	1	12.91	35.98
1	10	10	0	1	11.52	30.89
2	1	1	0	1	10.09	37.69
2	1	6	1	1	10.82	41.92
2	1	6	2	1	11.13	38.06
2	1	7	1	1	7.97	40.53
2	1	7	2	1	11.08	41.2
2	1	9	1	1	8.22	37.49
2	1	9	2	1	9.85	37.45
2	1	10	1	1	7.26	33.81
2	1	10	2	1	8.52	33.53
2	2	2	0	1	9.87	39.3
2	2	6	1	1	12.31	40.64
2	2	6	2	1	11.95	41.35
2	2	7	1	1	11.3	42.04
2	2	7	2	1	9.98	40.17
2	2	8	1	1	13.5	39.85
2	2	8	2	1	11.47	37.64
2	2	9	1	1	11.93	37.71
2	2	9	2	1	10.83	37.45
2	2	10	1	1	8.23	34.59
2	2	10	2	1	11.1	34.01
2	3	3	0	1	6.4	39.44
2	3	7	1	1	8	42.68
2	3	7	2	1	9.09	43.29
2	3	9	1	1	11.49	37.92
2	3	9	2	1	10.78	38.9
2	3	10	1	1	7.32	34.76
2	3	10	2	1	10.9	38.42
2	4	4	0	1	8.83	42.65
2	4	6	1	1	11.37	42.67
2	4	6	2	1	11.77	41.45
2	4	8	1	1	13.07	41.84
2	4	8	2	1	11.18	42.27
2	4	9	1	1	10.63	38.12
2	4	9	2	1	11.47	41.08
2	4	10	1	1	10.43	39.06
2	4	10	2	1	11.84	37.58
2	5	5	0	1	11.37	42.86
2	5	7	1	1	12.03	42.65
2	5	7	2	1	10.69	44.69
2	5	8	1	1	10.2	40.36
2	5	8	2	1	10.09	39.53

2	5	9	1	1	10.47	40.31
2	5	9	2	1	10.89	40.03
2	5	10	1	1	10.33	38.78
2	5	10	2	1	8.95	39.09
2	6	6	0	1	11.24	38.6
2	7	7	0	1	10.67	43.22
2	8	8	0	1	10.77	40.74
2	9	9	0	1	6.87	37.43
2	10	10	0	1	11.69	35.05

1. Run GENADE.EXE to create mating design matrix files and data for additive-dominance-epistasis (AD+AA) models. The data files (COTADAA.TXT) should have five columns: (1) environment, (2) maternal, (3) paternal, (4) generation, and (5) replication. There is a limitation (<100 traits) for the number of trait columns. An example of a data file is provided under the name COTADAA.TXT.

2. Run programs for variance and covariance analyses. Standard errors of estimates are calculated using jackknife procedures. If you have multiple blocks for your experiments, you can use GENVAR1R.EXE or GENCOV1R.EXE for jackknifing over blocks. Otherwise you can use GENVAR1C.EXE or GENCOV1C.EXE for jackknifing over cell means.

3. Run GENVAR1R.EXE or GENVAR1C.EXE for estimating variance components and predicting genetic effects before estimating covariance and correlation. The two programs in Step 2 will allow you to choose the parental type (inbred or outbred) and the prediction methods (LUP or AUP). You also need to input coefficients (1, 0, or −1) for conducting linear contrasts for genetic effects of parents.

4. After you finish variance analysis, you can run GENCOV1R.EXE or GENCOV1C.EXE for estimating covariance components and coefficients of correlation among all the traits analyzed.

5. If you want to predict heterosis and genotypic value for each F_1 or F_2 cross by an AD model, you can run GENHET1R.EXE or GENHET1C.EXE.

6. All results are automatically stored in text files for later use or printing. Examples of output files are provided with the names COTADAA.VAR for analysis of variance and genetic effects, COTADAA.PRE for heterosis, and COTADAA.COR for analysis of covariances and correlation.

Output 1 for Variance Analysis

```
Traits =, 2
Variance components = , 7
Degree of freedom = , 99
File name is COTADAA.VAR
Date and Time for Analysis: Fri Jun 23 08:33:02 2000
```

Variance Components Estimated by MINQUE(1) with GENVAR1R.EXE.
Jackknifing Over Block Conducted for Estimating S.E.
Predicting Genetic Effects by Adjusted Unbiased Prediction (AUP)
    Method.

NS = Not significant; S+ = Significant at 0.10 level.
S* = Significant at 0.05 level; S** = Significant at 0.01 level.

Linear Contrast Test:
+<1>   +<2>   +<3>   +<4>   +<5>   -<6>   -<7>   -<8>   -<9>   -<10>

Diallel Analysis of Trait, Bolls, for Public Users.

Var Comp	Estimate	S. E.	P-value	
(1): Additive Var	2.36714	0.474734	1.31e-006	S**
(2): Dominance Var	12.4508	2.25708	1.39e-007	S**
(3): Add.*Add. Var	3.48369	0.502654	1.9e-010	S**
(4): Add. * Env. Var	3.59761	0.745664	2.55e-006	S**
(5): Dom. * Env. Var	16.8931	2.83894	2.03e-008	S**
(6): (AA) * Env. Var	0	0	1	NS
(7): Residual Var	3.12779	0.712819	1.43e-005	S**
(8): Var(Pheno.)	41.9202	4.31614	2.55e-011	S**

Proportion of Var(G)/Var(T)	Estimate	S. E.	P-value	
(1): Additive Var/Vp	0.0564678	0.0211358	0.00441	S**
(2): Dominance Var/Vp	0.297013	0.0369701	2.44e-011	S**
(3): Add.*Add. Var/Vp	0.0831029	0.0214262	9.46e-005	S**
(4): Add. * Env. Var/Vp	0.0858205	0.0120405	5.86e-006	S**
(5): Dom. * Env. Var/Vp	0.402983	0.0341788	2.55e-011	S**
(6): (AA) * Env. Var/Vp	0	0	1	NS
(7): Residual Var/Vp	0.0746131	0.0151876	1.78e-006	S**

Heritability	Estimate	S. E.	P-value	
(8): Heritability(N)	0.139571	0.0266373	4.55e-007	S**
(9): Heritability(B)	0.436584	0.0348678	2.55e-011	S**
(10): Heritability(NE)	0.0858205	0.0120405	5.86e-006	S**
(11): Heritability(BE)	0.488803	0.03613	-2.55e-011	S**

Genetic Predictor, S.E., P-value for Two-tail t-test
(1): Random Effect is Additive Effects

A1	0.020391	1.046932	0.984	NS
A2	-0.172118	0.865219	0.843	NS
A3	0.243015	0.933404	0.795	NS
A4	0.024512	0.610581	0.968	NS
A5	-0.198413	0.284266	0.487	NS
A6	-0.045113	0.559257	0.936	NS
A7	-0.006029	0.525715	0.991	NS

```
A8                        -0.242755      0.336519   0.472        NS
A9                         0.177528      0.247130   0.474        NS
A10                        0.197105      0.780527   0.801        NS
Linear Contrast           -0.237404     11.14      0.983        NS

(2): Random Effect is Dominance Effects
D1*1                      -1.985631      1.605723   0.219        NS
D2*2                      -4.853828      3.095245   0.12         NS
D3*3                      -0.293169      0.924650   0.752        NS
D4*4                      -1.707263      1.061605   0.111        NS
D5*5                      -7.790652      3.617237   0.0337       S*
D6*6                      -2.681446      1.218021   0.03         S*
D7*7                      -4.154995      2.393013   0.0856       S+
D8*8                      -7.236830      3.889634   0.0658       S+
D9*9                      -3.026015      1.713898   0.0806       S+
D10*10                     1.364409      1.417177   0.338        NS
D1*6                       1.844893      1.020188   0.0736       S+
D1*7                       0.088293      2.692258   0.974        NS
D1*9                      -2.270995      1.155729   0.0522       S+
D1*10                      2.668295      1.548696   0.088        S+
D2*6                       1.351880      0.655828   0.0419       S*
D2*7                      -1.297874      1.003310   0.199        NS
D2*8                       9.815433      5.523754   0.0786       S+
D2*9                       5.188726      2.350358   0.0296       S*
D2*10                     -4.090970      1.895508   0.0333       S*
D3*7                       3.495850      2.000000   0.0836       S+
D3*9                       4.970930      2.262232   0.0303       S*
D3*10                     -9.112045      4.458072   0.0436       S*
D4*6                       2.946256      1.541041   0.0588       S+
D4*8                       5.140935      2.364873   0.0321       S*
D4*9                      -1.798652      0.660681   0.00766      S**
D4*10                     -1.958099      0.910923   0.034        S*
D5*7                       6.710816      3.253640   0.0418       S*
D5*8                      -0.368711      0.290930   0.208        NS
D5*9                       0.235750      0.576913   0.684        NS
D5*10                      8.804042      3.903617   0.0263       S*
Heterosis <Delta>          2.90056      12.8407     0.822        NS

(3): Random Effect is Add.*Add. Effects
AA1*1                     -1.827866      0.885620   0.0416       S*
AA2*2                      1.857440      0.692252   0.00855      S**
AA3*3                     -3.923112      1.819274   0.0335       S*
AA4*4                     -0.427958      0.372799   0.254        NS
AA5*5                      1.095843      0.431401   0.0126       S*
AA6*6                      1.074095      0.485307   0.0292       S*
AA7*7                      2.010456      0.866196   0.0223       S*
AA8*8                      0.866474      0.254778   0.00097      S**
AA9*9                     -2.618613      1.297275   0.0462       S*
AA10*10                   -1.434506      0.678498   0.037        S*
AA1*6                      1.127468      0.554236   0.0446       S*
AA1*7                      0.260405      0.142121   0.0699       S+
AA1*9                     -0.626353      0.353682   0.0796       S+
AA1*10                    -2.892538      1.389233   0.0399       S*
AA2*6                     -0.039485      0.150785   0.794        NS
AA2*7                      0.997226      0.482695   0.0414       S*
AA2*8                     -2.373975      1.062655   0.0277       S*
AA2*9                      0.916657      0.551207   0.0995       S+
```

```
AA2*10                  0.966337      0.475264    0.0447     S*
AA3*7                  -1.240906      0.602029    0.0419     S*
AA3*9                   0.736135      0.560053    0.192      NS
AA3*10                  3.882660      1.687257    0.0235     S*
AA4*6                  -0.333696      0.292460    0.257      NS
AA4*8                   0.232070      0.281352    0.411      NS
AA4*9                   3.758026      1.714850    0.0308     S*
AA4*10                  0.453800      0.251847    0.0746     S+
AA5*7                  -1.360470      0.556477    0.0163     S*
AA5*8                   0.698657      0.311950    0.0274     S*
AA5*9                   1.219364      0.571954    0.0355     S*
AA5*10                 -3.054544      1.399209    0.0314     S*
Heterosis <Delta>       1.12761       5.66678     0.843      NS


(4): Random Effect is Add. * Env. Effects
AE1  in E1             -1.871303      1.336265    0.165      NS
AE2  in E1              1.679535      1.500118    0.266      NS
AE3  in E1             -0.899398      0.974282    0.358      NS
AE4  in E1              0.355846      0.398988    0.375      NS
AE5  in E1              0.613201      0.505110    0.228      NS
AE6  in E1             -0.319194      0.433700    0.463      NS
AE7  in E1              3.154071      2.503520    0.211      NS
AE8  in E1             -1.709535      0.955588    0.0767     S+
AE9  in E1             -1.004626      0.705371    0.158      NS
AE10 in E1              0.000037      0.749113    1          NS
AE1  in E2             -0.687704      0.654509    0.296      NS
AE2  in E2              0.130732      0.339243    0.701      NS
AE3  in E2             -1.165105      0.703354    0.101      NS
AE4  in E2              1.212089      0.914817    0.188      NS
AE5  in E2             -0.818562      0.501433    0.106      NS
AE6  in E2              1.647312      1.146417    0.154      NS
AE7  in E2             -2.111514      1.652810    0.204      NS
AE8  in E2              1.666313      1.018155    0.105      NS
AE9  in E2              1.460007      0.882265    0.101      NS
AE10 in E2             -1.334250      0.692084    0.0567     S+
Linear Contrast       -0.000322198   0.00010415  0.00257    S**


(5): Random Effect is Dom. * Env. Effects
DE11 in E1             -7.339941      3.362180    0.0314     S*
DE22 in E1             -5.544965      3.293144    0.0954     S+
DE33 in E1             -1.685938      2.763209    0.543      NS
DE44 in E1             -1.592941      1.468080    0.281      NS
DE55 in E1             -6.919656      3.156330    0.0307     S*
DE66 in E1             -2.769618      1.534016    0.074      S+
DE77 in E1             -5.708889      3.470978    0.103      NS
DE88 in E1             -6.906281      3.149161    0.0306     S*
DE99 in E1             -3.155495      1.695257    0.0657     S+
DE1010 in E1           -4.448080      3.583114    0.217      NS
DE16 in E1              1.945276      1.444322    0.181      NS
DE17 in E1              6.019940      4.117258    0.147      NS
DE19 in E1              0.953192      0.789965    0.23       NS
DE110 in E1            5.306395      3.178059    0.0981     S+
DE26 in E1              0.018365      0.840712    0.983      NS
DE27 in E1             -4.879109      2.046507    0.019      S*
DE28 in E1             11.874683      6.855778    0.0864     S+
DE29 in E1              2.481694      2.213078    0.265      NS
DE210 in E1            1.704185      1.338270    0.206      NS
```

```
DE37  in E1          5.935123      3.679177    0.11        NS
DE39  in E1          2.916905      2.004578    0.149       NS
DE310 in E1         -5.481360      3.413565    0.112       NS
DE46  in E1          3.541911      2.501341    0.16        NS
DE48  in E1          1.163906      1.499863    0.44        NS
DE49  in E1         -1.462106      0.664229    0.03        S*
DE410 in E1          0.093319      0.547534    0.865       NS
DE57  in E1          5.091230      3.513362    0.15        NS
DE58  in E1          0.249648      0.907277    0.784       NS
DE59  in E1          1.192191      1.072943    0.269       NS
DE510 in E1          7.405446      4.610249    0.111       NS
DE11  in E2          6.233934      2.649984    0.0206      S*
DE22  in E2          0.070564      1.631085    0.966       NS
DE33  in E2          2.860760      1.834532    0.122       NS
DE44  in E2          0.496960      1.260403    0.694       NS
DE55  in E2          0.762586      0.976264    0.437       NS
DE66  in E2          0.750042      0.589078    0.206       NS
DE77  in E2          2.513161      1.959827    0.203       NS
DE88  in E2         -0.681558      1.133328    0.549       NS
DE99  in E2          1.396232      1.222769    0.256       NS
DE1010 in E2         6.754486      3.043240    0.0287      S*
DE16  in E2         -0.782136      0.571314    0.174       NS
DE17  in E2         -6.286620      3.932846    0.113       NS
DE19  in E2         -2.727444      1.812851    0.136       NS
DE110 in E2         -2.626815      1.660283    0.117       NS
DE26  in E2          0.500294      0.497259    0.317       NS
DE27  in E2          2.695619      1.378225    0.0533      S+
DE28  in E2          0.710618      1.939384    0.715       NS
DE29  in E2          1.408633      0.928007    0.132       NS
DE210 in E2         -5.416844      3.370826    0.111       NS
DE37  in E2         -2.775914      1.748225    0.116       NS
DE39  in E2          1.697015      1.049894    0.109       NS
DE310 in E2         -4.892908      3.552561    0.172       NS
DE46  in E2         -0.956358      0.906075    0.294       NS
DE48  in E2          2.794023      1.735292    0.111       NS
DE49  in E2         -1.011061      1.141284    0.378       NS
DE410 in E2         -1.635955      1.322434    0.219       NS
DE57  in E2          0.749728      1.094644    0.495       NS
DE58  in E2         -1.851171      0.743874    0.0145      S*
DE59  in E2         -1.672750      0.952113    0.082       S+
DE510 in E2          0.922233      1.377869    0.505       NS
```

(6): Random Effect is (AA) * Env. Effects
No Significant Effects.

Fixed Effect <1>, 15.345
Fixed Effect <2>, 10.3648

Results of Lint% are not presented.

Time Used (Hour) = 0.004722

Output 2 for Covariance Analysis

Traits =, 2

```
Covariance components = , 7
 Degree of freedom = , 99
 File name is COTADAA.COV
 Date and Time for Analysis: Fri Jun 23 08:33:35 2000

 Covariance Components Estimated by MINQUE(1) with GENCOV1C.EXE.
 Jackknifing Over Cell Mean Conducted for Estimating S.E.

NS = Not significant; S+ = Significant at 0.10 level.
S* = Significant at 0.05 level; S** = Significant at 0.01 level.

Covariances and Correlations Between, Bolls, , &, Lint%, for Public
    Users.:
```

Covariances	Estimates	S.E.	P-value	
Additive Cov	-0.165704	0.968417	0.864	NS
Dominance Cov	0.802175	2.31849	0.73	NS
Add.*Add. Cov	0.52236	0.79344	0.512	NS
Add. * Env. Cov	-0.585695	0.668664	0.383	NS
Dom. * Env. Cov	1.44656	2.09928	0.492	NS
(AA) * Env. Cov	-0.116658	1.07689	0.914	NS
Residual Cov	0.192467	0.376523	0.61	NS

Cov<1=Genotypic>				
Cov <2=Phenotypic>	Estimates	S.E.	P-value	
Cov 2	2.09551	1.35407	0.125	NS
Cov 1	1.90304	1.35435	0.163	NS

Correlation	Estimates	S.E.	P-value	
Additive Cor	-0.043057	0.0498808	0.39	NS
Dominance Cor	0.100186	0.0495757	0.046	S *
Add.*Add. Cor	0.174520	0.0516629	0.00104	S **
Add. * Env. Cor	-0.207088	0.0362769	1.19e-007	S **
Dom. * Env. Cor	0.000000	0	1	NS
(AA) * Env. Cor	0.000000	0	1	NS
Residual Cor	0.079717	0.0377952	0.0375	S *

Cor <1=Genotypic>				
Cor <2=Phenotypic>	Estimates	S.E.	P-value	
Cor 2	0.073183	0.0448333	0.106	NS
Cor 1	0.072636	0.0500355	0.15	NS

```
Time Used (Hour) = 0.003056
```

Output 3 for Heterosis Analysis

```
Traits =, 2
Variance components = , 7
Degree of freedom = , 99
File name is COTADJM.PRE
Date and Time for Analysis: Fri Jun 23 08:34:07 2000

Variance Components Estimated by MINQUE(1) with GENVAR1R.EXE.
Jackknifing Over Block Conducted for Estimating S.E.
```

Predicting Genetic Effects by Adjusted Unbiased Prediction (AUP)
    Method.
NS = Not significant; S+ = Significant at 0.10 level.
S* = Significant at 0.05 level; S** = Significant at 0.01 level.

Var Comp, Estimate, S. E. , P-value of One Tail t-test of, Bolls, for
    Public Users.

		Estimate	S. E.	P-value	Sig.
Additive	Var	2.36728	0.474749	1.31e-006	S **
Dominance	Var	12.4508	2.25708	1.39e-007	S **
Add.*Add.	Var	3.48381	0.502665	1.9e-010	S **
Add. * Env.	Var	3.59769	0.745673	2.54e-006	S **
Dom. * Env.	Var	16.893	2.83894	2.03e-008	S **
(AA) * Env.	Var	0	0	0.5	NS
Residual	Var	3.12783	0.712823	1.43e-005	S **

Heterosis Analysis of Trait, Bolls, for F_2 Seeds with total mean =,
    12.854884

No.	Cross		F1 (GE)	S.E.	P-value	Sig.	F2 (GE)	S.E.	P-value	Sig.
Cro 1 <E1>	<1 * 6>	-0.25	1.85	0.90	NS	-3.75	1.57	0.02	S *
Cro 2 <E1>	<1 * 7>	7.30	4.14	0.08	S +	1.03	1.39	0.46	NS
Cro 3 <E1>	<1 * 9>	-1.92	1.60	0.23	NS	-5.02	1.88	0.01	S **
Cro 4 <E1>	<1 * 10>	3.44	3.50	0.33	NS	-2.17	2.05	0.29	NS
Cro 5 <E1>	<2 * 6>	1.38	1.17	0.24	NS	-0.71	1.26	0.57	NS
Cro 6 <E1>	<2 * 7>	-0.04	3.21	0.99	NS	-0.42	2.55	0.87	NS
Cro 7 <E1>	<2 * 8>	11.84	6.84	0.09	S +	2.79	2.30	0.23	NS
Cro 8 <E1>	<2 * 9>	3.16	2.41	0.19	NS	-0.26	1.40	0.85	NS
Cro 9 <E1>	<2 * 10>	3.38	1.63	0.04	S *	0.03	1.43	0.98	NS
Cro 10 <E1>	<3 * 7>	8.19	4.09	0.05	S *	3.37	1.89	0.08	S +
Cro 11 <E1>	<3 * 9>	1.01	2.25	0.65	NS	-1.66	1.67	0.32	NS
Cro 12 <E1>	<3 * 10>	-6.38	3.52	0.07	S +	-5.17	2.04	0.01	S *
Cro 13 <E1>	<4 * 6>	3.58	2.53	0.16	NS	0.72	0.98	0.47	NS
Cro 14 <E1>	<4 * 8>	-0.19	1.73	0.91	NS	-2.90	1.47	0.05	S +
Cro 15 <E1>	<4 * 9>	-2.11	0.82	0.01	S *	-2.57	0.89	0.00	S **
Cro 16 <E1>	<4 * 10>	0.45	0.81	0.58	NS	-1.11	1.39	0.43	NS
Cro 17 <E1>	<5 * 7>	8.86	4.38	0.05	S *	3.16	2.50	0.21	NS
Cro 18 <E1>	<5 * 8>	-0.85	1.20	0.48	NS	-4.43	1.51	0.00	S **
Cro 19 <E1>	<5 * 9>	0.80	1.22	0.51	NS	-2.31	1.20	0.06	S +
Cro 20 <E1>	<5 * 10>	8.02	4.80	0.10	S +	1.47	1.73	0.40	NS
Cro 21 <E2>	<1 * 6>	0.18	0.77	0.82	NS	2.31	0.92	0.01	S *
Cro 22 <E2>	<1 * 7>	-9.09	4.23	0.03	S *	-3.76	1.82	0.04	S *
Cro 23 <E2>	<1 * 9>	-1.95	1.88	0.30	NS	1.32	0.89	0.14	NS
Cro 24 <E2>	<1 * 10>	-4.65	1.97	0.02	S *	-0.09	1.17	0.94	NS
Cro 25 <E2>	<2 * 6>	2.28	0.97	0.02	S *	2.23	1.05	0.04	S *
Cro 26 <E2>	<2 * 7>	0.72	1.80	0.69	NS	0.01	1.27	0.99	NS
Cro 27 <E2>	<2 * 8>	2.51	1.99	0.21	NS	2.00	1.12	0.08	S +
Cro 28 <E2>	<2 * 9>	3.00	1.18	0.01	S *	2.66	1.00	0.01	S **
Cro 29 <E2>	<2 * 10>	-6.62	3.59	0.07	S +	-2.21	1.30	0.09	S +
Cro 30 <E2>	<3 * 7>	-6.05	2.55	0.02	S *	-3.32	1.71	0.05	S +
Cro 31 <E2>	<3 * 9>	1.99	1.19	0.10	S +	2.21	0.94	0.02	S *
Cro 32 <E2>	<3 * 10>	-7.39	3.97	0.07	S +	-2.54	1.53	0.10	NS
Cro 33 <E2>	<4 * 6>	1.90	1.64	0.25	NS	2.69	1.44	0.06	S +
Cro 34 <E2>	<4 * 8>	5.67	2.21	0.01	S *	4.23	1.49	0.01	S **
Cro 35 <E2>	<4 * 9>	1.66	1.61	0.31	NS	2.64	1.29	0.04	S *
Cro 36 <E2>	<4 * 10>	-1.76	1.31	0.18	NS	0.87	0.94	0.35	NS
Cro 37 <E2>	<5 * 7>	-2.18	2.04	0.29	NS	-1.74	1.67	0.30	NS
Cro 38 <E2>	<5 * 8>	-1.00	0.91	0.27	NS	-0.06	0.76	0.94	NS
Cro 39 <E2>	<5 * 9>	-1.03	1.05	0.33	NS	0.35	0.65	0.60	NS

```
Cro 40  <E2>    <5 * 10> -1.23   1.67  0.46    NS    0.19  1.10   0.87     NS
```

```
Significance of F1 or F2 is over Population Mean 12.854884
Number          Cross    F1(G)  S.E.  P-value Sig. F2(G)  S.E.   P-value Sig.
Cro  1          <1 * 6>  16.18  1.01  0.00    S ** 14.09  0.98   0.21     NS
Cro  2          <1 * 7>  13.66  2.81  0.78    NS   12.08  1.30   0.55     NS
Cro  3          <1 * 9>   5.08  1.77  0.00    S **  4.96  1.64   0.00     S **
Cro  4          <1 * 10>  6.69  3.39  0.07    S +   5.20  2.95   0.01     S *
Cro  5          <2 * 6>  16.84  1.27  0.00    S ** 14.28  1.52   0.35     NS
Cro  6          <2 * 7>  17.24  1.85  0.02    S *  15.64  2.33   0.24     NS
Cro  7          <2 * 8>  20.23  5.09  0.15    NS   12.30  2.58   0.83     NS
Cro  8          <2 * 9>  19.12  2.07  0.00    S ** 14.56  1.07   0.12     NS
Cro  9          <2 * 10> 11.15  1.91  0.37    NS   12.32  1.42   0.71     NS
Cro 10          <3 * 7>  12.19  2.39  0.78    NS    9.33  1.62   0.03     S *
Cro 11          <3 * 9>  13.18  2.73  0.91    NS    9.86  1.89   0.12     NS
Cro 12          <3 * 10>  6.59  4.32  0.15    NS   11.41  2.81   0.61     NS
Cro 13          <4 * 6>  15.76  1.80  0.11    NS   13.19  1.32   0.80     NS
Cro 14          <4 * 8>  18.68  1.98  0.00    S ** 13.87  0.86   0.24     NS
Cro 15          <4 * 9>  15.73  1.49  0.06    S +  15.45  1.71   0.13     NS
Cro 16          <4 * 10> 10.16  0.76  0.00    S ** 11.06  0.67   0.01     S **
Cro 17          <5 * 7>  19.75  2.59  0.01    S ** 13.41  0.98   0.58     NS
Cro 18          <5 * 8>  15.40  0.84  0.00    S ** 11.83  1.97   0.61     NS
Cro 19          <5 * 9>  13.99  0.64  0.08    S +  11.16  1.12   0.13     NS
Cro 20          <5 * 10> 15.21  4.39  0.59    NS    9.20  2.55   0.16     NS
------------------------------------------------------------------------------
```

```
No.           Cross       H_pm(F_1) S.E. P-     Sig.  H_pm(F_2) S.E.  P-     Sig.
                          (GE)           value        (GE)           value
Cro  1 <E1> <1 * 6>        0.54    0.25 0.03   S *    0.27    0.12  0.03   S *
Cro  2 <E1> <1 * 7>        0.98    0.54 0.07   S +    0.49    0.27  0.07   S +
Cro  3 <E1> <1 * 9>        0.48    0.19 0.01   S *    0.24    0.10  0.01   S *
Cro  4 <E1> <1 * 10>       0.87    0.45 0.06   S +    0.44    0.23  0.06   S +
Cro  5 <E1> <2 * 6>        0.32    0.15 0.03   S *    0.16    0.07  0.03   S *
Cro  6 <E1> <2 * 7>        0.06    0.28 0.84   NS     0.03    0.14  0.84   NS
Cro  7 <E1> <2 * 8>        1.41    0.77 0.07   S +    0.70    0.38  0.07   S +
Cro  8 <E1> <2 * 9>        0.53    0.27 0.05   S +    0.27    0.13  0.05   S +
Cro  9 <E1> <2 * 10>       0.52    0.24 0.04   S *    0.26    0.12  0.04   S *
Cro 10 <E1> <3 * 7>        0.75    0.46 0.11   NS     0.37    0.23  0.11   NS
Cro 11 <E1> <3 * 9>        0.42    0.26 0.12   NS     0.21    0.13  0.12   NS
Cro 12 <E1> <3 * 10>      -0.19    0.43 0.67   NS    -0.09    0.22  0.67   NS
Cro 13 <E1> <4 * 6>        0.45    0.29 0.13   NS     0.22    0.15  0.13   NS
Cro 14 <E1> <4 * 8>        0.42    0.20 0.03   S *    0.21    0.10  0.03   S *
Cro 15 <E1> <4 * 9>        0.07    0.12 0.55   NS     0.04    0.06  0.55   NS
Cro 16 <E1> <4 * 10>       0.24    0.15 0.10   S +    0.12    0.07  0.10   S +
Cro 17 <E1> <5 * 7>        0.89    0.46 0.06   S +    0.44    0.23  0.06   S +
Cro 18 <E1> <5 * 8>        0.56    0.19 0.00   S **   0.28    0.10  0.00   S **
Cro 19 <E1> <5 * 9>        0.48    0.20 0.02   S *    0.24    0.10  0.02   S *
Cro 20 <E1> <5 * 10>       1.02    0.57 0.08   S +    0.51    0.29  0.08   S +

Cro 21 <E2> <1 * 6>       -0.33    0.13 0.01   S *   -0.17    0.06  0.01   S *
Cro 22 <E2> <1 * 7>       -0.83    0.46 0.08   S +   -0.41    0.23  0.08   S +
Cro 23 <E2> <1 * 9>       -0.51    0.24 0.04   S *   -0.25    0.12  0.04   S *
Cro 24 <E2> <1 * 10>      -0.71    0.30 0.02   S *   -0.35    0.15  0.02   S *
Cro 25 <E2> <2 * 6>        0.01    0.08 0.93   NS     0.00    0.04  0.93   NS
Cro 26 <E2> <2 * 7>        0.11    0.16 0.51   NS     0.05    0.08  0.51   NS
Cro 27 <E2> <2 * 8>        0.08    0.22 0.72   NS     0.04    0.11  0.72   NS
Cro 28 <E2> <2 * 9>        0.05    0.12 0.66   NS     0.03    0.06  0.66   NS
Cro 29 <E2> <2 * 10>      -0.69    0.39 0.08   S +   -0.34    0.19  0.08   S +
Cro 30 <E2> <3 * 7>       -0.42    0.23 0.06   S +   -0.21    0.11  0.06   S +
```

```
Cro 31 <E2> <3 * 9>   -0.03  0.12  0.77  NS   -0.02  0.06  0.77  NS
Cro 32 <E2> <3 * 10>  -0.75  0.43  0.08  S +  -0.38  0.22  0.08  S +
Cro 33 <E2> <4 * 6>   -0.12  0.12  0.29  NS   -0.06  0.06  0.29  NS
Cro 34 <E2> <4 * 8>    0.22  0.20  0.26  NS    0.11  0.10  0.26  NS
Cro 35 <E2> <4 * 9>   -0.15  0.14  0.29  NS   -0.08  0.07  0.29  NS
Cro 36 <E2> <4 * 10>  -0.41  0.20  0.04  S *  -0.20  0.10  0.04  S *
Cro 37 <E2> <5 * 7>   -0.07  0.13  0.59  NS   -0.03  0.06  0.59  NS
Cro 38 <E2> <5 * 8>   -0.15  0.09  0.09  S +  -0.07  0.04  0.09  S +
Cro 39 <E2> <5 * 9>   -0.21  0.12  0.08  S +  -0.11  0.06  0.08  S +
Cro 40 <E2> <5 * 10>  -0.22  0.20  0.26  NS   -0.11  0.10  0.26  NS
```

Significance of F1 or F2 is over Population Mean 12.854884

No.	Cro	$H_{pm}(F_1)$ (G)	S.E.	P-value	Sig.	$H_{pm}(F_2)$ (G)	S.E.	P-value	Sig.
Cro 1	<1 * 6>	0.56	0.12	0.00	S **	0.40	0.08	0.00	S **
Cro 2	<1 * 7>	0.27	0.31	0.39	NS	0.15	0.16	0.35	NS
Cro 3	<1 * 9>	0.27	0.10	0.01	S *	0.26	0.08	0.00	S **
Cro 4	<1 * 10>	0.04	0.19	0.85	NS	-0.08	0.12	0.50	NS
Cro 5	<2 * 6>	0.16	0.20	0.42	NS	-0.04	0.13	0.78	NS
Cro 6	<2 * 7>	0.10	0.18	0.56	NS	-0.02	0.10	0.84	NS
Cro 7	<2 * 8>	0.65	0.62	0.30	NS	0.04	0.37	0.92	NS
Cro 8	<2 * 9>	0.91	0.26	0.00	S **	0.56	0.13	0.00	S **
Cro 9	<2 * 10>	-0.07	0.17	0.70	NS	0.03	0.10	0.80	NS
Cro 10	<3 * 7>	0.40	0.21	0.06	S +	0.18	0.11	0.09	S +
Cro 11	<3 * 9>	1.14	0.21	0.00	S **	0.88	0.19	0.00	S **
Cro 12	<3 * 10>	0.27	0.52	0.61	NS	0.65	0.40	0.11	NS
Cro 13	<4 * 6>	0.30	0.18	0.10	NS	0.10	0.11	0.36	NS
Cro 14	<4 * 8>	0.75	0.28	0.01	S **	0.38	0.14	0.01	S *
Cro 15	<4 * 9>	0.87	0.24	0.00	S **	0.84	0.25	0.00	S **
Cro 16	<4 * 10>	0.08	0.12	0.53	NS	0.15	0.10	0.14	NS
Cro 17	<5 * 7>	0.53	0.43	0.22	NS	0.04	0.26	0.88	NS
Cro 18	<5 * 8>	0.51	0.23	0.03	S *	0.23	0.11	0.04	S *
Cro 19	<5 * 9>	0.75	0.16	0.00	S **	0.53	0.10	0.00	S **
Cro 20	<5 * 10>	0.49	0.42	0.25	NS	0.02	0.26	0.94	NS

No.	Cross	$H_{pb}(F_1)$ (GE)	S.E.	P-value	Sig.	$H_{pb}(F_2)$ (GE)	S.E.	P-value	Sig.
Cro 1 <E1>	<1 * 6>	0.25	0.23	0.28	NS	-0.03	0.13	0.85	NS
Cro 2 <E1>	<1 * 7>	0.52	0.56	0.35	NS	0.03	0.31	0.91	NS
Cro 3 <E1>	<1 * 9>	0.25	0.16	0.13	NS	0.01	0.10	0.92	NS
Cro 4 <E1>	<1 * 10>	0.61	0.45	0.17	NS	0.18	0.23	0.45	NS
Cro 5 <E1>	<2 * 6>	0.28	0.18	0.13	NS	0.11	0.15	0.44	NS
Cro 6 <E1>	<2 * 7>	-0.05	0.27	0.86	NS	-0.08	0.19	0.68	NS
Cro 7 <E1>	<2 * 8>	1.09	0.75	0.15	NS	0.39	0.37	0.30	NS
Cro 8 <E1>	<2 * 9>	0.42	0.31	0.18	NS	0.15	0.20	0.46	NS
Cro 9 <E1>	<2 * 10>	0.43	0.31	0.17	NS	0.17	0.23	0.45	NS
Cro 10 <E1>	<3 * 7>	0.59	0.50	0.24	NS	0.22	0.29	0.46	NS
Cro 11 <E1>	<3 * 9>	0.35	0.30	0.25	NS	0.14	0.19	0.45	NS
Cro 12 <E1>	<3 * 10>	-0.23	0.42	0.60	NS	-0.13	0.22	0.56	NS
Cro 13 <E1>	<4 * 6>	0.35	0.31	0.27	NS	0.12	0.17	0.46	NS
Cro 14 <E1>	<4 * 8>	0.05	0.22	0.81	NS	-0.16	0.17	0.36	NS
Cro 15 <E1>	<4 * 9>	-0.10	0.13	0.46	NS	-0.13	0.09	0.16	NS
Cro 16 <E1>	<4 * 10>	0.10	0.20	0.60	NS	-0.02	0.15	0.91	NS
Cro 17 <E1>	<5 * 7>	0.64	0.44	0.15	NS	0.20	0.22	0.38	NS
Cro 18 <E1>	<5 * 8>	0.38	0.23	0.11	NS	0.10	0.19	0.61	NS
Cro 19 <E1>	<5 * 9>	0.46	0.20	0.02	S *	0.22	0.13	0.08	S +
Cro 20 <E1>	<5 * 10>	0.97	0.59	0.11	NS	0.46	0.32	0.15	NS
Cro 21 <E2>	<1 * 6>	-0.36	0.19	0.06	S +	-0.20	0.14	0.16	NS

```
Cro 22 <E2><1 *  7>  -1.08  0.48 0.03  S *   -0.67 0.25 0.01  S **
Cro 23 <E2><1 *  9>  -0.53  0.27 0.05  S +   -0.28 0.17 0.10  S +
Cro 24 <E2><1 * 10>  -0.74  0.34 0.03  S *   -0.38 0.21 0.07  S +
Cro 25 <E2><2 *  6>  -0.14  0.11 0.21  NS    -0.14 0.08 0.08  S +
Cro 26 <E2><2 *  7>   0.03  0.23 0.90  NS    -0.02 0.16 0.88  NS
Cro 27 <E2><2 *  8>  -0.01  0.24 0.96  NS    -0.05 0.14 0.71  NS
Cro 28 <E2><2 *  9>  -0.10  0.14 0.48  NS    -0.13 0.10 0.19  NS
Cro 29 <E2><2 * 10>  -0.83  0.40 0.04  S *   -0.49 0.22 0.03  S *
Cro 30 <E2><3 *  7>  -0.51  0.26 0.05  S +   -0.30 0.16 0.06  S +
Cro 31 <E2><3 *  9>  -0.18  0.14 0.21  NS    -0.16 0.10 0.11  NS
Cro 32 <E2><3 * 10>  -0.89  0.44 0.05  S *   -0.52 0.23 0.03  S *
Cro 33 <E2><4 *  6>  -0.17  0.11 0.13  NS    -0.11 0.07 0.13  NS
Cro 34 <E2><4 *  8>   0.21  0.22 0.33  NS     0.10 0.13 0.43  NS
Cro 35 <E2><4 *  9>  -0.21  0.14 0.15  NS    -0.13 0.09 0.14  NS
Cro 36 <E2><4 * 10>  -0.45  0.22 0.04  S *   -0.25 0.14 0.07  S +
Cro 37 <E2><5 *  7>  -0.10  0.18 0.57  NS    -0.07 0.13 0.60  NS
Cro 38 <E2><5 *  8>  -0.28  0.12 0.02  S *   -0.21 0.09 0.03  S *
Cro 39 <E2><5 *  9>  -0.42  0.13 0.00  S **  -0.31 0.09 0.00  S **
Cro 40 <E2><5 * 10>  -0.41  0.24 0.09  S +   -0.30 0.15 0.05  S +
```

Significance of F1 or F2 is over Population Mean 12.854884

No.	Cro	$H_{pb}(F_1)$ (G)	S.E.	P-value	Sig.	$H_{pb}(F_2)$ (G)	S.E.	P-value	Sig.
Cro 1	<1 * 6>	0.14	0.17	0.43	NS	-0.02	0.15	0.88	NS
Cro 2	<1 * 7>	-0.24	0.31	0.44	NS	-0.36	0.18	0.05	S *
Cro 3	<1 * 9>	0.12	0.14	0.42	NS	0.11	0.13	0.42	NS
Cro 4	<1 * 10>	-0.17	0.21	0.42	NS	-0.29	0.15	0.05	S +
Cro 5	<2 * 6>	0.14	0.20	0.49	NS	-0.06	0.13	0.63	NS
Cro 6	<2 * 7>	0.04	0.20	0.85	NS	-0.08	0.13	0.53	NS
Cro 7	<2 * 8>	0.40	0.64	0.53	NS	-0.22	0.39	0.58	NS
Cro 8	<2 * 9>	0.31	0.24	0.19	NS	-0.04	0.16	0.80	NS
Cro 9	<2 * 10>	-0.31	0.25	0.22	NS	-0.22	0.20	0.28	NS
Cro 10	<3 * 7>	-0.35	0.23	0.14	NS	-0.58	0.20	0.00	S **
Cro 11	<3 * 9>	1.05	0.21	0.00	S **	0.79	0.21	0.00	S **
Cro 12	<3 * 10>	-0.18	0.53	0.74	NS	0.20	0.40	0.63	NS
Cro 13	<4 * 6>	0.11	0.18	0.55	NS	-0.09	0.11	0.41	NS
Cro 14	<4 * 8>	0.72	0.27	0.01	S **	0.34	0.15	0.03	S *
Cro 15	<4 * 9>	0.49	0.25	0.05	S +	0.46	0.26	0.08	S +
Cro 16	<4 * 10>	0.05	0.17	0.76	NS	0.12	0.16	0.44	NS
Cro 17	<5 * 7>	0.23	0.44	0.60	NS	-0.26	0.27	0.34	NS
Cro 18	<5 * 8>	0.49	0.23	0.04	S *	0.22	0.13	0.10	S +
Cro 19	<5 * 9>	0.38	0.19	0.05	S *	0.16	0.18	0.36	NS
Cro 20	<5 * 10>	0.48	0.39	0.23	NS	0.01	0.25	0.96	NS

Significance of Heterosis is over Population Mean 12.854884

```
Pre(F1)          13.987      0.730276    0.124       NS
Pre(F2)          12.6895     0.306025    0.59        NS
Hpm(F1)          0.147004    0.136824    0.285       NS
Hpm(F2)          0.0457101   0.0706352   0.519       NS
Hpb(F1)         -0.837894    0.143119    6.23e-008   S **
Hpb(F2)         -0.939187    0.0983169  -5.09e-011   S **
Generation n     0.569721    0.0534007  -5.09e-011   S **
0.5 Omiga(AA)    0.673029    0.0727739  -5.09e-011   S **
2Delta(AA)      -0.0555835   0.0447671   0.217       NS
```

Results of Lint% are not presented.
  Time Used (Hour) = 0.003056

Chapter 5

Diallel Analysis for an Animal Model with Sex-Linked and Maternal Effects Along with Genotype-by-Environment Interaction Effects

Jun Zhu

Purpose

To analyze balanced or unbalanced data of an animal genetic model for estimating components of variance, covariance, heritability, and selection response.

Definitions

Mating Design

A set of inbred lines is sampled from a reference population. Parents are used to produce F_1 crosses. Experiments with parents and their F_1s are conducted in multiple environments.

Genetic Model

The genetic model for the phenotypic mean (y_{ijsk}) of sex s in block k within environment h from the cross between maternal line i and paternal line j is

$$y_{hijsk} = \mu + E_h + G_{ijs} + GE_{hijs} + e_{hijsk}$$

where μ = population mean, E_h = environment effect, G_{ijs} = genotype effect, GE_{hijs} = genotype-environment effect, and e_{hijsk} = residual effect.

The total genotype effect G_{ijs} and genotype × environment interaction effect GE_{hijs} can be further partitioned into different components for heterogametic progeny (XY or ZW, $s = 1$) and for homogametic progeny (XX or ZZ, $s = 2$):

$$G_{ij1}^{XY} + GE_{hij1}^{XY} = A_i + A_j + D_{ij} + L_{i1} + M_i + AE_{hi} + AE_{hj} + DE_{hij} + LE_{hi1} + ME_{hi}$$

$$\text{or } G_{ij1}^{ZW} + GE_{hij1}^{ZW} = A_i + A_j + D_{ij} + L_{j1} + M_i + AE_{hi} + AE_{hj} + DE_{hij} + LE_{hj1} + ME_{hi}$$

$$G_{ij2}^{XX/ZZ} + GE_{hij2}^{XX/ZZ} = A_i + A_j + D_{ij} + \frac{1}{2}L_{i2} + \frac{1}{2}L_{j2} + M_i + AE_{hi} + AE_{hj} + DE_{hij} + \frac{1}{2}LE_{hi2}$$

$$+ \frac{1}{2}LE_{hj2} + ME_{hi}$$

where A_i (or A_j) $\sim (0, \sigma_A^2)$ is the additive effect of autosomal genes; $D_{ij} \sim (0, \sigma_D^2)$ is the dominance effect of autosomal genes; $L_i 1$ (or $L_j 1$) and $L_i 2$ (or $L_j 2$) $\sim (0, \sigma_L^2)$ is the additive effect of sex-linked genes; $M_i \sim (0, \sigma_M^2)$ is the maternal effect of dam i; AE_{hi} (or AE_{hj}) $\sim (0, \sigma_{AE}^2)$ is the additive × environment interaction effect of autosomal genes; $DE_{hij} \sim (0, \sigma_{DE}^2)$ is the dominance × environment interaction effect of autosomal genes; L_{i1} (or L_{j1}), L_{i2} (or L_{j2}) $\sim (0, \sigma_{LE}^2)$ is the sex-linked additive × environment interaction effect; and $ME_{hi} \sim (0, \sigma_M^2)$ is the maternal × environment interaction effect of dam i.

Analysis

Mixed Linear Model

The phenotypic mean of the genetic model can be expressed by a mixed linear model as

$$y = Xb + U_A e_A + U_D e_D + U_L e_L + U_M e_M + U_{AE} e_{AE} + U_{DE} e_{DE} + U_{LE} e_{LE}$$
$$+ U_{ME} e_{ME} + e_e$$
$$= Xb + \sum_u^9 U_u e_u$$

with variance-covariance matrix

$$\text{var}(y) = \sigma_A^2 U_A U_A^T + \sigma_D^2 U_D U_D^T + \sigma_L^2 U_L U_L^T + \sigma_M^2 U_M U_M^T + \sigma_{AE}^2 U_{AE} U_{AE}^T$$
$$+ \sigma_{DE}^2 U_{DE} U_{DE}^T + \sigma_{LE}^2 U_{LE} U_{LE}^T + \sigma_{ME}^2 U_{ME} U_{ME}^T + \sigma_e^2 I$$
$$= \sum_{u=1}^{9} \sigma_u^2 V_u.$$

Variance Components

Unbiased estimation of variances can be obtained by REML or MINQUE(1) approaches. When experimental variances are estimated, genetic variance components can be obtained by $V_A = 2\sigma_A^2$, $V_D = \sigma_D^2$, $V_L = \sigma_L^2$, $V_M = \sigma_M^2$, $V_{AE} = 2\sigma_{AE}^2$, $V_{DE} = \sigma_{DE}^2$, $V_{LE} = \sigma_{LE}^2$, $V_M = \sigma_M^2$, $V_e = \sigma_e^2$. The total phenotypic variance is $V_P = V_A + V_D + V_L + V_M + V_{AE} + V_{DE} + V_{LE} + V_{ME} + V_e$.

Covariance Components and Correlation

Unbiased estimation of covariances can be obtained by MINQUE(1) approaches (Zhu, 1997; Zhu and Weir, 1996). When experimental covariances are estimated, genetic covariance components can be obtained by $C_A = 2\sigma_{A/A}$, $C_D = \sigma_{D/D}$, $C_L = \sigma_{L/L}$, $C_M = \sigma_{M/M}$, $C_{AE} = \sigma_{AE/AE}$, $C_{DE} = \sigma_{DE/DE}$, $C_{LE} = \sigma_{LE/LE}$, $C_{ME} = \sigma_{ME/ME}$, $C_e = \sigma_{e/e}$. The total phenotypic covariance is $C_P = C_A + C_D + C_L + C_M + C_{AE} + C_{DE} + C_{LE} + C_{ME} + C_e$. For trait 1 and trait 2, correlation coefficients of genetic components can be estimated by

$$r_A = C_A / \sqrt{V_{A(1)} V_{A(2)}},$$
$$r_D = C_D / \sqrt{V D_{(1)} V_{D(2)}},$$
$$r_L = C_L / \sqrt{V_{L(1)} V_{L(2)}},$$
$$r_M = C_M / \sqrt{V_{M(1)} V_{M(2)}},$$
$$r_{AE} = C_{AE} / \sqrt{V_{AE(1)} V_{AE(2)}},$$
$$r_{LE} = r_{LE} = C_{LE} / \sqrt{V_{LE(1)} V_{LE(2)}},$$
$$r_{ME} = r_{ME} = C_{ME} / \sqrt{V_{ME(1)} V_{ME(2)}}, \text{ and}$$
$$r_{eE} = r_e = C_e / \sqrt{V_{e(1)} V_{e(2)}}.$$

Heritability Components

The total heritability (h^2) can be partitioned into two components ($h^2 = h_G^2 + h_{GE}^2$), where $h_G^2 = V_A / V_P$ is general heritability and $h_{GE}^2 = V_{AE} / V_P$ is interaction heritability (Zhu, 1997).

Selection Response

The total selection response ($R = ih^2 \sqrt{V_P}$) can be partitioned into two components (Zhu, 1997):

$$R = R_G + R_{GE}$$

where $R_G = ih_G^2 \sqrt{V_P}$ is general response and $R_{GE} = ih_{GE}^2 \sqrt{V_P}$ is interaction response.

Originators

Zhu, J. (1997). *Analysis Methods for Genetic Models.* Agricultural Publication House of China, Beijing.

Zhu, J. and Weir, B.S. (1996). Diallel analysis for sex-linked and maternal effects. *Theoretical and Applied Genetics,* 92(1):1-9.

Software Available

Zhu, J. (1997). GENSEX.EXE for constructing animal models, GENVAR1R.EXE or GENVAR1C.EXE for estimating components of variance and heritability, GENCOV1R.EXE or GENCOV1C.EXE for estimating components of covariance and correlation, GENHET1R.EXE or GENHET1C.EXE for predicting genetic effects and components of heterosis. *Analysis Methods for Genetic Models* (pp. 278-285), Agricultural Publication House of China, Beijing (program free of charge). Contact Dr. Jun Zhu, Department of Agronomy, Zhejiang University, Hangzhou, China. E-mail: <jzhu@zju.edu.cn>.

EXAMPLE

Balanced mice data (provided by William R. Atchley, Department of Genetics, North Carolina State University, Raleigh, NC) to be analyzed (Parent = 1, Year = 1, Sex = 1 & 2, Blk = 1):

Year	Fem	Male	Cross	Rep	Sex	35BW	35TL
1	1	1	0	1	1	20.23	79.78
1	1	1	0	1	2	17.71	78.93
1	1	2	1	1	1	22.01	84.79
1	1	2	1	1	2	19.44	82.87
1	1	3	1	1	1	22.48	93.66
1	1	3	1	1	2	18.34	88.74
1	1	4	1	1	1	22.80	85.48
1	1	4	1	1	2	20.41	85.09
1	1	5	1	1	1	22.57	82.83
1	1	5	1	1	2	19.25	81.83
1	1	6	1	1	1	25.11	86.36
1	1	6	1	1	2	21.79	84.54
1	1	7	1	1	1	22.67	89.44
1	1	7	1	1	2	19.80	87.06
1	2	1	1	1	1	22.91	88.60
1	2	1	1	1	2	19.14	86.39
1	2	2	0	1	1	20.94	83.13
1	2	2	0	1	2	18.50	82.40
1	2	3	1	1	1	22.09	91.83
1	2	3	1	1	2	18.28	88.57
1	2	4	1	1	1	22.37	81.91
1	2	4	1	1	2	20.30	82.42
1	2	5	1	1	1	23.61	86.13
1	2	5	1	1	2	20.16	83.73
1	2	6	1	1	1	26.45	88.73
1	2	6	1	1	2	22.01	86.77
1	2	7	1	1	1	22.86	87.86
1	2	7	1	1	2	19.85	86.45
1	3	1	1	1	1	23.73	85.75
1	3	1	1	1	2	19.86	84.80
1	3	2	1	1	1	24.18	84.48
1	3	2	1	1	2	19.75	82.55
1	3	3	0	1	1	23.72	87.41
1	3	3	0	1	2	19.09	84.93
1	3	4	1	1	1	25.36	87.27
1	3	4	1	1	2	20.00	85.20
1	3	5	1	1	1	21.98	79.03
1	3	5	1	1	2	18.77	77.21
1	3	6	1	1	1	26.48	85.66
1	3	6	1	1	2	21.85	83.52
1	3	7	1	1	1	24.99	86.89
1	3	7	1	1	2	20.41	85.06
1	4	1	1	1	1	23.33	84.48
1	4	1	1	1	2	20.77	83.14
1	4	2	1	1	1	23.18	81.61
1	4	2	1	1	2	19.47	79.41
1	4	3	1	1	1	22.50	88.10
1	4	3	1	1	2	18.90	84.24

1	4	4	0	1	1	24.24	85.97
1	4	4	0	1	2	20.91	84.16
1	4	5	1	1	1	23.22	83.22
1	4	5	1	1	2	19.33	82.19
1	4	6	1	1	1	24.01	83.07
1	4	6	1	1	2	20.62	81.25
1	4	7	1	1	1	24.86	86.73
1	4	7	1	1	2	20.70	85.19
1	5	1	1	1	1	22.07	87.14
1	5	1	1	1	2	19.37	87.90
1	5	2	1	1	1	21.05	82.04
1	5	2	1	1	2	18.78	82.88
1	5	3	1	1	1	21.32	84.22
1	5	3	1	1	2	18.19	82.25
1	5	4	1	1	1	23.31	90.07
1	5	4	1	1	2	20.18	89.91
1	5	5	0	1	1	23.79	84.50
1	5	5	0	1	2	20.48	84.69
1	5	6	1	1	1	24.48	87.04
1	5	6	1	1	2	21.15	87.06
1	5	7	1	1	1	21.41	81.79
1	5	7	1	1	2	19.18	82.03
1	6	1	1	1	1	22.28	87.39
1	6	1	1	1	2	18.81	85.55
1	6	2	1	1	1	18.86	75.66
1	6	2	1	1	2	15.75	74.44
1	6	3	1	1	1	21.68	87.52
1	6	3	1	1	2	16.24	83.10
1	6	4	1	1	1	23.01	85.38
1	6	4	1	1	2	18.64	85.04
1	6	5	1	1	1	22.97	84.62
1	6	5	1	1	2	18.69	82.70
1	6	6	0	1	1	25.60	84.43
1	6	6	0	1	2	20.88	83.36
1	6	7	1	1	1	22.91	84.57
1	6	7	1	1	2	18.81	82.43
1	7	1	1	1	1	22.59	91.29
1	7	1	1	1	2	17.91	88.00
1	7	2	1	1	1	22.48	86.97
1	7	2	1	1	2	17.50	83.04
1	7	3	1	1	1	21.71	86.41
1	7	3	1	1	2	17.54	83.27
1	7	4	1	1	1	24.23	92.60
1	7	4	1	1	2	19.88	89.00
1	7	5	1	1	1	23.79	86.83
1	7	5	1	1	2	18.93	84.83
1	7	6	1	1	1	25.07	89.44
1	7	6	1	1	2	20.21	87.38
1	7	7	0	1	1	24.31	90.48
1	7	7	0	1	2	19.81	86.65

1. Run GENSEX.EXE to create mating design matrix files and AD+ L+M model data. Before running this program, create a data file (MICEDATA.TXT) for your analysis with six design columns fol-

lowed by trait columns. The six design columns are (1) environment, (2) maternal, (3) paternal, (4) generation, (5) replication, and (6) sex. There is a limitation (<100 traits) for the number of trait columns. An example of the data file is provided with the name MICEDATA.TXT.

2. Run programs for variance and covariance analyses. Standard errors of estimates are calculated by the jackknife procedures. If you have multiple blocks for your experiments, you can use GENVAR1R.EXE or GENCOV1R.EXE for jackknifing over blocks. Otherwise you can use GENVAR1C.EXE or GENCOV1C.EXE for jackknifing over cell means.

3. Run GENVAR1R.EXE or GENVAR1C.EXE for estimating variance components and predicting genetic effects before estimating covariance and correlation. These two programs will allow you to choose the parental type (inbred or outbred) and the prediction methods (LUP or AUP). You also need to input coefficients (1, 0, or −1) for conducting linear contrasts for genetic effects of parents.

4. After finishing variance analysis, run GENCOV1R.EXE or GENCOV1C.EXE to estimate covariance components and coefficients of correlation among all analyzed traits.

5. Results will automatically be stored in text files for later use or printing.

Output 1 for Variance Analysis

```
Traits =, 2
Variance components = , 5
Degree of freedom = , 48
File name is micedata.VAR
Date and Time for Analysis: Sat Jun 24 20:03:15 2000

Variance Components Estimated by MINQUE(1) with GENVAR1R.EXE.
Jackknifing Over Block Conducted for Estimating S.E.
Predicting Genetic Effects by Adjusted Unbiased Prediction (AUP) Method.

NS = Not significant; S+ = Significant at 0.10 level.
S* = Significant at 0.05 level; S** = Significant at 0.01 level.

Linear Contrasts:
  +<1>   +<2>   +<3>   +<4>   −<5>   −<6>   −<7>

Diallel Analysis of Trait, 35BW, for Public Users.

Var Comp               Estimate      S. E.         P-value
(1): Additive Var      4.05678       0.914491      2.67e-005     S**
(2): Dominance Var     0.447741      0.114861      0.00015       S**
```

```
(3): Sex-linked Var         4.82177        0.332787     2.87e-017    S**
(4): Maternal Var           3.16826        0.912454     0.000551     S**
(5): Residual Var           0.979275       0.309302     0.00134      S**
(6): Var(Pheno.)            13.4738        1.91195      3.11e-009    S**

Proportion of Var(G)/Var(T)Estimate        S. E.        P-value
(1): Additive Var/Vp        0.301086       0.0251507    2.53e-016    S**
(2): Dominance Var/Vp       0.0332304      0.011576     0.00304      S**
(3): Sex-linked Var/Vp      0.357862       0.032024     2.98e-015    S**
(4): Maternal Var/Vp        0.235142       0.0202444    7.84e-016    S**
(5): Residual Var/Vp        0.0726798      0.0271762    0.0051       S**

Heritability                Estimate       S. E.        P-value
(6): Heritability(N)        0.301086       0.0251507    2.53e-016    S**
(7): Heritability(B)        0.334316       0.0236245    3.21e-017    S**

Genetic Predictor,  S. E. , P-value
(1): Random Effect is Additive Effects
A1                         -1.163629       0.383414     0.00388      S**
A2                         -1.622929       0.534681     0.00387      S**
A3                         -1.099644       0.273847     0.000208     S**
A4                          0.535614       0.274989     0.0573       S+
A5                          0.188242       0.412984     0.651        NS
A6                          2.638442       0.623674     0.000104     S**
A7                          0.517437       0.316664     0.109        NS
Linear Contrast            -6.21749        1.79317      0.00112      S**

(2): Random Effect is Dominance Effects
D1*1                       -1.575254       1.171944     0.185        NS
D2*2                       -0.823010       0.741883     0.273        NS
D3*3                        0.161857       0.293363     0.584        NS
D4*4                        0.306696       0.310885     0.329        NS
D5*5                        1.089396       0.790779     0.175        NS
D6*6                        1.449270       1.100055     0.194        NS
D7*7                        0.697693       0.513080     0.18         NS
D1*2                        0.719177       0.566455     0.21         NS
D1*3                        0.433285       0.376514     0.256        NS
D1*4                        0.532937       0.514004     0.305        NS
D1*5                        0.105613       0.398419     0.792        NS
D1*6                        0.617389       0.608657     0.316        NS
D1*7                       -0.008952       0.420072     0.983        NS
D2*3                        0.292272       0.710588     0.683        NS
D2*4                        0.047317       0.659437     0.943        NS
D2*5                        0.062535       0.567660     0.913        NS
D2*6                       -0.378167       2.358942     0.873        NS
D2*7                       -0.164429       0.355381     0.646        NS
D3*4                       -0.129561       0.270304     0.634        NS
D3*5                       -1.016032       0.950404     0.29         NS
D3*6                       -0.268319       0.703458     0.705        NS
D3*7                       -0.256718       0.609336     0.675        NS
D4*5                       -0.370931       0.502767     0.464        NS
D4*6                       -0.697666       0.733653     0.346        NS
D4*7                        0.261144       0.349548     0.459        NS
D5*6                       -0.182574       0.462678     0.695        NS
D5*7                       -0.609768       0.793060     0.446        NS
D6*7                       -0.298528       0.437983     0.499        NS
```

```
Heterosis <Delta>           -0.738068      1.8011       0.684        NS
(3): Random Effect is Sex-linked Effects
L1 for Sex1                 1.619670      0.268733     2.28e-007     S**
L2 for Sex1                 1.950941      0.409804     1.81e-005     S**
L3 for Sex1                 2.545023      0.396915     5.87e-008     S**
L4 for Sex1                 1.998426      0.326978     1.69e-007     S**
L5 for Sex1                 1.194290      0.283566     0.000111      S**
L6 for Sex1                 2.173739      0.465039     2.42e-005     S**
L7 for Sex1                 3.040606      0.367137     8.28e-011     S**
L1 for Sex2                -1.468479      0.283005     4.22e-006     S**
L2 for Sex2                -1.416539      0.315276     4.42e-005     S**
L3 for Sex2                -2.789509      0.420796     2.73e-008     S**
L4 for Sex2                -1.799656      0.408109     5.82e-005     S**
L5 for Sex2                -2.015034      0.390778     4.72e-006     S**
L6 for Sex2                -2.912965      0.350283     7.36e-011     S**
L7 for Sex2                -2.124706      0.288782     2.09e-009     S**
Linear Contrast             4.99948       0.147414     5.87e-017     S**

(4): Random Effect is Maternal Effects
M1                          0.532529      0.235722     0.0285        S*
M2                          1.445810      0.744384     0.058         S+
M3                          1.947175      0.459670     0.000102      S**
M4                          0.036886      0.341978     0.915         NS
M5                          0.198565      0.396981     0.619         NS
M6                         -3.214121      0.790225     0.000176      S**
M7                         -0.950092      0.338411     0.0072        S**
Linear Contrast             5.89251       1.83518      0.00236       S**

Fixed Effect <1>, 21.2871

Results of Tail Length are not presented.

Time Used (Hour) = 0.001389
```

Output 2 for Covariance Analysis

```
Traits =, 2
Covariance components = , 5
Degree of freedom = , 48
File name is micedata.COV
Date and Time for Analysis: Sat Jun 24 20:03:33 2000

Covariance Components Estimated by MINQUE(1) with GENCOV1C.EXE.
Jackknifing Over Cell Mean Conducted for Estimating S.E.

NS = Not significant; S+ = Significant at 0.10 level.
S* = Significant at 0.05 level; S** = Significant at 0.01 level.

Covariances and Correlations Between, 35BW, , &, 35TL, for Public
    Users.:

Covariances               Estimates      S.E.        P-value
Additive  Cov             1.2739         1.40446      0.369        NS
Dominance Cov            -0.279037       0.886415     0.754        NS
Sex-linked  Cov           2.07233        0.465324     5.04e-005    S**
```

| Maternal Cov | 0.848917 | 1.69412 | 0.619 | NS |
| Residual Cov | 1.76704 | 0.698536 | 0.0148 | S* |

Cov <1=Genotypic>				
Cov <2=Phenotypic>	Estimates	S.E.	P-value	
Cov 2	5.68315	2.84705	0.0516	S+
Cov 1	3.91611	2.85586	0.177	NS

Correlation	Estimates	S.E.	P-value	
Additive Cor	0.190211	0.07873	0.0195	S*
Dominance Cor	-0.169791	0.0696301	0.0185	S*
Sex-linked Cor	0.589663	0.0588193	2.33e-013	S**
Maternal Cor	0.138456	0.0704558	0.0552	S+
Residual Cor	0.664416	0.0763856	1.98e-011	S**

Cor <1=Genotypic>				
Cor <2=Phenotypic>	Estimates	S.E.	P-value	
Cor 2	0.248755	0.0830793	0.00434	S**
Cor 1	0.197347	0.0879154	0.0294	S*

Results of Tail Length are not presented.

Time Used (Hour) = 0.000556

Chapter 6

Generation Means Analysis

Michael M. Kenty
David S. Wofford

Importance

Development of elite varieties or germplasm often involves quantitatively inherited traits, such as insect resistance in soybeans. In these instances, it is advantageous for the breeder/geneticist to utilize methodology that allows not only the selection of desirable genotypes but also a determination of the underlying genetic effects contributing to the expression of the desired trait.

Definitions

Using Hayman's (1958) methodology and Gamble's (1962) notation, the models for the generation means analysis are as follows:

$$
\begin{aligned}
P_1 &= m + a - d/2 + aa - ad + dd/4 \\
P_2 &= m - a - d/2 + aa + ad + dd/4 \\
F_1 &= m + d/2 + dd/4 \\
F_2 &= m \\
BC_1 &= m + a/2 + aa/4 \\
BC_2 &= m - a/2 + aa/4 \\
F_3 &= m - d/4 + dd/16 \\
BS_1 &= m + a/2 - d/4 + aa/4 - ad/8 + dd/16 \\
BS_2 &= m - a/2 - d/4 + aa/4 + ad/8 + dd/16
\end{aligned}
$$

where m = overall mean, a = additive genetic effects, d = dominance genetic effects, aa = additive × additive genetic effects, ad = additive × dominance genetic effects, dd = dominance × dominance genetic effects.

Originators

Gamble, E.E. (1962). Gene effects in corn (*Zea mays* L.) I. Separation and relative importance of gene effects for yield. *Canadian Journal of Plant Science* 42:339-348.
Hayman, B.I. (1958). The separation of epistatic from additive and dominance variation in generation means. *Heredity* 12:371-390.

Software Available

Kenty, M.M. (1994). Inheritance of resistance to the soybean looper in soybean. Doctoral dissertation. University of Florida, Gainesville.

Key References

Kenty, M.M., Hinson, K., Quesenberry, K.H., and Wofford, D.S. (1996). Inheritance of resistance to the soybean looper in soybean. *Crop Science* 36:1532-1537.
Meredith, W.R., Jr. and Bridge, R.R. (1972). Heterosis and gene action in cotton, *Gossypium hirsutum* L. *Crop Science* 12:304-310.
Scott, G.E., Hallauer, A.R., and Dicke, F.F. (1964). Types of gene action conditioning resistance to European corn borer leaf feeding. *Crop Science* 4:603-605.

Contacts

Michael M. Kenty, 424 Quail Crest Drive, Collierville, TN 38017-1750, USA. E-mail: <mmkenty@aol.com>.
David S. Wofford, 2183 McCarty Hall, P.O. Box 110300, University of Florida, Gainesville, FL 32611-0300, USA. E-mail:< dsw@gnv.ifas.ufl.edu>.

EXAMPLE

The data are presented in this order: generation, plant no., rating 1(top 1/3 plant), rating 2 (middle 1/3 plant), and rating 3 (bottom 1/3 plant). Data to be analyzed (insect defoliation, 1 = 0-10%, 10 = 91-100%):

P1 1 1 2 1	P2 4 7 7 8	F2 2 6 7 8	F3 5 2 2 4
P1 2 1 3 1	P2 5 7 8 8	F2 3 4 5 5	BC1 1 1 1 2
P1 3 2 2 2	F1 1 2 3 3	F2 4 2 2 3	BC1 2 2 2 3
P1 4 1 2 1	F1 2 3 3 3	F2 5 5 6 6	BC1 3 1 2 1
P1 5 1 2 1	F1 3 2 3 4	F3 1 1 1 2	BC1 4 1 1 2
P2 1 7 8 8	F1 4 2 4 4	F3 2 3 3 4	BC1 5 2 3 3
P2 2 7 8 9	F1 5 3 3 4	F3 3 5 5 6	BC2 1 1 1 1
P2 3 8 8 9	F2 1 1 2 1	F3 4 2 3 3	BC2 2 1 2 2

BC2 3223	F3 10223	P1 12111	P2 19678
BC2 4222	BS1 6667	P1 13223	P2 20779
BC2 5333	BS1 7789	P1 14112	BC1 16123
BS1 1677	BS1 8889	P1 15223	BC1 17122
BS1 2789	BS1 9777	P2 11677	BC1 18112
BS1 3889	BS1 10677	P2 12789	BC1 19223
BS1 4788	BC1 6111	P2 13889	BC1 20233
BS1 5677	BC1 7223	P2 14778	BS1 16667
BS2 1667	BC1 8121	P2 15888	BS1 17889
BS2 2777	BC1 9122	BC1 11121	BS1 18777
BS2 3899	BC1 10223	BC1 12233	BS1 19678
BS2 4789	BC2 6112	BC1 13233	BS1 20889
BS2 5889	BC2 7223	BC1 14123	F1 16233
P2 6888	BC2 8222	BC1 15222	F1 17334
P2 7778	BC2 9333	BS1 11677	F1 18445
P2 8678	BC2 10123	BS1 12778	F1 19223
P2 9779	BS2 6677	BS1 13889	F1 20122
P2 10889	BS2 7778	BS1 14667	F2 16121
F1 6233	BS2 8889	BS1 15777	F2 17445
F1 7334	BS2 9789	BS2 11667	F2 18667
F1 8234	BS2 10678	BS2 12778	F2 19223
F1 9233	F1 11233	BS2 13888	F2 20123
F1 10333	F1 12222	BS2 14789	F3 16113
P1 6121	F1 13234	BS2 15677	F3 17122
P1 7122	F1 14334	BC2 11111	F3 18334
P1 8133	F1 15233	BC2 12223	F3 19556
P1 9123	F2 11121	BC2 13121	F3 20678
P1 10112	F2 12678	BC2 14112	BC2 16112
F2 6678	F2 13556	BC2 15123	BC2 17121
F2 7334	F2 14778	P1 16111	BC2 18223
F2 8223	F2 15223	P1 17212	BC2 19234
F2 9566	F3 11112	P1 18223	BC2 20122
F2 10112	F3 12334	P1 19122	BS2 16677
F3 6223	F3 13445	P1 20121	BS2 17778
F3 7112	F3 14567	P2 16678	BS2 18678
F3 8556	F3 15778	P2 17889	BS2 19899
F3 9667	P1 11121	P2 18899	BS2 20789

Program

```
/*This is a Generation Means Analysis based on Hayman's methodology,
    Heredity 12:371-390 */
DATA;
INFILE 'A:Cage1.dat';
/* The INPUT statement will vary according to the data set, you need
    generation ("GEN") and a dependent variable */
INPUT row plant gen $ fc $ rating1 rating2 rating3;
MEANRATE = (rating1+rating2+rating3)/3;
PROC SORT; BY GEN;
PROC MEANS; BY GEN; VAR MEANRATE;
OUTPUT OUT=NE MEAN=Y VAR=V STDERR=S;
DATA NEW; SET NE; RS=1/S; IF GEN='D' OR GEN='BD' THEN DELETE;
PROC PRINT;
/* You need a minimum of 6 generations to conduct this analysis, if
    the number of generations used are different than the amount (9)
```

in this example, refer to Gamble's paper (*Canadian Journal of Plant Science* 42:339-348) to obtain the proper coefficients. Another source of information is Jennings et al. (*Iowa State Journal of Research* 48:267-280) */

```
DATA COEFCNTS;

INPUT GEN $ X1 X2 X3 X4 X5;
CARDS;
BC1  0.5  0.0   0.25  0.0    0.0
BC2 -0.5  0.0   0.25  0.0    0.0
BS1  0.5 -0.25  0.25 -0.125  0.0625
BS2 -0.5 -0.25  0.25  0.125  0.0625
F1   0.0  0.5   0.0   0.0    0.25
F2   0.0  0.0   0.0   0.0    0.0
F3   0.0 -0.25  0.0   0.0    0.0625
P1   1.0 -0.5   1.0  -0.5    0.25
P2  -1.0 -0.5   1.0   0.5    0.25

DATA FINAL; MERGE NEW COEFCNTS;
PROC PRINT;
/* This model tests the significance of the additive (X1) and the dom-
    inance (X2) effects */
TITLE 'PROC REG WEIGHTED';
PROC REG;
MODEL Y = X1 X2;
WEIGHT RS;
/* The model must be weighted to account for the unequal population
    sizes among the generations. See Rowe and Alexander (Crop Science
    20:109-110) for further details. The next model tests for the
    epistatic effects, additive-additive (X3), additive-dominance
    (X4), and dominance-dominance (X5), as well as the additive (X1)
    and dominance (X2) effects. */
PRO REG;
MODEL Y = X1 X2 X3 X4 X5;
WEIGHT RS;
RUN;

/* This next set of models are tested with PROC GLM instead of PROC
    REG */
TITLE 'PROC GLM WEIGHTED';
PROC GLM;
MODEL Y = X1 X2;
WEIGHT RS;
PROC GLM;
MODEL Y = X1 X2 X3 X4 X5;
WEIGHT RS;
RUN;

/* The next series of models are the same as the previous models, ex-
    cept that they assume equal population size among the genera-
    tions, and are, therefore, not weighted. */
TITLE 'PROC REG NOWEIGHT';
PROC REG;
MODEL Y = X1 X2;
PROC REG;
MODEL Y = X1 X2 X3 X4 X5;
RUN;
```

```
TITLE 'PROC GLM NOWEIGHT';
MODEL Y = X1 X2;
PROC GLM; MODEL Y = X1 X2 X3 X4 X5;
RUN;
/* Depending on whether or not the population sizes among the genera-
    tions are equal, report the model that has the most significant
    effects from either PROC GLM or PROC REG. */
```

SAS Output File

The SAS System   15:21 Thursday, March 16, 2000   1

The MEANS Procedure
gen=BC1
Analysis Variable : MEANRATE

N	Mean	Std Dev	Minimum	Maximum
20	1.9166667	0.5606492	1.0000000	2.6666667

gen=BC2
Analysis Variable : MEANRATE

N	Mean	Std Dev	Minimum	Maximum
20	1.9166667	0.6386664	1.0000000	3.0000000

gen=BS1
Analysis Variable : MEANRATE

N	Mean	Std Dev	Minimum	Maximum
20	7.3166667	0.7683429	6.3333333	8.3333333

gen=BS2
Analysis Variable : MEANRATE

N	Mean	Std Dev	Minimum	Maximum
20	7.4833333	0.7606937	6.3333333	8.6666667

gen=F1
Analysis Variable : MEANRATE

N	Mean	Std Dev	Minimum	Maximum
20	2.9000000	0.5629912	1.6666667	4.3333333

gen=F2
Analysis Variable : MEANRATE

N	Mean	Std Dev	Minimum	Maximum
20	4.0166667	2.2412272	1.3333333	7.3333333

```
gen=F3
Analysis Variable : MEANRATE

N              Mean          Std Dev         Minimum          Maximum

20          3.7166667       2.0008039       1.3333333        7.3333333

gen=P1
Analysis Variable : MEANRATE

N              Mean          Std Dev         Minimum          Maximum

20          1.6333333       0.4445906       1.0000000        2.3333333

gen=P2
The MEANS Procedure

Analysis Variable : MEANRATE

N              Mean          Std Dev         Minimum          Maximum

20          7.7166667       0.5543554       6.6666667        8.6666667
```

Obs	gen	TYPE	FREQ	Y	V	S	RS
1	BC1	0	20	1.91667	0.31433	0.12536	7.9767
2	BC2	0	20	1.91667	0.40789	0.14281	7.0023
3	BS1	0	20	7.31667	0.59035	0.17181	5.8205
4	BS2	0	20	7.48333	0.57865	0.17010	5.8790
5	F1	0	20	2.90000	0.31696	0.12589	7.9435
6	F2	0	20	4.01667	5.02310	0.50115	1.9954
7	F3	0	20	3.71667	4.00322	0.44739	2.2352
8	P1	0	20	1.63333	0.19766	0.09941	10.0590
9	P2	0	20	7.71667	0.30731	0.12396	8.0673

```
PROC REG WEIGHTED
```

Obs	gen	TYPE	FREQ	Y	V	S	RS	X1	X2	X3	X4	X5
1	BC1	0	20	1.91667	0.31433	0.12536	7.9767	0.5	0.00	0.25	0.00	0.0000
2	BC2	0	20	1.91667	0.40789	0.14281	7.0023	-0.5	0.00	0.25	0.00	0.0000
3	BS1	0	20	7.31667	0.59035	0.17181	5.8205	0.5	-0.25	0.25	-0.25	0.0625
4	BS2	0	20	7.48333	0.57865	0.17010	5.8790	-0.5	-0.25	0.25	0.25	0.0625
5	F1	0	20	2.90000	0.31696	0.12589	7.9435	0.0	0.50	0.00	0.00	0.2500
6	F2	0	20	4.01667	5.02310	0.50115	1.9954	0.0	0.00	0.00	0.00	0.0000
7	F3	0	20	3.71667	4.00322	0.44739	2.2352	0.0	-0.25	0.00	0.00	0.0625
8	P1	0	20	1.63333	0.19766	0.09941	10.0590	1.0	-0.50	1.00	-1.00	0.2500
9	P2	0	20	7.71667	0.30731	0.12396	8.0673	-1.0	-0.50	1.00	1.00	0.2500

```
PROC REG WEIGHTED

The REG Procedure
Model: MODEL1
Dependent Variable: Y

Weight: RS

Analysis of Variance

                        Sum of          Mean
Source          DF      Squares         Square     F Value  Pr > F

Model           2       177.56093       88.78047   2.67     0.1485
```

```
Error              6      199.86954        33.31159
Corrected Total 8         377.43047
```

```
Root MSE              5.77162        R-Square  0.4704
Dependent Mean        4.09505        Adj R-Sq  0.2939
Coeff Var           140.94121
```

Parameter Estimates

Variable	DF	Parameter Estimate	Standard Error	t Value	Pr > |t|
Intercept	1	3.75399	0.84226	4.46	0.0043
X1	1	-2.32644	1.16302	-2.00	0.0924
X2	1	-2.93086	2.34026	-1.25	0.2570

PROC REG WEIGHTED

The REG Procedure
Model: MODEL1
Dependent Variable: Y

Weight: RS

Analysis of Variance

Source	DF	Sum of Squares	Mean Square	F Value	Pr > F
Model	5	301.20716	60.24143	2.37	0.2542
Error	3	76.22331	25.40777		
Corrected Total	8	377.43047			

```
Root MSE              5.04061        R-Square  0.7980
Dependent Mean        4.09505        Adj R-Sq  0.4615
Coeff Var           123.09023
```

Parameter Estimates

Variable	DF	Parameter Estimate	Standard Error	t Value	Pr > |t|
Intercept	1	4.42225	1.11945	3.95	0.0289
X1	1	0.38757	2.39420	0.16	0.8817
X2	1	-10.68980	4.64427	-2.30	0.1048
X3	1	-8.61894	4.64304	-1.86	0.1604
X4	1	3.35783	2.74842	1.22	0.3091
X5	1	14.76576	9.56113	1.54	0.2202

PROC GLM WEIGHTED

Number of observations   9
The GLM Procedure
Dependent Variable: Y

Weight: RS

Source	DF	Sum of Squares	Mean Square	F Value	Pr > F
Model	2	177.5609302	88.7804651	2.67	0.1485

```
Error         6              199.8695384    33.3115897
Corrected
Total         8              377.4304686
```

R-Square	Coeff Var	Root MSE	Y Mean
0.470447	140.9412	5.771619	4.095055

Source	DF	Type I SS	Mean Square	F Value	Pr > F
X1	1	125.3143256	125.3143256	3.76	0.1005
X2	1	52.2466046	52.2466046	1.57	0.2570

Source	DF	Type III SS	Mean Square	F Value	Pr > F
X1	1	133.2929249	133.2929249	4.00	0.0924
X2	1	52.2466046	52.2466046	1.57	0.2570

Parameter	Estimate	Standard Error	t Value	Pr > |t|
Intercept	3.753994272	0.84225532	4.46	0.0043
X1	-2.326444874	1.16301923	-2.00	0.0924
X2	-2.930859648	2.34025761	1.25	0.2570

PROC GLM WEIGHTED

The GLM Procedure
Number of observations   9
Dependent Variable: Y

Weight: RS

Source	DF	Sum of Squares	Mean Square	F Value	Pr > F
Model	5	301.2071567	60.2414313	2.37	0.2542
Error	3	76.2233119	25.4077706		
Corrected Total	8	377.4304686			

R-Square	Coeff Var	Root MSE	Y Mean
0.798047	123.0902	5.040612	4.095055

Source	DF	Type I SS	Mean Square	F Value	Pr > F
X1	1	125.3143256	125.3143256	4.93	0.1130
X2	1	52.2466046	52.2466046	2.06	0.2470
X3	1	29.6732521	29.6732521	1.17	0.3590
X4	1	33.3748169	33.3748169	1.31	0.3349
X5	1	60.5981574	60.5981574	2.39	0.2202

Source	DF	Type III SS	Mean Square	F Value	Pr > F
X1	1	0.6657960	0.6657960	0.03	0.8817
X2	1	134.6078535	134.6078535	5.30	0.1048
X3	1	87.5525268	87.5525268	3.45	0.1604
X4	1	37.9243276	37.9243276	1.49	0.3091
X5	1	60.5981574	60.5981574	2.39	0.2202

Parameter	Estimate	Standard Error	t Value	Pr > |t|
Intercept	4.42225240	1.11945357	3.95	0.0289
X1	0.38756832	2.39420301	0.16	0.8817
X2	-10.68980183	4.64427294	-2.30	0.1048
X3	-8.61894129	4.64304465	-1.86	0.1604
X4	3.35782810	2.74841807	1.22	0.3091
X5	14.76576243	9.56113494	1.54	0.2202

PROC REG NOWEIGHT

The REG Procedure
Model: MODEL1
Dependent Variable: Y

Analysis of Variance

Source	DF	Sum of Squares	Mean Square	F Value	Pr > F
Model	2	20.79725	10.39863	2.02	0.2141
Error	6	30.96170	5.16028		
Corrected Total	8	51.75895			

Root MSE	2.27163	R-Square	0.4018
Dependent Mean	4.29074	Adj R-Sq	0.2024
Coeff Var	52.94251		

Parameter Estimates

Variable	DF	Parameter Estimate	Standard Error	t Value	Pr > |t|
Intercept	1	3.83788	0.83885	4.58	0.0038
X1	1	-2.05556	1.31152	-1.57	0.1681
X2	1	-3.26061	2.59909	-1.25	0.2563

PROC REG NOWEIGHT

The REG Procedure
Model: MODEL1
Dependent Variable: Y

Analysis of Variance

Source	DF	Sum of Squares	Mean Square	F Value	Pr > F
Model	5	32.17318	6.43464	0.99	0.5404
Error	3	19.58577	6.52859		
Corrected Total	8	51.75895			

Root MSE	2.55511	R-Square	0.6216
Dependent Mean	4.29074	Adj R-Sq	-0.0091
Coeff Var	59.54940		

Parameter Estimates

Variable	DF	Parameter Estimate	Standard Error	t Value	Pr > |t|
Intercept	1	3.91960	1.23483	3.17	0.0503
X1	1	0.51587	3.25117	0.16	0.8840
X2	1	-7.22098	5.05505	-1.43	0.2485
X3	1	-4.82188	4.94994	-0.97	0.4018
X4	1	3.42857	3.86296	0.89	0.4402
X5	1	9.36501	12.09879	0.77	0.4953

PROC GLM NOWEIGHT

The REG Procedure
Model: MODEL2
Dependent Variable: Y

Analysis of Variance

Source	DF	Sum of Squares	Mean Square	F Value	Pr > F
Model	2	20.79725	10.39863	2.02	0.2141
Error	6	30.96170	5.16028		
Corrected Total	8	51.75895			

Root MSE	2.27163	R-Square	0.4018
Dependent Mean	4.29074	Adj R-Sq	0.2024
Coeff Var	52.94251		

Parameter Estimates

Variable	DF	Parameter Estimate	Standard Error	t Value	Pr > |t|
Intercept	1	3.83788	0.83885	4.58	0.0038
X1	1	-2.05556	1.31152	-1.57	0.1681
X2	1	-3.26061	2.59909	-1.25	0.2563

PROC GLM NOWEIGHT

The GLM Procedure
Number of observations   9
Dependent Variable: Y

Source	DF	Sum of Squares	Mean Square	F Value	Pr > F
Model	5	32.17318074	6.43463615	0.9	0.5404
Error	3	19.58576988	6.52858996		
Corrected Total	8	51.75895062			

R-Square	Coeff Var	Root MSE	Y Mean
0.621596	59.54940	2.555111	4.290741

Source	DF	Type I SS	Mean Square	F Value	Pr > F
X1	1	12.67592593	12.67592593	1.94	0.2578

X2	1	8.12132435	8.12132435	1.24	0.3460
X3	1	2.32149174	2.32149174	0.36	0.5930
X4	1	5.14285714	5.14285714	0.79	0.4402
X5	1	3.91158158	3.91158158	0.60	0.4953

Source	DF	Type III SS	Mean Square	F Value	Pr > F
X1	1	0.16437130	0.16437130	0.03	0.8840
X2	1	13.32175859	13.32175859	2.04	0.2485
X3	1	6.19514901	6.19514901	0.95	0.4018
X4	1	5.14285714	5.14285714	0.79	0.4402
X5	1	3.91158158	3.91158158	0.60	0.4953

Parameter	Estimate	Standard Error	t Value	Pr > |t|
Intercept	3.919597315	1.23482719	3.17	0.0503
X1	0.515873016	3.25116872	0.16	0.8840
X2	-7.220984340	5.05504842	-1.43	0.2485
X3	-4.821879195	4.94994238	-0.97	0.4018
X4	3.428571429	3.86296406	0.89	0.4402
X5	9.365011186	12.09878615	0.77	0.4953

Type-3 tests of estimates of fixed effects

Genetic effect	Numerator df	Denominator df	F value	Pr > F
Additive (a)	1	175	4.08	0.0450
Dominance (d)	1	175	7.49	0.0068
Epistatic effects				
aa	1	175	0.33	0.5693
dd	1	175	18.34	<.0001
ad	1	175	18.19	<.0001

Additive, dominance, additive x dominance, and dominance x dominance effects are significant.

Chapter 7

PATHSAS:
Path Coefficient Analysis
of Quantitative Traits

Christopher S. Cramer
Todd C. Wehner
Sandra B. Donaghy

Purpose

To calculate path coefficients (direct effects) and indirect effects between independent (x) and dependent (y) variables.

Definitions

Path coefficient analysis: the correlation between two traits is a function of the direct relationship between two traits and the indirect relationships of related traits (Wright, 1934).

$$r_{10} = \rho_{01} + \rho_{02}r_{12} + \rho_{03}r_{13} + \rho_{04}r_{14}$$

where r_{10} = the correlation between X_1 and Y; ρ_{01} = the path coefficient between X_1 and Y; ρ_{02} = the path coefficient between X_2 and Y; r_{12} = the correlation between X_1 and X_2; $\rho_{02}r_{12}$ = the indirect effect of X_2 on the correlation between X_1 and Y; ρ_{03} = the path coefficient between X_3 and Y; r_{13} = the correlation between X_1 and X_3; $\rho_{03}r_{13}$ = the indirect effect of X_3 on the correlation between X_1 and Y; ρ_{04} = the path coefficient between X_4 and Y; r_{14} = the correlation between X_1 and X_4; $\rho_{04}r_{14}$ = the indirect effect of X_4 on the correlation between X_1 and Y.

Originator

Wright, S. (1934). The method of path coefficients. *Annals of Mathematical Statistics* 5:161-215.

Software Available

Cramer, C.S., Wehner, T.C., and Donaghy, S.B. (1999). PATHSAS: A SAS computer program for path coefficient analysis of quantitative data. *Journal of Heredity* 90:260-262 (free of charge).

Some References Where the Software Has Been Used

Cramer, C.S. and Wehner, T.C. (1998). Fruit yield and yield component means and correlations of four slicing cucumber populations improved through six to ten cycles of recurrent selection. *Journal of American Society of Horticulture Science* 123:388-395.
Cramer, C.S. and Wehner, T.C. (1999). Little heterosis for yield and yield components in hybrids of six cucumber inbreds. *Euphytica* 110:101-110.
Cramer, C.S. and Wehner, T.C. (2000). Path analysis of the correlation between fruit number and plant traits of cucumber populations. *HortScience* 35(4):708-711.

Contact

Dr. Todd Wehner, Department of Horticultural Science, North Carolina State University, Raleigh, NC 27695-7609, USA. E-mail: <todd_wehner@ncsu.edu>.

EXAMPLE

Data to be analyzed:

Plot	Repli-cation	Cycle	Plant number	Pistillate flowers	Branch number	Leaf number	Total fruit number	Culled fruit number	Early fruit number
001	01	1	29	022	040	0240	01	00	00
002	02	1	21	017	034	0120	17	02	01
003	03	1	31	052	032	0440	29	15	03
004	04	1	30	049	077	0550	25	09	05
005	04	3	30	058	071	0810	30	10	05

006	03	3	23	039	040	0460	32	03	13
007	02	3	26	044	047	0460	27	07	08
008	01	3	22	023	027	0330	12	04	00
009	01	2	19	025	054	0510	27	05	04
010	02	2	27	035	038	0510	34	03	05
011	03	2	23	050	027	0290	01	00	00
012	04	2	32	035	055	0560	47	10	08
013	05	1	28	088	140	1315	58	13	27
014	06	1	28	162	105	0986	39	09	14
015	07	1	25	026	070	0741	36	05	10
016	08	1	31	074	125	0803	55	08	35
017	08	2	33	048	127	0870	57	11	13
018	07	2	25	069	038	0639	26	04	09
019	06	2	32	041	105	0878	31	04	13
020	05	2	30	021	098	0982	53	02	06
021	05	3	26	012	064	0622	31	03	09
022	06	3	31	024	111	1133	26	01	07
023	07	3	24	046	082	0879	33	04	04
024	08	3	28	048	161	1122	63	05	07

SAS Program

```
DATA DST1;
INPUT PLOT REP CYC PLANTNO PISTFLOW BRANCHNO LEAFNO TOTALNO CULLNO
     EARLYNO;
MARK=TOTALNO-CULLNO;
BRANPLAN=BRANCHNO/PLANTNO;
NODEBRAN=LEAFNO/(BRANCHNO+PLANTNO);
TOTFEMND=PISTFLOW+TOTALNO;
PERFENOD=(TOTFEMND/LEAFNO);
FRTSET=TOTALNO/PISTFLOW;
FRTPLANT=TOTALNO/PLANTNO;
MARKPLAN=MARK/PLANTNO;
EARPLAN=EARLYNO/PLANTNO;
CARDS;
001 01 1 29 022 040 0240 01 00 00
002 02 1 21 017 034 0120 17 02 01
003 03 1 31 052 032 0440 29 15 03
004 04 1 30 049 077 0550 25 09 05
005 04 3 30 058 071 0810 30 10 05
006 03 3 23 039 040 0460 32 03 13
007 02 3 26 044 047 0460 27 07 08
```

```
008 01 3 22 023 027 0330 12 04 00
009 01 2 19 025 054 0510 27 05 04
010 02 2 27 035 038 0510 34 03 05
011 03 2 23 050 027 0290 01 00 00
012 04 2 32 035 055 0560 47 10 08
013 05 1 28 088 140 1315 58 13 27
014 06 1 28 162 105 0986 39 09 14
015 07 1 25 026 070 0741 36 05 10
016 08 1 31 074 125 0803 55 08 35
017 08 2 33 048 127 0870 57 11 13
018 07 2 25 069 038 0639 26 04 09
019 06 2 32 041 105 0878 31 04 13
020 05 2 30 021 098 0982 53 02 06
021 05 3 26 012 064 0622 31 03 09
022 06 3 31 024 111 1133 26 01 07
023 07 3 24 046 082 0879 33 04 04
024 08 3 28 048 161 1122 63 05 07
;

%macro path(data,indep,dep0,dep,bylist,printreg,printout);
    /*
        Parameters to macro are:
        data =name of dataset to analyze
        indep=list of independent variables
        dep0=primary dependent variable
        dep=other dependent variables
        bylist=by variable list
        printreg=print regression? ( value is either yes or no)
        printout=print results(direct,indirect effects)?
                  (value is either yes or no)
    */

%local noind word nodep noby bylast printr;

    /*  create noind macro variable                          */
    /*  noind is the number of independent variables in &indep   */
    %let noind=0;
    %if &indep ne %then %do;
        %let word=%scan(&indep,1);
        %do %while (&word ne );
            %let noind=%eval(&noind+1);
            %let word=%scan(&indep,&noind+1);
            %end;
        %end;

    /*  create nodep macro variable                          */
    /*  nodep is the number of dependent variables in &dep     */
    %let nodep=0;
    %if &dep ne %then %do;
        %let word=%scan(&dep,1);
        %do %while (&word ne );
            %let nodep=%eval(&nodep+1);
            %let word=%scan(&dep,&nodep+1);
            %end;
        %end;

    /*  create noby  macro variable                          */
```

```
   /*   noby is the number of by variables in &bylist         */
   %let noby=0;
   %if &bylist ne %then %do;
      %let word=%scan(&bylist,1);
      %do %while (&word ne );
         %let noby=%eval(&noby+1);
         %let word=%scan(&bylist,&noby+1);
         %end;
      %end;
   %let bylast=%scan(&bylist,&noby);

   /*   create printr macro variable                          */
   /*   printr has a blank value or the value NOPRINT         */
   /*   specifies whether to print regression output or not   */
   %if %upcase(&printreg)=YES %then %let printr=;
      %else %let printr=noprint;

data data1; set &data;
   keep &bylist &dep0 &dep &indep;
   run;

proc sort data=data1;
   by &bylist;
proc standard data=data1 mean=0 std=1 out=sdata2;
      by &bylist;
      var &indep &dep0 &dep;
   run;

proc reg data=sdata2   &printr
         outsscp=sscp(keep=&bylist intercep _type_)
         outest=estdep(drop=_model_ _type_ _rmse_  intercep);
      by &bylist;
      model &dep0=&indep;
   run;

   /*
   _type_='N' is the number of obs in the dataset;
   nobs, number of obs., is created
   needed for checking that there are enough obs.
   if not, the reg. coefficients are biased, and need to set to miss-
      ing
   */
data sscp; set sscp;
   if _type_='N';
   rename intercep=nobs;
   drop _type_;

/*   if no. of obs. is <= the no. of indep. variables, then
     set the regression coefficients to missing              */
data estdep; merge sscp estdep;
   by &bylist;
   array v &indep;
   look='no ';
   if nobs<=&noind then do;
      look='yes';
      do over v;
         v=.;
```

```
            end;
        end;
    run;

proc print data=estdep;
    where look='yes';
    var &bylist nobs;
title3
'The following identification levels do not have enough obs. for anal-
    ysis';
title4 '    and the regression coefficients were set to missing
    ';
    run;
title3 ' ' ;

proc reg data=sdata2 &printr
    outest=estindep(drop=_model_  _type_  _rmse_  intercep);
    by &bylist;
    model &dep=&dep0;
    run;

data estind2; set estindep;
    by &bylist;
    array r   regc1-regc&nodep;
    retain    regc1-regc&nodep;
    if first.&bylast then _i_=0;
    _i_+1;
    r=&dep0;
    if last.&bylast then do;
        output;
        do over r;
            r=.;
            end;
        end;
    drop &dep0 &dep _depvar_;
    run;

proc corr data=data1  outp=corr  noprint;
    by &bylist;
    var &indep;
    run;
data corr; set corr;
    if _type_='CORR';
    drop _type_;
    run;

data estdep; set estdep;
    array reg &indep;
    array r2  reg1-reg&noind;
    do over reg;
        r2=reg;
        end;
    drop &indep;
    run;

data tog;
    merge corr estdep;
```

```
   by &bylist;
   array dir   &indep;
   array corr  &indep;
   array r2    reg1-reg&noind;
   if first.&bylast then do;
      totc=0;
      n=0;
      end;
   n+1;
   &dep0=.;
   do over dir;
      if n=_i_ then dir= r2;
         else dir=r2*corr;
       &dep0 + dir;
        end;
   drop n;
   keep &bylist--_name_ &indep   &dep0 _depvar_ nobs;
   format &indep &dep0 5.2;
   run;

data tog2; merge tog estind2; by &bylist;
   array r regc1-regc&nodep;
   array t &dep;
   do over r;
      t=&dep0 * r;
      end;
   format &dep &dep0 5.2;
   format regc1-regc&nodep 5.2;
 * drop regc1-regc&nodep;
   drop _depvar_;
   run;

%if %upcase(&printout)=YES %then
   %str(proc print data=tog2(drop=regc1-regc&nodep); run;);

%mend path;

%path(data=dst1,
      indep=branplan nodebran perfenod frtset,
      dep0=frtplant,
      dep=markplan earplan,
      bylist=cyc,
      printreg=no,
      printout=yes
      );

RUN;
```

SAS Output

_ N A	B R A N P	N O D E B	P E R F E	F R T	N	F R T P L	M A R K P	E A R P

```
  O  C   M          L      R      N      S   O   A      L      L
  B  Y   E          A      A      O      E   B   N      A      A
  S  C   _          N      N      D      T   S   T      N      N
001 1 BRANPLAN   0.72   0.18  -0.06   0.03  8   0.87   0.80   0.78
002 1 NODEBRAN   0.37   0.34  -0.10   0.04  8   0.65   0.61   0.59
003 1 PERFENOD  -0.17  -0.15   0.23   0.01  8  -0.08  -0.07  -0.07
004 1 FRTSET     0.07   0.05   0.01   0.30  8   0.42   0.39   0.38
005 2 BRANPLAN   0.72  -0.13  -0.41   0.42  8   0.61   0.54   0.31
006 2 NODEBRAN  -0.26   0.34   0.01  -0.01  8   0.08   0.07   0.04
007 2 PERFENOD  -0.58   0.01   0.50  -0.53  8  -0.60  -0.53  -0.30
008 2 FRTSET     0.40  -0.01  -0.34   0.77  8   0.82   0.72   0.41
009 3 BRANPLAN   1.06  -0.03  -0.37   0.12  8   0.78   0.75   0.28
010 3 NODEBRAN  -0.15   0.20  -0.31  -0.10  8  -0.36  -0.35  -0.13
011 3 PERFENOD  -0.46  -0.07   0.86  -0.22  8   0.10   0.09   0.04
012 3 FRTSET     0.28  -0.04  -0.42   0.46  8   0.28   0.26   0.10
```

Chapter 8

Restricted Maximum Likelihood Procedure to Estimate Additive and Dominance Genetic Variance Components

Agron Collaku

Purpose

To estimate narrow-sense heritability without any restriction in mating design.

Definitions

Estimation of genetic variance components for additive and dominance effects is important in plant breeding for estimating narrow-sense heritability and predicting the results of selection. Quantitative genetic methods used to estimate genetic variance components are based on strict mating designs with a number of restrictions for the way genotypes are produced. Many of the restrictions are untenable in common breeding programs. The restricted maximum likelihood (REML) method is advantageous because it provides genetic variance estimates without any restriction in mating design not only for balanced data but also for unbalanced data. REML estimates of additive and dominance variance components using a mixed-model approach are obtained based on the following formulas.

Additive Variance

The additive genetic variance component measures the expected mean genotypic effect in the selected material and can be estimated from the equation

$$COV_{HS} = 2r_{xy}\sigma^2_A$$

where COV_{HS} is the covariance among half-sib families, r_{xy} is the coancestry coefficient that measures the relationship among parents of half-sib families, and σ^2_A is additive genetic variance component.

Dominance Variance

The dominance genetic variance component measures deviations of genotypic values from their additive effects due to interaction between alleles at the same locus. It can be estimated from the equation

$$COV_{FS} = 2r_{xy}\sigma^2_A + u_{xy}\sigma^2_D$$

where COV_{FS} is the covariance among full-sib families, u_{xy} is the double coancestry coefficient that measures the relationship among parents of full-sib families, and σ^2_D is dominance genetic variance component.

The preceding equations assume no epistatic interaction of any kind.

The mixed model fitted to obtain additive and dominance genetic variances is

$$y = X\beta + Z_1\alpha + Z_2\gamma + Z_3\delta + \varepsilon$$

where y is a vector of $n \times 1$ observations; n is the number of observations for each entry in each year and each environment; X is the design matrix of fixed effects, and $\hat{a}$ is a $b \times 1$ vector of fixed effects, Z_1 is the design matrix of additive effects, and $\hat{a}$ is a $a \times 1$ vector; a is the number of populations or crosses in the genetic design; Z_2 is the design matrix of dominance effects, and $\tilde{a}$ is a $d \times 1$ vector, and d is the number of populations; Z_3 is the design matrix of entry-by-environment interaction (GE) effects, and $\ddot{a}$ is a $g \times 1$ vector; g is the cross combination of entries with environments; and $\mathring{a}$ is the vector of experimental error effects.

Random effects (additive, dominance, GE, and error) have the following variance-covariance matrix:

$$Var \begin{bmatrix} \alpha \\ \gamma \\ \delta \\ \varepsilon \end{bmatrix} = \begin{bmatrix} A\sigma^2_A & 0 & 0 & 0 \\ 0 & D\sigma^2_D & 0 & 0 \\ 0 & 0 & I\sigma^2_{GE} & 0 \\ 0 & 0 & 0 & I\sigma^2_\varepsilon \end{bmatrix}$$

where A is a matrix of $n \times n$. The diagonal elements of A are equal to 1 and the off-diagonal elements are equal to the coancestry coefficient times two $(2r_{xy})$ between n entries in the study. The matrix of coancestry coefficients can be obtained by using PROC INBREED of SAS (see example). D is a matrix of $n \times n$ with diagonal elements equal to 1/4, i.e., double coancestry coefficients within full-sib entries and off-diagonal elements are double coancestry coefficients among full-sib entries. I is an identity matrix.

Mixed model equations are used to obtain estimates of á, â, ã, and ä. REML estimates for genetic variance components (additive and dominance) as well as for other components are obtained by running PROC MIXED in SAS, in which A and D matrices are appended (see example).

References

These key references used maximum likelihood estimators to estimate additive and dominance genetic variance components:

Bernardo, R. (1994). Prediction of maize single-cross performance using RFLPs and information from related hybrids. *Crop Science* 34(1):20-25.

Collaku, A. (2000). Heritability of waterlogging tolerance in wheat (pp. 53-78). Doctoral dissertation, Louisiana State University, Baton Rouge, Louisiana.

Contact

A. Collaku, Department of Pharmacogenomics, Johnson & Johnson Research and Development, Raritan, New Jersey 08869.

EXAMPLE

Four F_2 populations (crosses) of soft red winter wheat were studied:

Cross No. 1 - Tchere/savannah//GA 85240
Cross No. 2 - Tchere/savannah//PION 2643
Cross No. 3 - Tchere/DS 2368//GA 85240
Cross No. 4 - Tchere/DS 2368//PION 2691

Five random full-sib families from each cross totaling twenty entries were studied in a randomized complete block design with three replications. The experiment was conducted for two years. Double coancestry coefficients between entries 6-15=0.0625. Diagonal elements of D matrix = 0.25 which represented double coancestry coefficients of within full-sib families. All other elements of D matrix were zero. Based on these elements, a matrix was constructed and appended to the data analyzed using PROC MIXED.

Data to be analyzed (Yield):

LINE	YEAR	REP	CROSS	YIELD
1	1	1	1	150.0
1	1	2	1	130.0
1	1	3	1	141.3
2	1	1	1	178.0
2	1	2	1	172.6
2	1	3	1	167.0
3	1	1	1	148.0
3	1	2	1	120.0
3	1	3	1	.
4	1	1	1	108.5
4	1	2	1	113.0
4	1	3	1	123.0
5	1	1	1	89.0
5	1	2	1	90.0
5	1	3	1	92.0
6	1	1	2	93.8
6	1	2	2	121.5
6	1	3	2	86.0
7	1	1	2	113.0
7	1	2	2	107.0
7	1	3	2	.
8	1	1	2	113.0
8	1	2	2	111.4
8	1	3	2	109.8
9	1	1	2	68.0

9	1	2	2	47.0
9	1	3	2	56.0
10	1	1	2	100.0
10	1	2	2	68.0
10	1	3	2	94.0
11	1	1	3	159.0
11	1	2	3	151.3
11	1	3	3	.
12	1	1	3	172.0
12	1	2	3	139.0
12	1	3	3	.
13	1	1	3	127.0
13	1	2	3	.
13	1	3	3	92.0
14	1	1	3	109.0
14	1	2	3	138.0
14	1	3	3	119.2
15	1	1	3	159.0
15	1	2	3	152.0
15	1	3	3	137.0
16	1	1	4	152.6
16	1	2	4	147.0
16	1	3	4	.
17	1	1	4	134.0
17	1	2	4	184.3
17	1	3	4	125.0
18	1	1	4	134.0
18	1	2	4	146.3
18	1	3	4	186.4
19	1	1	4	107.0
19	1	2	4	90.0
19	1	3	4	124.0
20	1	1	4	77.0

LINE	YEAR	REP	CROSS	YIELD
20	1	2	4	113.0
20	1	3	4	96.7

LINE	YEAR	REP	CROSS	YIELD
1	2	1	1	118.6
1	2	2	1	98.2
1	2	3	1	108.9
2	2	1	1	80.9
2	2	2	1	119.5
2	2	3	1	76.0
3	2	1	1	112.6
3	2	2	1	96.5
3	2	3	1	99.3
4	2	1	1	80.3
4	2	2	1	73.2
4	2	3	1	72.2
5	2	1	1	65.3
5	2	2	1	54.3
5	2	3	1	58.4
6	2	1	2	61.7
6	2	2	2	81.3
6	2	3	2	66.0
7	2	1	2	65.5
7	2	2	2	95.8
7	2	3	2	88.2
8	2	1	2	.
8	2	2	2	103.0
8	2	3	2	103.9
9	2	1	2	60.3
9	2	2	2	79.8
9	2	3	2	70.2
10	2	1	2	137.8

10	2	2	2	128.4
10	2	3	2	.
11	2	1	3	82.5
11	2	2	3	74.3
11	2	3	3	78.5
12	2	1	3	132.4
12	2	2	3	128.7
12	2	3	3	116.1
13	2	1	3	87.0
13	2	2	3	88.9
13	2	3	3	61.8
14	2	1	3	121.9
14	2	2	3	98.4
14	2	3	3	112.3
15	2	1	3	84.7
15	2	2	3	105.0
15	2	3	3	98.2
16	2	1	4	113.4
16	2	2	4	105.7
16	2	3	4	101.3
17	2	1	4	114.8
17	2	2	4	131.2
17	2	3	4	117.7
18	2	1	4	71.3
18	2	2	4	41.8
18	2	3	4	45.2
19	2	1	4	.
19	2	2	4	100.6
19	2	3	4	115.0
20	2	1	4	129.1
20	2	2	4	113.6
20	2	3	4	139.7

data c;

```
input F $ M $ Cross $ Line $ Gene;
datalines;
```

TCH.	SAV	P1	P1	1
P1	GA	1	1	2
P1	GA	1	2	2
P1	GA	1	3	2
P1	GA	1	4	2
P1	GA	1	5	2
P1	P.2643	2	6	2
P1	P.2643	2	7	2
P1	P.2643	2	8	2
P1	P.2643	2	9	2
P1	P.2643	2	10	2
TCH.	DS	P2	P2	1
P2	GA	3	11	2
P2	GA	3	12	2
P2	GA	3	13	2
P2	GA	3	14	2
P2	GA	3	15	2
P2	P.2691	4	16	2
P2	P.2691	4	17	2
P2	P.2691	4	18	2
P2	P.2691	4	19	2
P2	P.2691	4	20	2

```
;
proc inbreed data=c covar outcov=matrix;
var line f m;
data matrix;
set matrix;
if substr(line,1,1)>0;
drop OBS _TYPE_ _PANEL_ LINE F M;
proc print data=matrix;
run;
data two;
```

```
retain row (0);
parm=1;
set matrix(drop=_col_ col1-col4 col10 col16 col17 col23);
if substr(line,1,1)>0;
drop _type_ _panel_ line f m;
array old col5--col28;
array new ncol1-ncol20;
do over old;
  new=old;
end;
drop col5--col28;
row+1;
run;
data two;
set two;
array old ncol1-ncol20;
array new col1-col20;
do over old;
  new=old;
end;
drop ncol1-ncol20;
run;
data three;
row=1;
parm=2;
array dmat{20} col1-col20;
do j=1 to 20;
  dmat{j}=0;
end;
do i=1 to 5;
  do j=1 to 5;
    dmat{j}=0.25;
  end;
  parm=2;
  row=i;
  output;
end;
do j=1 to 20;
  dmat{j}=0;
end;
do i=6 to 10;
```

```
  do j=6 to 10;
    dmat{j}=0.25;
    dmat{j+5}=0.0625;
  end;
  parm=2;
  row=i;
  output;
end;
do j=1 to 20;
  dmat{j}=0;
end;
do i=11 to 15;
  do j=11 to 15;
    dmat{j}=0.0625;
    dmat{j+5}=0.25;
  end;
  parm=2;
  row=i;
  output;
end;
do j=1 to 20;
  dmat{j}=0;
end;
do i=16 to 20;
  do j=16 to 20;
    dmat{j}=0.25;
  end;
  parm=2;
  row=i;
  output;
end;
drop i j;
run;
data mat20;
set two three;
run;
data varcomp;
input line year rep cross yield;
cards;
```

1	1	1	1	150.0
1	1	2	1	130.0

1	1	3	1	141.3
2	1	1	1	178.0
2	1	2	1	172.6
.	.	.	.	.
.	.	.	.	.
18	2	1	4	71.3
18	2	2	4	41.8
18	2	3	4	45.2
19	2	1	4	.
19	2	2	4	100.6
19	2	3	4	115.0
20	2	1	4	129.1
20	2	2	4	113.6
20	2	3	4	139.7

```
;
Proc Mixed covtest;
class I yr rep cr;
model yield = yr rep(yr);
random I/type=Lin(2) Idata=mat80 ;
random yr*I(cr);
parms (110) (100) (700) (200);
run;
```

SAS Output

Covariance Parameter Estimates (REML)

Cov Parm	Estimate	Std Error	Z	Pr>|Z|
LIN (1)	111.66689661	19.28580197	5.79	0.0001
LIN(2)	386.10778285	592.07749876	0.65	0.5143
ENTRY*YEAR(CROSS)	577.41756955	156.55363443	3.69	0.0002
Residual	203.03072111	35.06509448	5.79	0.0001

Tests of Fixed Effects

Source	NDF	DDF	Type III	F Pr>F
YEAR	1	19	11.91	0.0027
REP(YEAR)	4	67	0.31	0.8707

In the previous output,

Lin(1) is $\hat{\sigma}_A^2$ – the estimate of additive variance component, and
Lin(2) is $\hat{\sigma}_D^2$ – the estimate of dominance variance component.

Changes needed in this program:

- Calculate double coancestry coefficients among a group of full-sib families and then construct matrix D as shown in the program.
- When appending matrices A and D, use the number of columns corresponding to each set of full-sib families included in the study.

Chapter 9

Calculating Additive Genetic Correlation Using ANOVA and the Sum Method of Estimating Covariance

Blair L. Waldron

Importance

Genetic correlations allow breeders to predict the correlated response due to pleiotropy and/or linkage in unselected traits. For this reason, they are often required in many selection indices.

Definitions

Additive Genetic Correlations

Estimated as

$$r_{A(xy)} = \sigma_{A(xy)} / \sqrt{(\sigma^2_{A(x)}\sigma^2_{A(y)})}$$

where $\sigma_{A(xy)}$ is the additive genetic covariance of means for traits x and y, and $\sigma_{A(x)}$ and $\sigma_{A(y)}$ are the additive genetic standard deviations for traits x and y, respectively. Approximate standard errors for genetic correlations can be calculated as described by Falconer (1989). This assumes an appropriate family structure exists within the population of interest such that additive genetic variance (σ^2_A) can be derived. Most often, half-sib families (HSFs) are used where it is assumed that the variance among HSFs = $\frac{1}{4}\sigma^2_A$.

Sum Method of Estimating Covariance

The sum method is based on the statistical property of the sum of two random variables, which states:

$$Var(X+Y) = Var(X) + Var(Y) + 2Cov(X,Y)$$

This can be rearranged and written as

$$Cov(X,Y) = \left[Var(X+Y) - Var(X) - Var(Y)\right]/2.$$

In this case, X and Y refer to two different traits evaluated within the same population.

Using ANOVA we get an estimate (mean square) for $Var(X)$, $Var(Y)$, and $Var(X+Y)$ (most often referred to as the mean cross product, or MCP). The $Var(X+Y)$ (e.g., MCP) is obtained by running ANOVA on a new variable created by summing the plot mean values for trait X and trait Y for each level of observation within each HSF ($X_{ij}+Y_{ij}$).

Additive Genetic Variance and Covariance

Standard procedures for isolating appropriate variance components, based on expected mean squares, are used to estimate additive genetic variances and covariances where, as previously stated, it is assumed that

$$\sigma^2_{HSF(x)} = \tfrac{1}{4}\sigma^2_{A(x)}; \ \sigma^2_{HSF(y)} = \tfrac{1}{4}\sigma^2_{A(y)}; \ \text{and}$$

$\sigma_{HSF(xy)}$ (e.g., covariance among HSFs for x and y) $= \tfrac{1}{4}\sigma_{A/(xy)}$ (e.g., additive genetic covariance between traits x and y).

The appropriate linear combination of mean squares or mean cross products is used to solve for $\sigma^2_{HSF(x)}$, $\sigma^2_{HSF(y)}$, and $\sigma^2_{HSF(xy)}$, respectively (see the following example).

Important Considerations

a. Because the additive genetic correlation is a function of two mathematical equations (i.e., the linear combination of mean squares or mean cross products to solve for variance components, and the statistical property that $Cov(X,Y) = [Var(X+Y) - Var(X) - Var(Y)]/2$), all components will contribute to the standard error of the correlation. In some cases, this can lead to large standard errors for the correlation estimate, resulting in correlations greater than 1.

b. As is the case with the estimation of all genetic parameters, consideration should be given as to how many HSFs are needed to obtain reliable estimates. In general, the more heritable a character, the fewer HSFs needed to obtain good genetic correlation estimates. Fifteen to twenty HSFs is the minimal acceptable range that can be used.

c. Traits X and Y may differ widely in overall scale due to differences in concentration and/or units of measurement. For example, for a particular forage quality study, mean neutral detergent fiber (NDF) may equal 508 $g \cdot kg^{-1}$ (range: 478 to 539), whereas mean magnesium concentration may equal 2.18 $g \cdot kg^{-1}$ (range: 1.56 to 2.69). The sum of the average NDF and magnesium concentrations would equal 510.18 and is heavily weighted toward NDF. In such a case, the sum method could produce misleading results. When these types of data are used, they should first be standardized, such that traits X and Y are both on the same scale. One standardization method that effectively cancels scaling differences is to divide each trait's raw data by its mean standard deviation prior to creating the $X + Y$ variable (Frey and Horner, 1957).

Originator

The sum method of estimating additive genetic covariances originated at North Carolina State University during the theoretical development and application of mating designs I and II. This method was taught by Dr. Robert E. Stucker (a North Carolina State University graduate) at the University of Minnesota in the Statistical Topics in Plant Sciences course.

Key References Using the Formula

Waldron, B.L., Ehlke, N.J., Wyse, D.L., and Vellekson, D.J. (1998). Genetic variation and predicted gain from selection for winterhardiness and turf quality in a perennial ryegrass topcross population. *Crop Science* 38:817-822.

Contact

Dr. Blair L. Waldron, USDA-ARS Forage and Range Research Laboratory, Utah State University, Logan, UT 84322-6300, USA. E-mail: <blw@cc.usu.edu>.

EXAMPLE

Traits X and Y are visual scores where the range is 1-9; HSFs were evaluated in multiple locations for one year using a randomized complete block (RCB) design. All variables are assumed to be random.

Step 1. Creating $X + Y$ Variable

HSF	Trait X	Trait Y	Trait $X + Y$
1	9	8	17
2	8	3	11
3	5	4	9
.	.	.	.
.	.	.	.
.	.	.	.
n	X_n	Y_n	$X_n + Y_n$

Step 2. ANOVA for Traits X and Y

Source	DF	Mean Square	Expected Mean Square
Location	$l-1$	$MS_{L(x \text{ or } y)}$	$\sigma^2_{E(x \text{ or } y)} + r\,\sigma^2_{H L(x \text{ or } y)} + h\,\sigma^2_{R/L(x \text{ or } y)} + rh\,\sigma^2_{L(x \text{ or } y)}$
Rep in Loc	$l(r-1)$	$MS_{R/L(x \text{ or } y)}$	$\sigma^2_{E(x \text{ or } y)} + h\,\sigma^2_{R/L(x \text{ or } y)}$
HSF	$h-1$	$MS_{H(x \text{ or } y)}$	$\sigma^2_{E(x \text{ or } y)} + r\,\sigma^2_{H L(x \text{ or } y)} + rl\,\sigma^2_{H(x \text{ or } y)}$
HSF × Loc	$(h-1)(l-1)$	$MS_{H L(x \text{ or } y)}$	$\sigma^2_{E(x \text{ or } y)} + r\,\sigma^2_{H L(x \text{ or } y)}$
Error	$l(r-1)(h-1)$	$MS_{E(x \text{ or } y)}$	$s^2_{E(x \text{ or } y)}$

Step 3. ANOVA for Traits $X + Y$

Source	DF	MCP	Expected Mean Cross Product
Location	$l-1$	$MCP_{L(x+y)}$	$\sigma_{E(x+y)} + r\,\sigma_{H L(x+y)} + h\,\sigma_{R/L(x+y)} + rh\,\sigma_{L(x+y)}$
Rep in Loc	$l(r-1)$	$MCP_{R/L(x+y)}$	$\sigma_{E(x+y)} + h\,\sigma_{R/L(x+y)}$
HSF	$h-1$	$MCP_{H(x+y)}$	$\sigma_{(x+y)} + r\,\sigma_{H L(x+y)} + rl\,\sigma_{H(x+y)}$

HSF × Loc $(h{-}1)(l{-}1)$ $MCP_{HL(x+y)}$ $\sigma_{E(x+y)} + r\,\sigma_{HL(x+y)}$

Error $l(r{-}1)(h{-}1)$ $MCP_{E(x+y)}$ $\sigma_{E(x+y)}$

MCP is the the mean square value resulting from ANOVA on the new variable $X+Y$.

Step 4. Solving for Appropriate Variance Components

$$\sigma^2_{A(x)} = ((MS_{H(x)} - MS_{H\,L(x)}) \,/\, rl)$$
$$\sigma^2_{A(y)} = ((MS_{H(y)} - MS_{H\,L(y)}) \,/\, rl)$$
$$\sigma^2_{A(x+y)} = ((MCP_{H(x+y)} - MCP_{H\,L(x+y)}) \,/\, rl)$$

Step 5. Solving for Additive Genetic Covariance Between Traits X and Y

$$\sigma_{A(xy)} = (\sigma^2_{A(x+y)} - \sigma^2_{A(x)} - \sigma^2_{A(y)}) / 2$$

Step 6. Calculating Additive Genetic Correlation Between Traits X and Y

$$r_{A(xy)} = \sigma_{A(xy)} \,/\, \sqrt{(\sigma^2_{A(x)}\sigma^2_{A(y)})}$$

REFERENCES

Falconer, D.S. (1989). *Introduction to quantitative genetics,* Third edition. John Wiley & Sons, Inc., New York, p. 317.

Frey, K.J. and Horner, T. (1957). Heritability in standard units. *Agronomy Journal* 49: 59-62.

Chapter 10

Developmental Analysis
for Quantitative Traits

Jun Zhu

Purpose

To analyze developmental quantitative traits.

Definitions

Genetic Model

For time-dependent traits, the phenotypic data observed at time t ($t = 1$, 2, ...) have the following mixed linear model:

$$y_{(t)} = Xb_{(t)} + \sum_{u=1}^{m} U_u e_{u(t)}$$

$$\sim N(Xb_{(t)}, V_{(t)} = \sum_{u=1}^{m} \sigma_{u(t)}^2 U_u U_u^T)$$

Variance at time t, $\sigma_{u(t)}^2$, can measure genetic variation accumulated from the initial time to time t. Given the observed phenotype vector $y_{(t-1)}$ measured at time $(t-1)$, the conditional random variables of $y_{(t)} \mid y_{(t-1)}$ at time t have conditional distribution:

$$y_{(t)} \mid y_{(t-1)} = Xb_{(t|t-1)} + \sum_{u=1}^{m} U_u e_{u(t|t-1)}$$

$$\sim N(Xb_{(t|t-1)}, V_{(t|t-1)} = \sum_{u=1}^{m} \sigma_{u(t|t-1)}^2 U_u U_u^T)$$

Since conditional $y_{(t)} \mid y_{(t-1)}$ is independent of $y_{(t-1)}$, conditional random effects, $e_{(t|t-1)}$, and conditional variance components, $\sigma^2_{u(t|t-1)}$ contain extra variation from time $t-1$ to time t, which is not explainable by the accumulated effects of the initial time to time $t-1$.

Analysis

With observed phenotypic data at time $t-1$ $(y_{(t-1)})$ and time $t(y_{(t)})$, a new random vector $y_{(*)}$ can be obtained using mixed model approaches (Zhu, 1995):

$$y_{(*)} = y_{(t)} - C_{(t-1,t)} V^{-1}_{(t-1)} (y_{(t-1)} - Xb_{(t-1)})$$

The new random vector has variance,

$$\mathrm{var}(y_{(*)}) = V_{(t)} - C_{(t-1,t)} V^{-1}_{(t-1)} C_{(t,t-1)},$$

which is identical to the conditional variance-covariance matrix of $V_{(t|t-1)}$. It can be proved that $y_{(*)}$ is independent of $y_{(t-1)}$.

When the new data $(y_{(*)})$ are used to fit the genetic model,

$$y_{(*)} = Xb_{(*)} + \sum_{u-1}^{m} \sigma^2_{u(*)} U_u U_u^T$$

$$\sim N(Xb_{(*)}, V_{(*)} = \sum_{u=1}^{m} \sigma^2_{u(*)} U_u U_u^T)$$

unbiased estimation of variances, $\sigma^2_{u(*)}$, can be obtained by REML or MINQUE(1) approaches (Zhu, 1995). Prediction of random effects, $e_{u(*)}$, can be obtained by the linear unbiased prediction (LUP) method (Zhu, 1992; Zhu and Weir, 1996) or the adjusted unbiased prediction (AUP) method (Zhu, 1993; Zhu and Weir, 1996). Since $\sigma^2_{u(*)}$ is equivalent to the conditional variance $\sigma^2_{u(t|t-1)}$, genetic effects $e_{u(*)}$ also have an equivalency to the conditional genetic effects $e_{u(t|t-1)}$.

Originator

Zhu, J. (1992). Mixed model approaches for estimating genetic variances and covariances. *Journal of Biomathematics* 7(1):1-11.

Zhu, J. (1993). Methods of predicting genotype value and heterosis for offspring of hybrids. *Journal of Biomathematics* 8(1):32-44.

Zhu, J. (1995). Analysis of conditional effects and variance components in developmental genetics *Genetics* 141(4):1633-1639.

Zhu, J. and Weir, B.S. (1996). Diallel analysis for sex-linked and maternal effects. *Theoretical and Applied Genetics* 92(1):1-9.

Software Available

Zhu, J. (1997). GENCOND1.EXE a computer software for calculating conditional phenotypic data. *Analysis Methods for Genetic Models* (pp. 278-285), Agricultural Publication House of China, Beijing (program free of charge). Contact Dr. Jun Zhu, Department of Agronomy, Zhejiang University, Hangzhou, China. E-mail: <jzhu@zju.edu.cn>.

EXAMPLE

Unconditional data (BOL8/4 and BOL8/9) to be analyzed (file: COTBOLM.TXT) (Parent = 4, Year = 2, Blk = 1):

Env	Fem	Male	Cross	BLK	BOL8/4	BOL8/9
1	1	1	0	1	6.46	8.14
1	1	2	1	1	5.77	7.85
1	1	3	1	1	8.64	9.01
1	1	4	1	1	8.33	10.30
1	2	1	1	1	6.70	8.74
1	2	2	0	1	5.65	7.90
1	2	3	1	1	7.94	9.13
1	2	4	1	1	8.47	11.24
1	3	1	1	1	8.72	9.29
1	3	2	1	1	9.32	10.36
1	3	3	0	1	4.98	5.35
1	3	4	1	1	8.90	10.14
1	4	1	1	1	7.58	9.74
1	4	2	1	1	8.74	11.08
1	4	3	1	1	9.34	11.49
1	4	4	0	1	7.02	8.90
2	1	1	0	1	8.06	11.63
2	1	2	1	1	11.36	15.18
2	1	3	1	1	9.31	10.58
2	1	4	1	1	13.30	15.76
2	2	2	0	1	8.09	12.39

2	2	3	1	1	10.87	13.50
2	2	4	1	1	15.60	20.45
2	3	3	0	1	5.05	5.78
2	3	4	1	1	12.76	14.26
2	4	4	0	1	12.29	15.86

Conditional data (BOL8/9|BOL8/4) produced and to be analyzed (Parent = 4, Year = 2, Blk = 1):

Year	Fem	Male	Cross	Blk	BOL8/9|BOL8/4
1	1	1	0	1	9.75974
1	1	2	1	1	10.0186
1	1	3	1	1	8.54954
1	1	4	1	1	9.42577
1	2	1	1	1	9.98357
1	2	2	0	1	10.3079
1	2	3	1	1	9.30607
1	2	4	1	1	10.1621
1	3	1	1	1	8.74997
1	3	2	1	1	9.16348
1	3	3	0	1	8.62587
1	3	4	1	1	8.8201
1	4	1	1	1	9.61174
1	4	2	1	1	9.73354
1	4	3	1	1	9.73246
1	4	4	0	1	8.86123
2	1	1	0	1	14.6502
2	1	2	1	1	14.8385
2	1	3	1	1	12.6994
2	1	4	1	1	12.6991
2	2	2	0	1	14.9987
2	2	3	1	1	13.9182
2	2	4	1	1	14.949
2	3	3	0	1	12.5327
2	3	4	1	1	12.0607
2	4	4	0	1	12.8132

1. Run GENAD.EXE to create mating design matrix files and unconditional data for the additive-dominance (AD) model. Before running these programs, create a data file (e.g., COTBOLM.TXT) for your analysis of unconditional data with five design columns followed by trait columns, which are (1) environment, (2) maternal, (3) paternal, (4) generation, and (5) replication. There is a limitation (<100 traits)

for the number of trait columns. The data file COTBOLM.TXT contains phenotypic data of two traits (BOL8/4 and BOL8/9).

2. Run the program GENCOND1.EXE for constructing conditional data. The conditional data will have five design columns and will be stored in a file with the name COTBOLM.CON. Afterward, run GENAD.EXE again using the conditional data file COTBOLM. CON to create files for mating design matrix and conditional data by the AD model.

3. Conditional variances and conditional genetic effects can be obtained by running programs for variance analyses. Standard errors of estimates are calculated by jackknife procedures. If you have multiple blocks for your experiments, you can use GENVAR1R.EXE for jackknifing over blocks. Otherwise, you can use GENVAR1C.EXE or GENCOV1C.EXE for jackknifing over cell means. These two programs will allow you to choose the parental type (inbred or outbred) and the prediction methods (LUP or AUP). You also need to input coefficients (1, 0, or −1) for conducting linear contrasts for genetic effects of parents.

4. The results will be automatically stored in text files for later use or printing. An example of results is provided in a file named COTBOLM.VAR (output 1) for analysis of conditional variance and conditional genetic effects.

5. Developmental genetic analysis can also be conducted for other genetic models, such as GENADM.EXE for additive, dominance, and maternal models with $G = A + D + M$; GENADE.EXE for additive, dominance, and epistatic models with $G = A + D + AA$; GENSEX. EXE for additive, dominance, sex-linked, and maternal models with $G = A + D + L + M$; GENDIPLD.EXE for traits of diploid seeds or animals; GENTRIPL.EXE for traits of triploid endosperm.

Output 1 for Conditional Variance Analysis

```
Traits =, 1
Variance components = , 5
Degree of freedom = , 25
File name is cotbolm.VAR
Date and Time for Analysis: Sat Jun 24 19:07:06 2000

Variance Components Estimated by MINQUE(1) with GENVAR1R.EXE.
Jackknifing Over Block Conducted for Estimating S.E.
Predicting Genetic Effects by Adjusted Unbiased Prediction (AUP)
    Method.
```

```
NS = Not significant; S+ = Significant at 0.10 level.
S* = Significant at 0.05 level; S** = Significant at 0.01 level.
Linear Contrast Test:
 +<1> +<2> -<3> +<4>
```

Diallel Analysis of Trait, BOL8/9|BOL8/4, for Public Users.

Var Comp	Estimate	S. E.	P-value	
(1): Additive Var	0.665074	0.120759	5.04e-006	S**
(2): Dominance Var	0.180163	0.0462009	0.00032	S**
(3): Add. * Env. Var	0.193749	0.0588614	0.00148	S**
(4): Dom. * Env. Var	0.331579	0.0779674	0.000129	S**
(5): Residual Var	0.189768	0.0819921	0.0146	S*
(6): Var(Pheno.)	1.56033	0.209525	4.22e-008	S**

Proportion of Var(G)/Var(T)	Estimate	S. E.	P-value	
(1): Additive Var/Vp	0.426239	0.0429252	1.59e-010	S**
(2): Dominance Var/Vp	0.115464	0.0393502	0.00353	S**
(3): Add. * Env. Var/Vp	0.124172	0.0301349	0.000182	S**
(4): Dom. * Env. Var/Vp	0.212505	0.0386894	5.24e-006	S**
(5): Residual Var/Vp	0.12162	0.0341315	0.000753	S**

Heritability	Estimate	S. E.	P-value	
(6): Heritability(N)	0.426239	0.0429252	1.59e-010	S**
(7): Heritability(B)	0.541703	0.0394694	2.53e-011	S**
(8): Heritability(NE)	0.124172	0.0301349	0.000182	S**
(9): Heritability(BE)	0.336677	0.0399954	4.56e-009	S**

Genetic Predictor, S. E. , P-value
(1): Random Effect is Additive Effects

A1,	0.223513,	0.155049,	0.162,	NS
A2,	0.677601,	0.121236,	8.19e-006,	S**
A3,	-0.562930,	0.091853,	2.09e-006,	S**
A4,	-0.338327,	0.119019,	0.00878,	S**
Linear				
Contrast,	1.95226,	0.371664,	1.94e-005,	S**

(2): Random Effect is Dominance Effects

D1*1	0.798935	0.436678	0.0793	S+
D2*2	0.018615	0.066023	0.78	NS
D3*3	0.072842	0.101742	0.481	NS
D4*4	-0.412087	0.328823	0.222	NS
D1*2	0.425023	0.229382	0.0757	S+
D1*3	-1.004059	0.598641	0.106	NS
D1*4	-0.661568	0.375175	0.0901	S+
D2*3	-0.076713	0.091950	0.412	NS
D2*4	1.092527	0.629749	0.0951	S+
D3*4	-0.253545	0.321869	0.438	NS
Heterosis <Delta>	-0.563434	0.767809	0.47	NS

(3): Random Effect is Add. * Env. Effects

AE1 in E1	-0.022308	0.102366	0.829	NS
AE2 in E1	0.036740	0.088970	0.683	NS
AE3 in E1	-0.011993	0.081400	0.884	NS
AE4 in E1	-0.002428	0.153184	0.987	NS
AE1 in E2	0.158196	0.194168	0.423	NS
AE2 in E2	0.357913	0.255052	0.173	NS

```
AE3 in E2                        -0.305599     0.174689      0.0925    S+
AE4 in E2                        -0.210610     0.188787      0.275     NS
Linear Contrast                   1.58114e-005 1.77345e-005  0.381     NS

(4): Random Effect is Dom. * Env. Effects
DE1 in E1                        -0.320161     0.196316      0.115     NS
DE2 in E1                         0.368241     0.179784      0.0512    S+
DE3 in E1                        -0.184366     0.161795      0.265     NS
DE4 in E1                        -0.541939     0.380790      0.167     NS
DE1 in E2                        -0.053143     0.150489      0.727     NS
DE2 in E2                        -0.057755     0.290609      0.844     NS
DE3 in E2                         0.687891     0.349814      0.0604    S+
DE4 in E2                        -0.303718     0.189203      0.121     NS
DE1 in E3                        -0.340375     0.372988      0.37      NS
DE2 in E3                         0.745340     0.610413      0.233     NS
DE3 in E3                         0.927552     0.534879      0.0952    S+
DE4 in E3                        -0.569171     0.246751      0.0296    S*
DE1 in E4                         0.343324     0.197179      0.0939    S+
DE2 in E4                         0.301441     0.324563      0.362     NS
DE3 in E4                         0.139931     0.334570      0.679     NS
DE4 in E4                        -0.560693     0.344209      0.116     NS
DE1 in E5                        -1.175256     0.700765      0.106     NS
DE2 in E5                         0.311932     0.239635      0.205     NS
DE3 in E5                         1.167883     0.778234      0.146     NS
DE4 in E5                        -0.887039     0.664529      0.194     NS
Heterosis <Delta>                 0            0            1         NS

Fixed Effect <1>, 9.42573
Fixed Effect <2>, 13.616

Time Used (Hour) = 0.000556
```

Chapter 11

Ecovalence and Stability Variance

Manjit S. Kang

Purpose

To identify and select genotypes with consistent (stable) performance across diverse environments (broad adaptation).

Definitions

Ecovalence

Ecovalence is the sum of squares contributed by a genotype to a genotype-by-environment interaction:

$$W_i = \sum_j (u_{ij} - \bar{u}_i.)^2$$

where W_i = *ecovalence* for ith genotype, $u_{ij} = x_{ij} - \bar{x}._j$, x_{ij} = observed trait value for the ith genotype in jth environment, $\bar{x}._j$ = mean of all genotypes in jth environment; $\bar{u}_i. = \sum_j u_{ij} / s$, s = number of environments.

Originator

Wricke, G. (1962). Über eine Methode zur Erfassung der ökologischen Streubreite. *Zeitschrift für Pflanzenzüchtung* 47:92-96.

Stability Variance

Stability variance measures the consistency of performance of a genotype across a set of diverse environments. The smaller the value, the greater the stability:

$$\sigma_i^2 = \left[1/(s-1)(t-1)(t-2)\right] \times \left[t(t-1)\sum_j (u_{ij} - \bar{u}_i.)^2 - \sum_i \sum_j (u_{ij} - \bar{u}_i.)^2\right],$$

where, σ_i^2 = stability variance for the ith genotype, and t = total number of genotypes evaluated.

When genotype-by-environment interaction is significant, it is desirable to know the factor(s) responsible for the interaction. Technically, factors are used as covariates and their linear effects, which represent heterogeneity or nonadditivity, are removed. Following the removal of heterogeneity, the remainder of the genotype-by-environment interaction is examined for significance. If heterogeneity is significant, the contribution of each genotype (s_i^2) to the residual genotype-by-environment interaction can be determined using the following formula provided by Shukla (1972):

$$s_i^2 = \left[t/(t-2)(s-2)(s-2)\right] \times \left[S_i - \sum_i S_i / t(t-1)\right],$$

where $S_i = \sum_j (u_{ij} - \bar{u}_i. - b_i Z_j)^2$,   and $b_i = \sum_i \left[(u_{ij} - \bar{u}_i.)Z_j / \sum_j Z_j,\right]$ and
$Z_j = \bar{x}._j - \bar{x}..$

Originator

Shukla, G. K. (1972). Some statistical aspects of partitioning genotype-environmental components of variability. *Heredity* 29:237-245.

Software Available

Kang, M.S. (1989). A new SAS program for calculating stability-variance parameters. *Journal of Heredity* 80:415. (software is free of charge).

Key Reference(s) Using the Concept/Software/Formula

Kang, M.S. (1993). Simultaneous selection for yield and stability in crop performance trials: Consequences for growers. *Agronomy Journal* 85:754-757.
Pazdernick, D.L., Hardman, L.L., and Orf, J.H. (1997). Agronomic performance and stability of soybean varieties grown in three maturity zones of Minnesota. *Journal of Production Agriculture* 10:425-430.

Contact

Dr. M.S. Kang, 105 Sturgis Hall, Louisiana State University, Baton Rouge, LA 70803-2110, USA. E-mail: mKang@agctr.lsu.edu.

EXAMPLE

Data to be analyzed (Yield):

Environments	Env1	Env2	Env3	Env4	Env5	Env6
Genotypes						
Genotype1	161.7†	247.0	185.4	218.7	165.3	154.6
Genotype2	187.7	257.5	182.4	183.3	138.9	143.8
Genotype3	200.1	262.9	194.9	220.2	165.8	146.3
Genotype4	196.9	339.2	271.2	266.3	151.2	193.6
Genotype5	182.5	253.8	219.2	200.5	184.4	190.1
Zj	-16.41	69.93	8.43	15.63	-41.07	-36.51

†Each cell represents summation of six observations (six replications).

Case 1: Using Totals Across Replications

Since each cell in the above data represents a "total" of six observations, the value for N in the computer program provided in this chapter will be 6, and that for *REP* will be 1 ($N = 6$; *REP* = 1); For the given data set, $\bar{x}.. = 202.18$ (grand mean), and

$$Z_j = \bar{x}._j - \bar{x}.. = -16.41 \; 69.93 \; 8.43 \; 15.63 \; -41.07 \; -36.51$$

Note: Z_j represents a covariate.

Program Listing for Case 1

```
DATA GENETIC;
INPUT X1 - X6;
CARDS;
161.7   247.0   185.4   218.7   165.3   154.6
187.7   257.5   182.4   183.3   138.9   143.8
200.1   262.9   194.9   220.2   165.8   146.3
196.9   339.2   271.2   266.3   151.2   193.6
182.5   253.8   219.2   200.5   184.4   190.1
;
PROC IML; USE GENETIC;
READ ALL VAR _ALL_ INTO X;
N=6; REP=1;
ZJ={-16.41 69.93 8.43 15.63 -41.07 -36.51};
P=NROW(X); Q=NCOL(X);
```

```
CMEAN=X(|+,|)/P;
GENO=J(P,Q); START;
DO I=1 TO P; GENO(|I,|)=CMEAN(|1,1:Q|); END;
FINISH; RUN;
U=X - GENO; UM=U/Q;
ENV=J(P,Q); START;
DO K=1 TO Q;
ENV(|,K|)=UM(|,+|); END;
FINISH; RUN;
DIFF=U-ENV; SSDIFF=(DIFF#DIFF)(|,+|);
SUMSS=SUM(SSDIFF); ECOV=SSDIFF/N;
L=P*(P-1); E=(Q-1)*(P-1)*(P-2);
LSSDIFF=(SSDIFF*L)/N;
D=J(P,1,(SUMSS/N));
SIG=LSSDIFF-D; SIGMA=SIG/E;
SUMSQZJ=SUM(ZJ#ZJ); HAT=J(P,Q);
START; DO R=1 TO P;
HAT(|R,|)=ZJ(|1,1:Q|); END;
FINISH; RUN;
NEW=DIFF#HAT; BETA=(NEW/SUMSQZJ)(|,+|);
GP=J(P,Q); START;
DO C=1 TO Q; GP(|,C|)=BETA(|1:P,1|); END;
FINISH; RUN;
BIZJ=HAT#GP; NEWDIFF=(DIFF-BIZJ);
SI=(NEWDIFF#NEWDIFF)(|,+|); TS=P/((P-2)*(Q-2));
TOTSI=SUM(SI)/L; SP=((SI-TOTSI)*TS)/N;
F=D(|1,1|); START;
IF N=1 THEN DO; ECOV=ECOV*REP;
F=F*REP; SIGMA=SIGMA*REP; SP=SP*REP; END;
FINISH; RUN;
TITLE 'STABILITY-VARIANCE';
TITLE2 'X MATRIX REPRESENTS INPUT DATA';
TITLE3 'ECOV MATRIX REPRESENTS GXE SS FOR EACH GENOTYPE';
TITLE4 'F MATRIX REPRESENTS TOTAL GXE SS';
TITLE5 'SIGMA MATRIX REPRESENTS STABILITY VARIANCE FOR EACH GENOTYPE';
TITLE6 'SP MATRIX REPRESENTS SMALL S-SQUARE SUB-I';
PRINT X, ECOV, F, SIGMA, SP; RUN;
```

Case 1 Output

```
X MATRIX REPRESENTS INPUT DATA
ECOV MATRIX REPRESENTS GXE SS FOR EACH GENOTYPE
TITLE4 'F MATRIX REPRESENTS TOTAL GXE SS';
SIGMA MATRIX REPRESENTS STABILITY VARIANCE FOR EACH GENOTYPE
SP MATRIX REPRESENTS SMALL S-SQUARE SUB-I
```

X	COL1	COL2	COL3	COL4	COL5	COL6
ROW1	161.7	247.0	185.4	218.7	165.3	154.6
ROW2	187.7	257.5	182.4	183.3	138.9	143.8
ROW3	200.1	262.9	194.9	220.2	165.8	146.3
ROW4	196.9	339.2	271.2	266.3	151.2	193.6
ROW5	182.5	253.8	219.2	200.5	184.4	190.1

ECOV	COL1
ROW1	151.5

```
ROW2                132.7
ROW3                142.1
ROW4                750.5
ROW5                300.9

F                   COL1
ROW1                1477.7

SIGMA               COL1
ROW1                25.8791
ROW2                19.5998
ROW3                22.7305
ROW4               225.5
ROW5                75.6800

SP                  COL1
ROW1                34.1065
ROW2                40.4848
ROW3                42.8052
ROW4                79.6618
ROW5                23.2671

NOTE: EXIT FROM IML
```

Case 2: Using Means Across Replications

Differences when running the program using means across replications:

- After the *CARDS* statement, enter the means instead of totals across replications.
- Calculate *ZJ* using means instead of totals.
- $N = 1$; $REP = 6$.

Chapter 12

Genotype-by-Environment Interaction Variance

Robert Magari
Manjit S. Kang

Purpose

To estimate genotype-by-environment interaction and evaluate performance across a range of environments.

Definitions

Genotype-by-environment variance (GE variance) program is a restricted maximum likelihood estimator of Shukla's (1972) stability variance.

Originator

Shukla, G.K. (1972). Some statistical aspects of partitioning genotype-environment components of variability. *Heredity* 29:237-245.

Software Available

Magari, R. and Kang, M.S. (1997). SAS_STABLE: Analysis of balanced and unbalanced data. *Agronomy Journal* 89:929-932. The software is provided free of charge.

Key References

Kang, M.S. and Magari, R. (1996). New developments in selecting for phenotypic stability in crop breeding. In Kang, M.S. and Gauch, H.G. (Eds.), *Genotype by Environment Interaction* (pp. 1-14). CRC Press, Boca Raton, FL.

Magari, R., Kang, M.S., and Zhang, Y. (1997). Genotype by environment interaction for ear moisture loss in corn. *Crop Science* 37:774-779.

Contact

Dr. M.S. Kang, 105 Sturgis Hall, Louisiana State University, Baton Rouge, LA 70803-2110, USA. E-mail: <mKang@agctr.lsu.edu>.

EXAMPLE

Replicated data of several genotypes in different environments are entered in SAS format. The solutions to the parameters are obtained as follows:

$$
\begin{pmatrix} \hat{\beta} \\ \hat{\alpha} \\ \hat{\gamma} \\ \cdot \\ \hat{\delta}\hat{K} \\ \cdot \end{pmatrix} = \begin{pmatrix} X'X & X'Z_1 & X'Z_2 & \cdot & X'Z_{3k} & \cdot \\ Z_1'X & Z_1'Z_1 + I\frac{\hat{\sigma}^2\delta}{\hat{\sigma}^2 E} & Z_1'R^{-1}Z_2 & \cdot & Z_1'Z_{3k} & \cdot \\ Z_2'X & Z_2'Z_1 & Z_2'Z_2 + I\frac{\hat{\sigma}^2\epsilon}{\hat{\sigma}^2 E} & \cdot & Z_2'Z_{3k} & \cdot \\ \cdot & \cdot & \cdot & \cdot & \cdot & \cdot \\ Z_{3k}'X & \cdot & Z_{3k}'Z_2 & \cdot & Z_{3k}'Z_{3k} + I\frac{\hat{\sigma}^2\epsilon}{\hat{\sigma}^2_{EG(k)}} & \cdot \\ \cdot & \cdot & \cdot & \cdot & \cdot & \cdot \end{pmatrix}^{-1} \begin{pmatrix} X'y \\ Z_1'y \\ Z_2'y \\ \cdot \\ Z_{3k}'y \\ \cdot \end{pmatrix}
$$

where **y** is vector of observations, $\hat{\beta}, \hat{\alpha}, \hat{\gamma}$, and $\hat{\delta}_k$ are the vector of estimates for genotypes, environments, replicates within environments, and each GEI, respectively. **X** is the design matrix of the fixed effects (genotypes), $\mathbf{Z}_1$ is the design matrix for environments, $\mathbf{Z}_2$ is the design matrix for replications-within-environments, and $\mathbf{Z}_{3k}$ is the design matrix for GEI of the *k*th genotype.

Variance components are defined and calculated as follows:

Environmental variance

$$\hat{\sigma}^2{}_E = \frac{\hat{a}'I\hat{a}+tr\left(C_{22}\right)\hat{\sigma}^2{}_\varepsilon}{a}$$

Replications-within-environment variance

$$\hat{\sigma}^2{}_{R/E} = \frac{\hat{\gamma}'I\hat{\gamma}+tr\left(C_{33}\right)\hat{\sigma}^2{}_\varepsilon}{b}$$

Genotype-by-environment interaction variance

$$\hat{\sigma}^2{}_{GE(k)} = \frac{\left\lfloor \hat{\delta}'_k I\hat{\delta}_k +tr\left(C_{(3+k)(3+k)}\right)\hat{\sigma}^2{}_\varepsilon \right\rfloor}{\text{no. columns of}\, Z_{3k}}$$

Experimental error variance

$$\hat{\sigma}^2{}_\varepsilon = \frac{y'y-\hat{\beta}'X'y-\hat{u}'Z'y}{n-c}$$

where I represents identity matrix, C represents corresponding blocks of inverse of the aforementioned (large) matrix, where solutions of parameters are obtained, a is the number of environments, b is the number of replicates, c is the number of genotypes, and n is the dimension of the y vector.

Program Listing

```
proc iml;
use early;

/* Read data into vectors*/;
read all var{earwt} into y;
read all var{hybrid} into X;
read all var{date} into ZE;
read all var{rep} into r;

/* Set up design matrices */;
R=design(r);
X=design(X);
ZE=design(ZE);
U=hdir(X,ZE);
RE=hdir(R,ZE);
nr=ncol(RE);
W=X||ZE||U||RE;
dim=ncol(W);
t=nrow(y);
a=ncol(X);
b=ncol(ZE);
```

```
yy=y`*y;
wy=W`*y;
ww=W`*W;
bb=ncol(U);
/* Starting values for iterations */;
lambda=0.4;
lambda1=0.4;
lambda2=0.4;

/* Set up matrices and start iterations */;
block1=j(a,a,0);
do iter=1 to 1000;
block2=I(b)*lambda;
block3=I(bb)*lambda1;
block4=I(nr)*lambda2;
addon=block(block1,block2,block3,block4);
M=ww+addon;
invM=inv(M);

/* Solutions for vector of effects (BLUE and BLUP) */;
solution=invM*wy;

/* Variance components */;
sigmae=(yy-solution`*wy)/(t-a);
ue=solution[a+1:a+b];
ue1=solution[a+1+b:a+2*b];
ue2=solution[a+1+2*b:a+3*b];
ue3=solution[a+1+3*b:a+4*b];
ue4=solution[a+1+4*b:a+5*b];
ue5=solution[a+1+5*b:a+6*b];
ue6=solution[a+1+6*b:a+7*b];
ue7=solution[a+1+7*b:a+8*b];
ue8=solution[a+1+8*b:a+9*b];
ur=solution[a+1+9*b:a+9*b+nr];

true=trace(invM[a+1:a+b,a+1:a+b]);
sigmaue=(ue`*ue+true*sigmae)/b;
true1=trace(invM[a+1+b:a+2*b,a+1+b:a+2*b]);
s1=(ue1`*ue1+true1*sigmae)/b;
true2=trace(invM[a+1+2*b:a+3*b,a+1+2*b:a+3*b]);
s2=(ue2`*ue2+true2*sigmae)/b;
true3=trace(invM[a+1+3*b:a+4*b,a+1+3*b:a+4*b]);
s3=(ue3`*ue3+true3*sigmae)/b;
true4=trace(invM[a+1+4*b:a+5*b,a+1+4*b:a+5*b]);
s4=(ue4`*ue4+true4*sigmae)/b;
true5=trace(invM[a+1+5*b:a+6*b,a+1+5*b:a+6*b]);
s5=(ue5`*ue5+true5*sigmae)/b;
true6=trace(invM[a+1+6*b:a+7*b,a+1+6*b:a+7*b]);
s6=(ue6`*ue6+true6*sigmae)/b;
true7=trace(invM[a+1+7*b:a+8*b,a+1+7*b:a+8*b]);
s7=(ue7`*ue7+true7*sigmae)/b;
true8=trace(invM[a+1+8*b:a+9*b,a+1+8*b:a+9*b]);
s8=(ue8`*ue8+true8*sigmae)/b;
truer=trace(invM[a+b+bb+1:a+b+bb+nr, a+b+bb+1:a+b+bb+nr]);
sigmar=(ur`*ur+truer*sigmae)/nr;

if(mod(iter,10)=0) then print iter sigmae sigmaue sigmar;
```

```
if(mod(iter,10)=0) then print iter s1 s2 s3 s4 s5 s6 s7 s8;
sig=(s1+s2+s3+s4+s5+s6+s7+s8)/a;
lambda=sigmae/sigmaue;
lambda1=sigmae/sig;
lambda2=sigmae/sigmar;
if(mod(iter,10)=0) then print iter sig;
end;

/* Set up of Fisher's information matrix */;
e1={1,0,0,0,0,0,0,0};
e2={0,1,0,0,0,0,0,0};
e3={0,0,1,0,0,0,0,0};
e4={0,0,0,1,0,0,0,0};
e5={0,0,0,0,1,0,0,0};
e6={1,0,0,0,0,1,0,0};
e7={1,0,0,0,0,0,1,0};
e8={1,0,0,0,0,0,0,1};
a=ncol(X);
k=I(b);
ee1=e1@k;
ee2=e2@k;
ee3=e3@k;
ee4=e4@k;
ee5=e5@k;
ee6=e6@k;
ee7=e7@k;
ee8=e8@k;
Z1=U*ee1;
Z2=U*ee2;
Z3=U*ee3;
Z4=U*ee4;
Z5=U*ee5;
Z6=U*ee6;
Z7=U*ee7;
Z8=U*ee8;
fi=j(11,11,0);
V=ZE*ZE`*sigmaue+RE*RE`*sigmar+U*U`*sig+I(t)*sigmae;
vi=inv(V);
p=vi-vi*X*inv(X`*vi*X)*X`*vi;
fi[1,1]=0.5*trace(p*ZE*ZE`*p*ZE*ZE`);
fi[1,2]=0.5*trace(p*ZE*ZE`*p*RE*RE`);
fi[1,3]=0.5*trace(p*ZE*ZE`*p*Z1*Z1`);
fi[1,4]=0.5*trace(p*ZE*ZE`*p*Z2*Z2`);
fi[1,5]=0.5*trace(p*ZE*ZE`*p*Z3*Z3`);
fi[1,6]=0.5*trace(p*ZE*ZE`*p*Z4*Z4`);
fi[1,7]=0.5*trace(p*ZE*ZE`*p*Z5*Z5`);
fi[1,8]=0.5*trace(p*ZE*ZE`*p*Z6*Z6`);
fi[1,9]=0.5*trace(p*ZE*ZE`*p*Z7*Z7`);
fi[1,10]=0.5*trace(p*ZE*ZE`*p*Z8*Z8`);
fi[1,11]=0.5*trace(p*ZE*ZE`*p);

fi[2,1]=fi[1,2]; fi[2,2]=0.5*trace(p*RE*RE`*p*RE*RE`);
fi[2,3]=0.5*trace(p*RE*RE`*p*Z1*Z1`);
fi[2,4]=0.5*trace(p*RE*RE`*p*Z2*Z2`);
fi[2,5]=0.5*trace(p*RE*RE`*p*Z3*Z3`);
fi[2,6]=0.5*trace(p*RE*RE`*p*Z4*Z4`);
```

```
fi[2,7]=0.5*trace(p*RE*RE`*p*Z5*Z5`);
fi[2,8]=0.5*trace(p*RE*RE`*p*Z6*Z6`);
fi[2,9]=0.5*trace(p*RE*RE`*p*Z7*Z7`);
fi[2,10]=0.5*trace(p*RE*RE`*p*Z8*Z8`);
fi[2,11]=0.5*trace(p*RE*RE`*p);
fi[3,1]=fi[1,3]; fi[3,2]=fi[2,3];
fi[3,3]=0.5*trace(p*Z1*Z1`*p*Z1*Z1`);
fi[3,4]=0.5*trace(p*Z1*Z1`*p*Z2*Z2`);
fi[3,5]=0.5*trace(p*Z1*Z1`*p*Z3*Z3`);
fi[3,6]=0.5*trace(p*Z1*Z1`*p*Z4*Z4`);
fi[3,7]=0.5*trace(p*Z1*Z1`*p*Z5*Z5`);
fi[3,8]=0.5*trace(p*Z1*Z1`*p*Z6*Z6`);
fi[3,9]=0.5*trace(p*Z1*Z1`*p*Z7*Z7`);
fi[3,10]=0.5*trace(p*Z1*Z1`*p*Z8*Z8`);
fi[3,11]=0.5*trace(p*Z1*Z1`*p);

fi[4,1]=fi[1,4]; fi[4,2]=fi[2,4];
fi[4,3]=fi[3,4]; fi[4,4]=0.5*trace(p*Z2*Z2`*p*Z2*Z2`);
fi[4,5]=0.5*trace(p*Z2*Z2`*p*Z3*Z3`);
fi[4,6]=0.5*trace(p*Z2*Z2`*p*Z4*Z4`);
fi[4,7]=0.5*trace(p*Z2*Z2`*p*Z5*Z5`);
fi[4,8]=0.5*trace(p*Z2*Z2`*p*Z6*Z6`);
fi[4,9]=0.5*trace(p*Z2*Z2`*p*Z7*Z7`);
fi[4,10]=0.5*trace(p*Z2*Z2`*p*Z8*Z8`);
fi[4,11]=0.5*trace(p*Z2*Z2`*p);

fi[5,1]=fi[1,5]; fi[5,2]=fi[2,5];
fi[5,3]=fi[3,5]; fi[5,4]=fi[4,5];
fi[5,5]=0.5*trace(p*Z3*Z3`*p*Z3*Z3`);
fi[5,6]=0.5*trace(p*Z3*Z3`*p*Z4*Z4`);
fi[5,7]=0.5*trace(p*Z3*Z3`*p*Z5*Z5`);
fi[5,8]=0.5*trace(p*Z3*Z3`*p*Z6*Z6`);
fi[5,9]=0.5*trace(p*Z3*Z3`*p*Z7*Z7`);
fi[5,10]=0.5*trace(p*Z3*Z3`*p*Z8*Z8`);
fi[5,11]=0.5*trace(p*Z3*Z3`*p);

fi[6,1]=fi[1,6]; fi[6,2]=fi[2,6];
fi[6,3]=fi[3,6]; fi[6,4]=fi[4,6];
fi[6,5]=fi[5,6]; fi[6,6]=0.5*trace(p*Z4*Z4`*p*Z4*Z4`);
fi[6,7]=0.5*trace(p*Z4*Z4`*p*Z5*Z5`);
fi[6,8]=0.5*trace(p*Z4*Z4`*p*Z6*Z6`);
fi[6,9]=0.5*trace(p*Z4*Z4`*p*Z7*Z7`);
fi[6,10]=0.5*trace(p*Z4*Z4`*p*Z8*Z8`);
fi[6,11]=0.5*trace(p*Z4*Z4`*p);

fi[7,1]=fi[1,7]; fi[7,2]=fi[2,7];
fi[7,3]=fi[3,7]; fi[7,4]=fi[4,7];
fi[7,5]=fi[5,7]; fi[7,6]=fi[6,7];
fi[7,7]=0.5*trace(p*Z5*Z5`*p*Z5*Z5`);
fi[7,8]=0.5*trace(p*Z5*Z5`*p*Z6*Z6`);
fi[7,9]=0.5*trace(p*Z5*Z5`*p*Z7*Z7`);
fi[7,10]=0.5*trace(p*Z5*Z5`*p*Z8*Z8`);
fi[7,11]=0.5*trace(p*Z5*Z5`*p);

fi[8,1]=fi[1,8]; fi[8,2]=fi[2,8];
fi[8,3]=fi[3,8]; fi[8,4]=fi[4,8];
fi[8,5]=fi[5,8]; fi[8,6]=fi[6,8];
```

```
fi[8,7]=fi[7,8]; fi[8,8]=0.5*trace(p*Z6*Z6`*p*Z6*Z6`);
fi[8,9]=0.5*trace(p*Z6*Z6`*p*Z7*Z7`);
fi[8,10]=0.5*trace(p*Z6*Z6`*p*Z8*Z8`);
fi[8,11]=0.5*trace(p*Z6*Z6`*p);
fi[9,1]=fi[1,9]; fi[9,2]=fi[2,9];
fi[9,3]=fi[3,9]; fi[9,4]=fi[4,9];
fi[9,5]=fi[5,9]; fi[9,6]=fi[6,9];
fi[9,7]=fi[7,9]; fi[9,8]=fi[8,9];
fi[9,9]=0.5*trace(p*Z7*Z7`*p*Z7*Z7`);
fi[9,10]=0.5*trace(p*Z7*Z7`*p*Z8*Z8`);
fi[9,11]=0.5*trace(p*Z7*Z7`*p);

fi[10,1]=fi[1,10]; fi[10,2]=fi[2,10];
fi[10,3]=fi[3,10]; fi[10,4]=fi[4,10];
fi[10,5]=fi[5,10]; fi[10,6]=fi[6,10];
fi[10,7]=fi[7,10]; fi[10,8]=fi[8,10];
fi[10,9]=fi[9,10]; fi[10,10]=0.5*trace(p*Z8*Z8`*p*Z8*Z8`);
fi[10,11]=0.5*trace(p*Z8*Z8`*p);

fi[11,1]=fi[1,11]; fi[11,2]=fi[2,11];
fi[11,3]=fi[3,11]; fi[11,4]=fi[4,11];
fi[11,5]=fi[5,11]; fi[11,6]=fi[6,11];
fi[11,7]=fi[7,11]; fi[11,8]=fi[8,11];
fi[11,9]=fi[9,11]; fi[11,10]=fi[10,11];
fi[11,11]=0.5*trace(p*p);

/* Inverse of Fisher's information matrix */;
asvc=inv(fi);

/* Standard errors */;
errue1=j(a,1,0);
errue=sqrt(asvc[1,1]);
errar=sqrt(asvc[2,2]);
errue1[1]=sqrt(asvc[3,3]);
errue1[2]=sqrt(asvc[4,4]);
errue1[3]=sqrt(asvc[5,5]);
errue1[4]=sqrt(asvc[6,6]);
errue1[5]=sqrt(asvc[7,7]);
errue1[6]=sqrt(asvc[8,8]);
errue1[7]=sqrt(asvc[9,9]);
errue1[8]=sqrt(asvc[10,10]);
errae=sqrt(asvc[11,11]);

/* Testing */;
zerror=sigmae/errae;

zenv=sigmaue/errue;
zrepenv=sigmar/errar;
perror=(1-probnorm(zerror))*2;
penv=(1-probnorm(zenv))*2;
prepenv=(1-probnorm(zrepenv))*2;
s=j(a,1,0);
z=j(a,1,0);
pge=j(a,1,0);

s[1]=s1;
s[2]=s2;
```

```
s[3]=s3;
s[4]=s4;
s[5]=s5;
s[6]=s6;
s[7]=s7;
s[8]=s8;
ii=j(a,1,1);
z=s/errue1;
pge=(ii-probnorm(z))*2;
gen=j(a,1,0);
do i=1 to a;
gen[i]=i;
end;
print'Fisher's information matrix';
print fi;
print' Inverse of Fisher's information matrix';
print asvc;
print'Individual GxE variance components';
print  gen      s     errue1    z        pge;
print'Error';
print  sigmae     errae  perror;
print'Environment';
print  sigmaue     errue  penv;
print'Replications within environment';
print  sigmar      errar  prepenv;
```

Output

```
ITERATION HISTORY OF STABILITY VARIANCES

ITER          S
  10 0.0034962
     0.0030041
     0.003348
     0.0044955

ITER          S
  20 0.0027947
     0.0024548
     0.0026893
     0.0035634
ITER          S
 100 0.002327
     0.0020955
     0.0022605
     0.0029363
```

SOURCE	GENOTYPE	MEAN	VARIANCE	ERROR	Z	PROB
ENV	.	.	0.025184	0.021807	1.15486	0.24815
GXE	1	4.15499	0.002327	0.001933	1.20366	0.22872
GXE	2	4.16907	0.002095	0.001933	1.08388	0.27842
GXE	3	4.14078	0.002261	0.001880	1.20265	0.22911
GXE	4	4.18702	0.002936	0.001880	1.56219	0.11824
REP/ENV	.	.	0.000643	0.001527	0.42118	0.67363

ERROR   .   .   0.008739   0.002464   3.54657   0.00039

Chapter 13

Code for Simulating Degrees of Freedom for the Items in a Principal Components Analysis of Variance

Walter T. Federer
Russell D. Wolfinger

Purpose

To provide simulations required to approximate degrees of freedom for such items as principal components, autoregressions, smoothing, kriging, and the like.

Data

The experiment design is a balanced lattice square in which $v = 16$ insecticide treatments and $r = 5$ replicates (complete blocks). The measurement y is the mean of three counts of plants infected with boll weevil. The variable grad in the input statement of the following program listing is the linear polynomial regression coefficients of count on column order in each row of a replicate.

For each simulation using random unit normal deviates, the randomization plan of the experiment for which degrees of freedom are being estimated, is utilized. Then, the sum of squares for a line in the analysis of variance is an estimate of the degrees of freedom since the expected value of each mean square is one.

Originator

Cochran, W.G. and Cox, G.M. (1957). *Experimental Designs,* John Wiley & Sons, New York.

Contact

Dr. Walter T. Federer, Department of Biometrics, Cornell University, Ithaca, New York.
E-mail: <wtf1@Cornell.edu>.

```
/* Here the data are included, but an infile statement may be used to
    input the plan of the experiment to be simulated. */
data original;
input y rep row col grad treat;
   label row='incomplete block';
   datalines;
 9.0  1 1 1 -3 10
20.3  1 1 2 -1 12
17.7  1 1 3  1  9
26.3  1 1 4  3 11
 4.7  1 2 1 -3  2
 9.0  1 2 2 -1  4
 7.3  1 2 3  1  1
 8.3  1 2 4  3  3
 9.0  1 3 1 -3 14
 6.7  1 3 2 -1 16
11.7  1 3 3  1 13
 4.3  1 3 4  3 15
 4.0  1 4 1 -3  6
 5.0  1 4 2 -1  8
 5.7  1 4 3  1  5
14.3  1 4 4  3  7
19.0  2 1 1 -3  5
 8.7  2 1 2 -1 12
13.0  2 1 3  1 15
15.7  2 1 4  3  2
12.0  2 2 1 -3 10
 6.0  2 2 2 -1  7
15.3  2 2 3  1  4
12.0  2 2 4  3 13
12.7  2 3 1 -3 16
 6.3  2 3 2 -1  1
 1.7  2 3 3  1  6
13.0  2 3 4  3 11
 3.7  2 4 1 -3  3
 3.7  2 4 2 -1 14
 8.0  2 4 3  1  9
13.3  2 4 4  3  8
17.0  3 1 1 -3 10
 7.0  3 1 2 -1 15
10.3  3 1 3  1  8
 1.3  3 1 4  3  1
11.3  3 2 1 -3  9
12.3  3 2 2 -1 16
 3.0  3 2 3  1  7
 5.3  3 2 4  3  2
12.3  3 3 1 -3 12
 8.7  3 3 2 -1 13
 8.0  3 3 3  1  6
 9.3  3 3 4  3  3
```

```
30.3   3  4  1 -3 11
22.3   3  4  2 -1 14
11.0   3  4  3  1  5
12.7   3  4  4  3  4
 5.0   4  1  1 -3 16
10.3   4  1  2 -1 12
 5.7   4  1  3  1  8
12.7   4  1  4  3  4
 2.7   4  2  1 -3 11
 6.7   4  2  2 -1 15
10.3   4  2  3  1  3
 5.7   4  2  4  3  7
 1.0   4  3  1 -3  1
10.3   4  3  2 -1  5
11.3   4  3  3  1  9
11.7   4  3  4  3 13
11.0   4  4  1 -3  6
19.0   4  4  2 -1  2
20.7   4  4  3  1 14
29.7   4  4  4  3 10
 2.0   5  1  1 -3  3
 5.0   5  1  2 -1 16
 4.0   5  1  3  1  5
13.7   5  1  4  3 10
 9.3   5  2  1 -3  6
 1.7   5  2  2 -1  9
 6.3   5  2  3  1  4
12.3   5  2  4  3 15
16.7   5  3  1 -3 12
 4.3   5  3  2 -1  7
18.7   5  3  3  1 14
 8.7   5  3  4  3  1
16.7   5  4  1 -3 13
30.0   5  4  2 -1  2
25.7   5  4  3  1 11
14.0   5  4  4  3  8
run;
/* data sets for the pc analysis */
proc sort data=original;
   by rep row col;
run;
%let nsim=2;    /* nsim=2 is for 2 simulations.  Usually nsim will be
     large. */
%let seed=2834701;   /* Any random seed may be specified. */
data sim;
   set original;
   do k=1 to &nsim;
     y = rannor(&seed); /* This statement says that unit normal ran-
     dom deviates are to be used in the simulation. */
      output;
   end;
run;
/* principal component analysis, by k, i. e. for each simulated analy-
     sis, and rep */
proc sort data=sim;
   by k rep col row;
proc transpose data=sim prefix=row out=simr(drop=_name_);
```

```
    by k rep col;
    var y;
proc princomp data=simr prefix=rpc n=2 out=rowvar noprint;
    by k rep;
    var row1-row4; /* Four rows in the design. */
proc sort data=sim;
    by k rep row col;
proc transpose data=sim prefix=col out=simc(drop=_name_);
    by k rep row;
    var y;
proc princomp data=simc prefix=cpc n=2 out=colvar noprint;
    by k rep;
    var col1-col4; /* Four columns in the design. */
/* expand data sets and merge */
data cc;
    set colvar;
    array colv{4} col1-col4;
    do col = 1 to 4;
        y = colv{col};
        output;
    end;
    drop col1-col4;
data rr;
    set rowvar;
    array rowv{4} row1-row4;
    do row = 1 to 4;
        y = rowv{row};
        output;
    end;
    drop row1-row4;
proc sort data=rr;
    by k rep row col;
data ana;
    merge sim cc rr;
    by k rep row col;
/* analysis of variance using the principal components, non-nested */
proc glm data=ana outstat=o1 noprint;
    by k;
    class rep treat;
    model y=rep treat cpc1 cpc2 rpc1 rpc2 cpc1*rpc1 cpc1*rpc2
        cpc2*rpc1 cpc2*rpc2 ;
proc print data=o1;
run;
/* using the principal components, nested */
proc glm data=ana outstat=o2 noprint;
    by k;
    class rep treat;
    model y=rep treat cpc1(rep) cpc2(rep) rpc1(rep) rpc2(rep)
        cpc1*rpc1(rep) cpc1*rpc2(rep) cpc2*rpc1(rep) cpc2*rpc2(rep) ;
proc print data=o2;
run;
/* using the textbook analysis of the design as in Cochran and Cox
    (1957), page 493, and as given above.  This provides a check on
    the simulations as the sums of squares are the degrees of free-
    dom. */
proc glm data=ana outstat=o3 noprint;
    by k;
```

```
   class rep row col treat;
   model y=rep treat row(rep) col(rep);
   lsmean treat;
proc print data=o3;
run;
```

Output from this program follows. SS1 is type I sum of squares; SS3 is type III sum of squares; and the sum of squares is the degrees of freedom as the expected value of each mean square in the table is one.

```
Unnested PCTA ANOVA - run 1
OBS  K   _NAME_   _SOURCE_    _TYPE_   DF      SS         F        PROB
 1   1     Y      ERROR       ERROR    52   41.0582       .         .
 2   1     Y      REP         SS1       4   10.6315     3.3662    0.01594
 3   1     Y      TREAT       SS1      15   16.8648     1.4239    0.17144
 4   1     Y      CPC1        SS1       1    9.5108    12.0454    0.00105
 5   1     Y      CPC2        SS1       1    0.0890     0.1127    0.73848
 6   1     Y      RPC1        SS1       1    5.9200     7.4976    0.00844
 7   1     Y      RPC2        SS1       1    1.1131     1.4097    0.24049
 8   1     Y      CPC1*RPC1   SS1       1    0.4601     0.5828    0.44869
 9   1     Y      CPC1*RPC2   SS1       1    2.9147     3.6915    0.06018
10   1     Y      CPC2*RPC1   SS1       1    0.0440     0.0558    0.81427
11   1     Y      CPC2*RPC2   SS1       1    0.0642     0.0813    0.77664
12   1     Y      REP         SS3       4   10.6315     3.3662    0.01594
13   1     Y      TREAT       SS3      15    5.6731     0.4790    0.94084
14   1     Y      CPC1        SS3       1    9.0891    11.5113    0.00133
15   1     Y      CPC2        SS3       1    0.3867     0.4898    0.48715
16   1     Y      RPC1        SS3       1    6.1529     7.79260   0.00732
17   1     Y      RPC2        SS3       1    0.9345     1.18358   0.28165
18   1     Y      CPC1*RPC1   SS3       1    0.6489     0.82188   0.36881
19   1     Y      CPC1*RPC2   SS3       1    2.9887     3.78517   0.05712
20   1     Y      CPC2*RPC1   SS3       1    0.0470     0.05948   0.80827
21   1     Y      CPC2*RPC2   SS3       1    0.0642     0.08133   0.77664

Unnested PCTA ANOVA - run 2
22   2     Y      ERROR       ERROR    52   38.1947       .         .
23   2     Y      REP         SS1       4    1.9787     0.67346   0.61338
24   2     Y      TREAT       SS1      15   19.1134     1.73479   0.07252
25   2     Y      CPC1        SS1       1    1.7968     2.44618   0.12388
26   2     Y      CPC2        SS1       1    1.8594     2.53141   0.11766
27   2     Y      RPC1        SS1       1    6.0489     8.23524   0.00593
28   2     Y      RPC2        SS1       1    1.1192     1.52375   0.22260
29   2     Y      CPC1*RPC1   SS1       1    5.9688     8.12616   0.00624
30   2     Y      CPC1*RPC2   SS1       1    0.0026     0.00354   0.95277
31   2     Y      CPC2*RPC1   SS1       1    0.09327    0.12699   0.72302
32   2     Y      CPC2*RPC2   SS1       1    0.74168    1.00975   0.31962
33   2     Y      REP         SS3       4    1.97867    0.67346   0.61338
34   2     Y      TREAT       SS3      15    4.41711    0.40091   0.97267
35   2     Y      CPC1        SS3       1    3.54125    4.82123   0.03260
36   2     Y      CPC2        SS3       1    1.62114    2.20710   0.14341
37   2     Y      RPC1        SS3       1    6.56799    8.94197   0.00425
38   2     Y      RPC2        SS3       1    1.24960    1.70127   0.19787
39   2     Y      CPC1*RPC1   SS3       1    5.87599    7.99984   0.00663
40   2     Y      CPC1*RPC2   SS3       1    0.01225    0.01668   0.89773
```

```
41  2     Y     CPC2*RPC1     SS3     1     0.06838   0.09310   0.76149
42  2     Y     CPC2*RPC2     SS3     1     0.74168   1.00975   0.31962
```

Nested PCTA ANOVA - run 1

OBS	K	_NAME_	_SOURCE_	_TYPE_	DF	SS	F	PROB
1	1	Y	ERROR	ERROR	20	4.7282	.	.
2	1	Y	REP	SS1	4	10.6315	11.2427	0.00006
3	1	Y	TREAT	SS1	15	16.8648	4.7558	0.00076
4	1	Y	CPC1(REP)	SS1	5	15.5141	13.1248	0.00001
5	1	Y	CPC2(REP)	SS1	5	2.7152	2.2970	0.08381
6	1	Y	RPC1(REP)	SS1	5	10.0044	8.4637	0.00020
7	1	Y	RPC2(REP)	SS1	5	4.7658	4.0318	0.01080
8	1	Y	CPC1*RPC1(REP)	SS1	5	7.5124	6.3554	0.00110
9	1	Y	CPC1*RPC2(REP)	SS1	5	4.6495	3.9335	0.01203
10	1	Y	CPC2*RPC1(REP)	SS1	5	8.2553	6.9839	0.00064
11	1	Y	CPC2*RPC2(REP)	SS1	5	3.0294	2.5628	0.06004
12	1	Y	REP	SS3	4	10.6315	11.2427	0.00006
13	1	Y	TREAT	SS3	15	3.1950	0.9010	0.57497
14	1	Y	CPC1(REP)	SS3	5	13.1371	11.1139	0.00003
15	1	Y	CPC2(REP)	SS3	5	2.6241	2.2200	0.09243
16	1	Y	RPC1(REP)	SS3	5	8.5463	7.2301	0.00052
17	1	Y	RPC2(REP)	SS3	5	4.3279	3.6614	0.01630
18	1	Y	CPC1*RPC1(REP)	SS3	5	3.5137	2.9726	0.03638
19	1	Y	CPC1*RPC2(REP)	SS3	5	3.0198	2.5547	0.06065
20	1	Y	CPC2*RPC1(REP)	SS3	5	5.8240	4.9271	0.00423
21	1	Y	CPC2*RPC2(REP)	SS3	5	3.0294	2.5628	0.06004

Nested PCTA ANOVA - run 2

22	2	Y	ERROR	ERROR	20	4.3221	.	.
23	2	Y	REP	SS1	4	1.9787	2.2890	0.09551
24	2	Y	TREAT	SS1	15	19.1134	5.8963	0.00018
25	2	Y	CPC1(REP)	SS1	5	9.1962	8.5108	0.00019
26	2	Y	CPC2(REP)	SS1	5	3.9972	3.6993	0.01562
27	2	Y	RPC1(REP)	SS1	5	8.7444	8.0927	0.00026
28	2	Y	RPC2(REP)	SS1	5	0.8850	0.8190	0.55043
29	2	Y	CPC1*RPC1(REP)	SS1	5	15.2570	14.1200	0.00001
30	2	Y	CPC1*RPC2(REP)	SS1	5	6.2267	5.7626	0.00188
31	2	Y	CPC2*RPC1(REP)	SS1	5	3.81879	3.53420	0.01883
32	2	Y	CPC2*RPC2(REP)	SS1	5	3.37785	3.12611	0.03029
33	2	Y	REP	SS3	4	1.97867	2.28901	0.09551
34	2	Y	TREAT	SS3	15	1.79833	0.55477	0.87591
35	2	Y	CPC1(REP)	SS3	5	6.68211	6.18412	0.00128
36	2	Y	CPC2(REP)	SS3	5	3.52047	3.25810	0.02592
37	2	Y	RPC1(REP)	SS3	5	8.15456	7.54683	0.00040
38	2	Y	RPC2(REP)	SS3	5	1.92512	1.78165	0.16248
39	2	Y	CPC1*RPC1(REP)	SS3	5	8.04745	7.44771	0.00043
40	2	Y	CPC1*RPC2(REP)	SS3	5	7.22510	6.68665	0.00082
41	2	Y	CPC2*RPC1(REP)	SS3	5	4.65856	4.31137	0.00799
42	2	Y	CPC2*RPC2(REP)	SS3	5	3.37785	3.12611	0.03029

Textbook ANOVA - run 1

OBS	K	_NAME_	SOURCE_	_TYPE_	DF	SS	F	PROB
1	1	Y	ERROR	ERROR	30	25.6177	.	.
2	1	Y	REP	SS1	4	10.6315	3.11254	0.02958
3	1	Y	TREAT	SS1	15	16.8648	1.31665	0.25255
4	1	Y	ROW(REP)	SS1	15	19.6950	1.53760	0.15374
5	1	Y	COL(REP)	SS1	15	15.8615	1.23832	0.29886

```
  6    1    Y    REP        SS3      4   10.6315   3.11254   0.02958
  7    1    Y    TREAT      SS3     15    9.8877   0.77194   0.69613
  8    1    Y    ROW(REP)   SS3     15   15.7929   1.23297   0.30227
  9    1    Y    COL(REP)   SS3     15   15.8615   1.23832   0.29886

Textbook ANOVA - run 2
 10    2    Y    ERROR      ERROR   30   29.5471      .         .
 11    2    Y    REP        SS1      4    1.9787   0.50225   0.73428
 12    2    Y    TREAT      SS1     15   19.1134   1.29376   0.26542
 13    2    Y    ROW(REP)   SS1     15   16.4413   1.11289   0.38676
 14    2    Y    COL(REP)   SS1     15    9.8368   0.66584   0.79576
 15    2    Y    REP        SS3      4    1.9787   0.50225   0.73428
 16    2    Y    TREAT      SS3     15    9.3513   0.63298   0.82440
 17    2    Y    ROW(REP)   SS3     15   13.8467   0.93726   0.53691
 18    2    Y    COL(REP)   SS3     15    9.8368   0.66584   0.79576
```

Chapter 14

Principal Components (PC) and Additive Main Effects and Multiplicative Interaction (AMMI) Trend Analyses for Incomplete Block and Lattice Rectangle-Designed Experiments

Walter T. Federer
Russell D. Wolfinger
José Crossa

Importance

A principal component (PC) is a linear combination of data that has a maximum sum of squares. No other linear combination can be associated with a larger sum of squares. Therefore, the analyses outlined in this chapter could prove useful when describing spatial variation found in field experiments. PC and additive main effects and multiplicative interaction (AMMI) analyses have been used in genotype-by-environment studies. The problem is that the degrees of freedom for these linear combinations need to be obtained via simulations. A program for doing this is given in Chapter 13 of this book. The SAS code allocates a single degree of freedom for each PC, but this is not correct. In such a case, if the F-value associated with a PC is less than the F-value at the 25 percent level, the PC sum of squares is pooled with the residual sum of squares. Rather than applying this rule, one may use a SAS/MIXED procedure to eliminate all effects from the model that has variance components estimated as zero; however, the two procedures do not give the same result in general. Since the properties of this procedure have not been established, it is not recommended.

References

Federer, W.T., Crossa, J., and Franco, J. (1998). *Forms of spatial analyses with mixed model effects and exploratory model selection.* BU-1406-M, Technical Report, Department of Biometrics, Cornell University, Ithaca, NY.

Gauch, H.G. (1988). Model selection and validation for yield trials with interaction. *Biometrics* 44:705-715.

Moreno-Gonzales, J. and Crossa, J. (1998). Combining environments, genotypes, and attribute variables in regression models for predicting the cell-means of multi-environment trials. *Theoretical and Applied Genetics* 96:803-811.

Zobel, R.W. (1990). A powerful statistical tool for understanding genotype-by-environment interaction. In Kang, M.S. (Ed.), *Genotype-by-Environment Interaction and Plant Breeding* (pp. 126-140). Louisiana State University, Baton Rouge, Louisiana.

Contact

Dr. Walter T. Federer, Department of Biometrics, Cornell University, Ithaca, New York. E-mail: <wtfl@Cornell.edu>.

Data

The example used to illustrate the SAS code is a balanced lattice square-designed experiment. By altering the program appropriately, the code can also be used for incomplete block and row-column-designed experiments. Also, PCs can be computed using the correlation matrix or the variance-covariance matrix. The SAS default uses the correlation matrix.

Originator

Cochran, W.G. and Cox, G.M. (1957). *Experimental Designs,* John Wiley & Sons, New York.

```
/* Here the data are included, but an infile statement may be used to
    input data. */
data original;
   input y rep row col grad treat; /* grad is the linear regression
   coefficient on column order. */
   label row='incomplete block';
   datalines;
 9.0  1 1 1 -3 10
20.3  1 1 2 -1 12
17.7  1 1 3  1  9
26.3  1 1 4  3 11
 4.7  1 2 1 -3  2
 9.0  1 2 2 -1  4
```

```
 7.3   1 2 3   1   1
 8.3   1 2 4   3   3
 9.0   1 3 1  -3  14
 6.7   1 3 2  -1  16
11.7   1 3 3   1  13
 4.3   1 3 4   3  15
 4.0   1 4 1  -3   6
 5.0   1 4 2  -1   8
 5.7   1 4 3   1   5
14.3   1 4 4   3   7
19.0   2 1 1  -3   5
 8.7   2 1 2  -1  12
13.0   2 1 3   1  15
15.7   2 1 4   3   2
12.0   2 2 1  -3  10
 6.0   2 2 2  -1   7
15.3   2 2 3   1   4
12.0   2 2 4   3  13
12.7   2 3 1  -3  16
 6.3   2 3 2  -1   1
 1.7   2 3 3   1   6
13.0   2 3 4   3  11
 3.7   2 4 1  -3   3
 3.7   2 4 2  -1  14
 8.0   2 4 3   1   9
13.3   2 4 4   3   8
17.0   3 1 1  -3  10
 7.0   3 1 2  -1  15
10.3   3 1 3   1   8
 1.3   3 1 4   3   1
11.3   3 2 1  -3   9
12.3   3 2 2  -1  16
 3.0   3 2 3   1   7
 5.3   3 2 4   3   2
12.3   3 3 1  -3  12
 8.7   3 3 2  -1  13
 8.0   3 3 3   1   6
 9.3   3 3 4   3   3
30.3   3 4 1  -3  11
22.3   3 4 2  -1  14
11.0   3 4 3   1   5
12.7   3 4 4   3   4
 5.0   4 1 1  -3  16
10.3   4 1 2  -1  12
 5.7   4 1 3   1   8
12.7   4 1 4   3   4
 2.7   4 2 1  -3  11
 6.7   4 2 2  -1  15
10.3   4 2 3   1   3
 5.7   4 2 4   3   7
 1.0   4 3 1  -3   1
10.3   4 3 2  -1   5
11.3   4 3 3   1   9
11.7   4 3 4   3  13
11.0   4 4 1  -3   6
19.0   4 4 2  -1   2
20.7   4 4 3   1  14
```

```
29.7   4 4 4   3 10
 2.0   5 1 1  -3  3
 5.0   5 1 2  -1 16
 4.0   5 1 3   1  5
13.7   5 1 4   3 10
 9.3   5 2 1  -3  6
 1.7   5 2 2  -1  9
 6.3   5 2 3   1  4
12.3   5 2 4   3 15
16.7   5 3 1  -3 12
 4.3   5 3 2  -1  7
18.7   5 3 3   1 14
 8.7   5 3 4   3  1
16.7   5 4 1  -3 13
30.0   5 4 2  -1  2
25.7   5 4 3   1 11
14.0   5 4 4   3  8
run;

/* principal component analysis.  */
proc sort data=original;
   by  rep col row;
proc transpose data=original prefix=row out=origr(drop=_name_);
   by  rep col;
   var y;
/* The SAS default option is the correlation matrix.  If it is desired
      to use the variance-covariance matrix, simply add COV at the end
      of the following statement and also in the next PROC PRINCOMP
      statement.*/
proc princomp data=origr prefix=rpc out=rowvar noprint;
   by  rep;
   var row1-row4; /* Four rows in the design. */
proc sort data=original;
   by  rep row col;
proc transpose data=original prefix=col out=origc(drop=_name_);
   by  rep row;
   var y;
proc princomp data=origc prefix=cpc  out=colvar noprint;
   by  rep;
   var col1-col4; /* Four columns in the design. */

/* expand data sets and merge */
data cc;
   set colvar;
   array colv{4} col1-col4;
   do col = 1 to 4;
      y = colv{col};
      output;
   end;
   drop col1-col4;
data rr;
   set rowvar;
   array rowv{4} row1-row4;
   do row = 1 to 4;
      y = rowv{row};
      output;
   end;
```

```
    drop row1-row4;
proc sort data=rr;
   by  rep row col;
data ana;
   merge original cc rr;
   by  rep row col;

/* analysis of variance, fixed principal component effects, non-nested
   */
proc glm data=ana;
   class rep treat;
   model y=rep treat cpc1 cpc2 rpc1 rpc2 cpc1*rpc1 cpc1*rpc2
      cpc2*rpc1 cpc2*rpc2 ;
run;

/* fixed  principal component effects, nested */
proc glm data=ana;
   class rep treat;
   model y=rep treat cpc1(rep) cpc2(rep) rpc1(rep) cpc1*rpc1(rep)
      cpc1*rpc2(rep)
   cpc2*rpc1(rep) cpc2*rpc2(rep) ;
run;

/*  random principal component effects, nested */
proc mixed data = ana;
   class rep treat row col;
   model y = treat;
   random rep cpc1(rep) cpc2(rep) rpc1(rep) cpc1*rpc1(rep)
      cpc1*rpc2(rep)
   cpc2*rpc1(rep) cpc2*rpc2(rep);
   lsmeans treat;
run;

/* fixed effects textbook analysis of the design as in Cochran and Cox
     (1957), page 493. */
proc glm data=ana;
   class rep row col treat;
   model y=rep treat row(rep) col(rep);
run;

/* fixed AMMI trend analysis, PC within row within replicate */
proc glm data = ana;
   class rep treat row col;
   model y = rep treat row(rep) rpc1*row(rep);
run;
/* random AMMI effect within row and random row and rep effects */
proc mixed data = ana;
   class rep treat row col;
   model y = treat;
   random rep row(rep) rpc1*row(rep);
   lsmeans treat;
run;
```

An abbreviated form of the output from this program is given here.

```
/*fixed effect un-nested PC analysis */
```

Dependent Variable: Y

Source	DF	Sum of Squares	Mean Square	F Value	Pr > F
Model	27	2723.926541	100.886168	5.93	0.0001
Error	52	884.611459	17.011759		
Corrected Total	79	3608.538000			

R-Square	C.V.	Root MSE	Y Mean
0.754856	37.82239	4.124531	10.90500

Dependent Variable: Y

Source	DF	Type I SS	Mean Square	F Value	Pr > F
REP	4	31.563000	7.890750	0.46	0.7619
TREAT	15	1244.202000	82.946800	4.88	0.0001
CPC1	1	937.165951	937.165951	55.09	0.0001
CPC2	1	28.780965	28.780965	1.69	0.1991
RPC1	1	465.277940	465.277940	27.35	0.0001
RPC2	1	7.694324	7.694324	0.45	0.5042
CPC1*RPC1	1	1.186538	1.186538	0.07	0.7927
CPC1*RPC2	1	7.642424	7.642424	0.45	0.5057
CPC2*RPC1	1	0.325597	0.325597	0.02	0.8905
CPC2*RPC2	1	0.087801	0.087801	0.01	0.9430

Source	DF	Type III SS	Mean Square	F Value	Pr > F
REP	4	31.5630000	7.8907500	0.46	0.7619
TREAT	15	371.7330707	24.7822047	1.46	0.1571
CPC1	1	913.7729632	913.7729632	53.71	0.0001
CPC2	1	75.9641822	75.9641822	4.47	0.0394
RPC1	1	466.8813930	466.8813930	27.44	0.0001
RPC2	1	8.7776829	8.7776829	0.52	0.4758
CPC1*RPC1	1	1.8655460	1.8655460	0.11	0.7419
CPC1*RPC2	1	7.7122489	7.7122489	0.45	0.5037
CPC2*RPC1	1	0.3476938	0.3476938	0.02	0.8869
CPC2*RPC2	1	0.0878007	0.0878007	0.01	0.9430

/* Random PC effects, nested analysis*/
Dependent Variable: Y

Source	DF	Sum of Squares	Mean Square	F Value	Pr > F
Model	54	3503.080701	64.871865	15.38	0.0001
Error	25	105.457299	4.218292		
Corrected Total	79	3608.538000			

R-Square	C.V.	Root MSE	Y Mean
0.970776	18.83400	2.053848	10.90500

General Linear Models Procedure

Dependent Variable: Y

Source	DF	Type I SS	Mean Square	F Value	Pr > F
REP	4	31.563000	7.890750	1.87	0.1470
TREAT	15	1244.202000	82.946800	19.66	0.0001
CPC1(REP)	5	1034.531257	206.906251	49.05	0.0001
CPC2(REP)	5	46.108785	9.221757	2.19	0.0879
RPC1(REP)	5	480.204652	96.040930	22.77	0.0001

```
CPC1*RPC1(REP)       5      262.322233      52.464447      12.44      0.0001
CPC1*RPC2(REP)       5       71.654226      14.330845       3.40      0.0177
CPC2*RPC1(REP)       5      117.938209      23.587642       5.59      0.0014
CPC2*RPC2(REP)       5      214.556340      42.911268      10.17      0.0001
```

```
Source              DF      Type III SS     Mean Square  F Value      Pr > F
REP                  4       31.563000       7.890750       1.87      0.1470
TREAT               15       84.006127       5.600408       1.33      0.2576
CPC1(REP)            5     1145.198035     229.039607      54.30      0.0001
CPC2(REP)            5       56.156010      11.231202       2.66      0.0462
RPC1(REP)            5      401.390987      80.278197      19.03      0.0001
CPC1*RPC1(REP)       5      197.143931      39.428786       9.35      0.0001
CPC1*RPC2(REP)       5       67.960365      13.592073       3.22      0.0222
CPC2*RPC1(REP)       5      138.518519      27.703704       6.57      0.0005
CPC2*RPC2(REP)       5      214.556340      42.911268      10.17      0.0001
```

The MIXED Procedure

Least Squares Means

```
Effect   TREAT        LSMEAN     Std Error   DF       t   Pr > |t|
TREAT     1        7.54548158   1.72555784   29    4.37    0.0001
TREAT     2       10.37683195   1.75725118   29    5.91    0.0001
TREAT     3        9.37048456   1.56493910   29    5.99    0.0001
TREAT     4       12.34206915   1.57505850   29    7.84    0.0001
TREAT     5       11.45961200   1.61696971   29    7.09    0.0001
TREAT     6        9.11047013   1.69326123   29    5.38    0.0001
TREAT     7        7.86631476   1.75974765   29    4.47    0.0001
TREAT     8       11.11913803   1.67931870   29    6.62    0.0001
TREAT     9       12.18761611   1.64388978   29    7.41    0.0001
TREAT    10       14.08160555   1.92299755   29    7.32    0.0001
TREAT    11       13.13295404   1.92869896   29    6.81    0.0001
TREAT    12       11.40375822   1.67381535   29    6.81    0.0001
TREAT    13       10.59779026   1.54426804   29    6.86    0.0001
TREAT    14       12.13599166   1.70542148   29    7.12    0.0001
TREAT    15        9.18610407   1.62649961   29    5.65    0.0001
TREAT    16       12.56377795   1.63547347   29    7.68    0.0001
```

/* textbook analysis, fixed effects */
Dependent Variable: Y

```
                         Sum of         Mean
Source              DF   Squares        Square       F Value     Pr > F
Model               49   2928.370083    59.762655      2.64      0.0029
Error               30    680.167917    22.672264
Corrected Total     79   3608.538000
```

```
           R-Square        C.V.        Root MSE          Y Mean
           0.811511      43.66382      4.761540         10.90500
```

Dependent Variable: Y
```
Source              DF    Type I SS     Mean Square   F Value     Pr > F
REP                  4     31.563000      7.890750      0.35      0.8433
TREAT               15   1244.202000     82.946800      3.66      0.0012
ROW(REP)            15   1093.015500     72.867700      3.21      0.0032
COL(REP)            15    559.589583     37.305972      1.65      0.1197
```

```
Source              DF    Type III SS    Mean Square   F Value     Pr > F
```

```
REP                4        31.563000        7.890750      0.35      0.8433
TREAT             15       319.452083       21.296806      0.94      0.5350
ROW(REP)          15      1026.755833       68.450389      3.02      0.0049
COL(REP)          15       559.589583       37.305972      1.65      0.1197
```

... ...

```
/* treatment means from random effects AMMI analysis */
    Effect   TREAT          LSMEAN       Std Error    DF       t    Pr > |t|
    TREAT      1         6.37311825     2.38292075    25    2.67      0.0130
    TREAT      2        10.61497350     2.15761108    25    4.92      0.0001
    TREAT      3         7.93436160     2.06059560    25    3.85      0.0007
    TREAT      4        12.28827249     1.99475730    25    6.16      0.0001
    TREAT      5        10.76991952     1.97476997    25    5.45      0.0001
    TREAT      6         8.24325882     2.17029859    25    3.80      0.0008
    TREAT      7         5.71709277     2.38681041    25    2.40      0.0244
    TREAT      8        10.24396596     2.09650099    25    4.89      0.0001
    TREAT      9        12.25845651     2.11969951    25    5.78      0.0001
    TREAT     10        14.74535577     2.35708086    25    6.26      0.0001
    TREAT     11        15.30490826     2.42463576    25    6.31      0.0001
    TREAT     12        13.57092699     2.14533617    25    6.33      0.0001
    TREAT     13        11.45714987     2.01082282    25    5.70      0.0001
    TREAT     14        13.94971794     2.10372824    25    6.63      0.0001
    TREAT     15         8.62267164     2.09619203    25    4.11      0.0004
    TREAT     16        12.38585010     2.14469670    25    5.78      0.0001
```

Chapter 15

A Method for Classifying Observations Using Categorical and Continuous Variables

Jorge Franco
José Crossa

Purpose

Classifying observations into homogeneous subpopulations or groups using categorical and continuous variables is important in various fields of research, such as genetic resource conservation, genetics, plant breeding, biotechnology, agronomy, and ecology.

The program outlined in this chapter uses a statistical method for classifying observations into homogeneous groups.

Definitions

The objective is to classify n observations using the statistical technique known as the mixture of a finite number of distributions. The model assumes a statistical distribution for variables. The probability of membership of each observation in each subpopulation or group is computed. The program allows use of continuous variables (Gaussian Model, GM) (McLachlan and Basford, 1988) or of continuous and categorical variables (Modified Location Model, MLM) (Franco et al., 1998). Homogeneity of variance-covariance matrices within subpopulations is also assumed.

The GM model assumes that each vector y_j ($j = 1,...,n$), formed with p continuous variables is distributed as a mixture of g multivariate, a multivariate normal with p variables, each corresponding to a subpopulation. Thus, assuming homogeneity of variance-covariance matrices within subpopulations, its probability density function (PDF) is

$$f(y_j;\Theta)=\sum_{i=1}^{g}\alpha_i(2\pi)^{-p/2}|\Sigma|^{-1/2}\exp\left[-(1/2)(y_j-\mu_i)'\Sigma^{-1}(y_j-\mu_i)\right]$$

where the vector Θ contains the parameters of the model; α_i ($i = 1,2,...,g$) is the proportion of observations in each subpopulation (cluster) of the mixture; Σ is the common variance-covariance matrix within a subpopulation; and μ_i represents vectors of means of the ith subpopulation.

The MLM model transforms the vector formed with p continuous and q categorical variables into a $p + 1$ vector in which all the categorical values are transformed into a unique multinomial variable W that takes values $s = 1,2,...,m$, where m is the number of combinations observed or multinomial cells. The vector of $p + 1$ variables (x_{sj}) is assumed to be distributed as a mixture of the product of the multinomial and multinormal variables. The model assumes that the dispersion matrices and mean vectors are equal for all of the multinomial cells within each subpopulation; thus, its probability density function is

$$f(x_{sj};\Theta)=\sum_{i=1}^{g}\alpha_i p_{is}(2\pi)^{-p/2}|\Sigma|^{-1/2}\exp\left[-(1/2)(y_{sj}-\mu_i)'\Sigma^{-1}(y_{sj}-\mu_i)\right]$$

where the vector Θ contains the parameters of the model; α_i ($i = 1,2,...,g$) is the proportion of observations in each subpopulation (cluster) of the mixture; p_{is} is the proportion of observations in the sth multinomial cell of the ith subpopulation; Σ is the common variance-covariance matrix within a subpopulation; and μ_i are the vectors of means of the ith subpopulation.

The program uses the expectation maximization (EM) algorithm (Dempster et al., 1977) to estimate parameters (maximization) and to calculate the probability of membership for each observation (expectation).

The likelihood function corresponding to the matrix of the whole sample data, X, is the objective function for the maximization. For the MLM model, this function is

$$L(\Theta;X)=\prod_{s=1}^{m}\prod_{j=1}^{n_s}f(x_{sj};\Theta)$$

$$=\prod_{s=1}^{m}\prod_{j=1}^{n_s}\sum_{i=1}^{g}\alpha_i p_{is}(2\pi)^{-p/2}|\Sigma|^{-1/2}\exp\left[-(1/2)(y_{sj}-\mu_i)'\Sigma^{-1}(y_{sj}-\mu_i)\right].$$

Originator

Franco, J., Crossa, J., Villaseñor, J., Taba, S., and Eberhart, S.A. (1998). Classifying genetic resources by categorical and continuous variables. *Crop Science* 38(6):1688-1696.

Software Available

Franco, J. and Crossa, J. (2001). SAS program for classifying observations using categorical and continuous attributes. Centro Internacional de Mejoramiento de Maiz y Trigo (CIMMYT, INT), access online: <http://www.cimmyt.org/biometrics>.

Key References

Dempster, A.P., Laird, N.M., and Rubin, D.B. (1977). Maximum likelihood from incomplete data via the EM algorithm. *Journal of Royal Society,* Series B, 39:1-38.

Franco, J., Crossa, J., Ribaut, J.M., Betran, J., Warbuton, M.L., and Khairallah, M. (2001). A method for combining molecular markers and phenotypic attributes for classifying plant genotypes. *Theoretical and Applied Genetics* 103(6/7):944-952.

Franco, J., Crossa, J., Villaseñor, J., Taba, S., and Eberhart, S.A. (1998). Classifying genetic resources by categorical and continuous variables. *Crop Science* 38(6):1688-1696.

McLachlan, G.J. and Basford, K.E. (1988). *Mixture models: Inference and applications to clustering.* Marcel Dekker, New York.

Contacts

José Crossa, Biometrics and Statistics Unit, Centro Internacional de Mejoramiento de Maiz y Trigo (CIMMYT), Apdo. Postal 6-641, 0600, Mexico DF, Mexico. E-mail: <j.crossa@cgiar.org>.

Jorge Franco, Departamento de Biometría, Estadística y Computación, Facultad de Agronomía, Universidad de la República, Ave. Garzón 780, 12900, Montevideo, Uruguay. E-mail: <jfranco@fagro.edu.uy>.

Modifications to the Program

Lines that can be modified begin with /*, end with */, and are in bold. The program starts with a data file called DATA0 (this name can be changed) with the following characteristics:

1. Observations are rows; variables are columns.

2. The variable showing initial subpopulations must be called CLASS0 (CLASS zero); the discrete variables must be called Q1, Q2,..., Qq. If the GM model (only continuous variables) is required, use a unique discrete variable (Q1) with all values equal to 1.
3. Names for continuous variables can be any valid SAS name no longer than eight characters and should not begin with a number.
4. Any other variable (not included in the analysis) must be dropped from line L2.

The next lines that can be modified are:

1. L3: The TABLES statement must be followed by a list of discrete variables Q1*Q2*...*Qq joined by an asterisk (*).
2. L4, L5, L6: The statements must be Q1-Qq, where q is the subindex for the last discrete variable.
3. L8, L9: These lines can be modified to allow more than fifty iterations or a lesser value of convergence.
4. L10, L11, L12: In these lines you must write the names of continuous variables.

The program produces a file (SAS file) called FINAL that contains the following information:

1. All the initial information plus the number of group to which each observation was assigned (named FINGROUP)
2. Membership probabilities for each observation in each group (named GROUP1,...,GROUPg)
3. Starting group (called INIGROUP)
4. Results from the canonical analysis

The program performs a canonical analysis on the continuous variables to observe the separation of the groups on the first two canonical variables and to allow characterization of the groups relative to the continuous variables.

EXAMPLE

The example comes from Franco et al. (2001). Fifteen maize genotypes are classified using five continuous variables (days to anthesis and silking, plant and ear height, and grain weight) and fifteen discrete variables (restriction fragment length polymorphism [RFLP] markers). There are five initial subpopulations (or groups). Note that categorical variables can be binary, ordinal, or multistate. The data T0 is created from the original and five initial groups are defined to form data set DATA0. Lines for forming the data set DATA0 are shown in bold.

SAS Program

```
OPTIONS LS=132 PS=9000 NODATE NOCENTER;
TITLE Modified Location Model, Franco et al. 1998;
TITLE2 Example, Table 6, Franco et al. 2001;
DATA T0;INPUT NOBS ad sd ph eh grw Q1-Q15;
LABEL NOBS='ENTRY';
CARDS;
    1    87.61 91.37  97.00 36.36 286.07 0 1 1 0 0 1 0 1 0 0 1 0 1 1 1
    2    89.96 96.20 110.00 38.42 292.08 0 0 1 0 0 0 0 1 0 1 0 1 0 0 1
    3    87.49 94.91 119.33 56.36 169.67 1 1 1 0 0 1 0 1 0 0 1 0 1 1 0
    4    92.97 96.77 100.33 35.77 290.78 0 0 0 1 0 0 1 0 1 0 0 1 1 0 0
    5    90.73 91.36 120.33 50.48 702.64 1 0 1 1 1 0 1 0 1 0 0 1 0 0 1
    6    90.85 94.33 117.33 50.16 427.10 1 0 1 1 1 1 1 0 0 1 0 1 0 0 0
    7    83.85 88.02 101.67 41.91 263.24 0 0 1 0 1 0 1 0 1 0 1 0 1 1 1
    8    84.93 87.09 113.67 41.53 498.19 0 0 0 1 0 0 0 0 1 0 1 0 0 1 0
    9    91.30 92.32 119.33 37.26 522.76 1 0 1 1 1 1 1 0 0 1 0 1 0 0 0
   10    87.95 88.04 115.00 53.76 505.95 0 0 0 0 1 0 0 0 1 0 1 0 1 0 1
   11    91.51 92.37 106.67 36.64 451.09 0 0 0 1 0 1 0 1 0 1 0 1 1 1 1
   12    83.33 84.72 103.67 31.66 316.84 1 1 0 1 0 0 0 0 1 0 0 1 0 1
   13    89.43 91.59 113.67 42.55 572.25 0 1 0 1 1 0 0 1 0 1 0 1 1 0 1
   14    86.19 87.82  97.33 40.15 581.17 0 1 0 0 0 0 0 1 0 1 0 1 1 0 1
   15    86.88 87.75 126.67 47.99 518.48 1 0 1 1 1 1 0 1 0 0 1 0 0 0
DATA T6;INPUT G6 @@;CARDS;
 1  2  1  3  4  4  5  3  4  5  5  6  2  2  4
;
DATA T5;INPUT G5 @@;CARDS;
 1  2  1  3  4  4  5  3  4  5  5  2  2  2  4
;
DATA T4;INPUT G4 @@;CARDS;
 1  2  1  3  4  4  3  3  4  3  3  2  2  2  4
;
DATA T3;INPUT G3 @@;CARDS;
 1  1  1  2  3  3  2  2  3  2  2  1  1  1  3
;
DATA T2;INPUT G2 @@;CARDS;
 1  1  1  1  2  2  1  1  2  1  1  1  1  1  2
;
/*L1*/ DATA DATA0; MERGE T0 T2 T3 T4 T5 T6;
/*L2*/ DATA DATA1;SET DATA0; CLAS0 = G5; DROP G2-G6;
```

```
/*L3*/ PROC FREQ; TABLES
    Q1*Q2*Q3*Q4*Q5*Q6*Q7*Q8*Q9*Q10*Q11*Q12*Q13*Q14*Q15 / LIST;
/*L4*/ PROC SORT DATA=DATA1;BY Q1-Q15 NOBS;
/*L5*/ DATA T1;SET DATA1;DROP Q1-Q15;
/*L6*/ DATA T2;SET DATA1;KEEP Q1-Q15;

PROC IML;                                     /* GENERATING VARIABLE W */
REMOVE _ALL_;
USE T1;
NAM1=CONTENTS(T1);
READ ALL INTO A;
READ ALL VAR{CLAS0} INTO CLAS0;
G=CLAS0[];
VARCUA=CONTENTS(T2);
USE DATA1;
READ ALL VAR VARCUA INTO Q;
N=NROW(Q);
NQ=NCOL(Q);
P=NCOL(A)-2;
W=J(N,1,1);
DO I=2 TO N;
  IF Q[I,] = Q[I-1,] THEN W[I]=W[I-1];
  ELSE W[I] = W[I-1] + 1;
END;
M=W[];
D1=A||W;
NAM2=NAM1`||{W};
CREATE D1 FROM D1 [COLNAME=NAM2];
APPEND FROM D1;
STORE N NQ P G M;
QUIT;
PROC SORT DATA=D1; BY W CLAS0 NOBS;
PROC IML;                                     /* I-MATRIX CREATION */
LOAD N P G M;
NAM1=CONTENTS (D1);
USE D1;
READ ALL INTO D1;
READ ALL VAR {W CLAS0} INTO M0;
MI=J(M,G,0); M1=J(1,2,0);
DO S=1 TO M;
  DO I=1 TO G;
    M1[1,1]=S;M1[1,2]=I;
    DO J=1 TO N;
      IF M0[J,] = M1 THEN MI[S,I] = 1;
    END;
  END;
END;
STORE MI;
QUIT;

DATA T1;SET D1;DROP NOBS W CLAS0;
DATA T2;SET D1;KEEP NOBS W CLAS0;

PROC IML;
LOAD N P G M MI;

START CERO;                                   /* NAMES AND DIMENSIONS */
```

```
VARCLAS=CONTENTS(T2);
VARCONT=CONTENTS(T1);
T={GROUP};
IG=INT(G/10);
VARTAO=CHAR(J(G,1,0));
XX=1;
DO X=0 TO IG;
  DO Y=0 TO 9;
    IF (0 < (10*X+Y) & (10*X+Y) <= G) THEN DO;
        VARTAO[XX]=CONCAT(T,CHAR(X,1),CHAR(Y,1));
        XX=XX+1;
      END;
  END;
END;
ENE=J(M,G,0);
USE D1;
CONT=0;
DO S=1 TO M;
  DO I=1 TO G;
    IF MI[S,I] = 1 THEN DO;
      READ ALL WHERE(W=S & CLAS0=I) VAR VARCONT INTO B;
      ENE[S,I]=NROW(B);
    END;
  END;
END;
ENEI=ENE[+,]`; ENES=ENE[,+]; N=SUM(ENE);
PRINT VARCLAS VARCONT M G P VARTAO;
PRINT ENE N ENEI ENES [FORMAT=5.0];
FINISH CERO;

START UNO;                                    /* INITIAL ESTIMATIONS */
                            /* MEANS AND SUM OF SQUARES BY GROUP */
USE D1;
MEDI=J(G,P,0); SC=J(P,P,0);
DO I=1 TO G;
  READ ALL WHERE(CLAS0=I) VAR VARCONT INTO B;
  MEDI[I,]=J(1,NROW(B),1/NROW(B))*B;
  SC=SC+B`*(I(NROW(B))-J(NROW(B),NROW(B),1/NROW(B)))*B;
END;
V=SC/N;
SINV=INV(V);
DETS=DET(V);
/* ALPHA AND P ESTIMATION */
ALFA=ENEI/N;
PE=J(M,G,0);
DO S=1 TO M;
  DO I=1 TO G;
    IF MI[S,I] = 1 THEN PE[S,I] = ENE[S,I] / ENEI[I];
    ELSE PE[S,I] = 1E-04;
  END;
END;
DO I=1 TO G;
  NCERO=0;
  DO S=1 TO M;
    IF PE[S,I] <= 1E-04 THEN NCERO=NCERO+1;
  END;
  DO S=1 TO M;
```

```
      IF PE[S,I] > 1E-04 THEN PE[S,I] =
         PE[S,I] - NCERO*1E-04 / (M - NCERO);
   END;
END;
PRINT V [FORMAT=9.4];
PRINT DETS;
PRINT MEDI [FORMAT=9.4];
PRINT ALFA [FORMAT=7.5];
PRINT PE [FORMAT=7.5];
PRINT "LOG LIKELIHOOD :";
FINISH UNO;

START DOS;         /* LIKELIHOOD AND POSTERIOR PROBABILITY ESTIMATION*/
CONT3=0; TAO=J(N,G,0); L=J(N,G+1,0);ID=J(N,3,0);B1=J(N,P,0);
DO S=1 TO M;
  DO I=1 TO G;
    IF MI[S,I] = 1 THEN DO;
       READ ALL WHERE(W=S & CLAS0=I) VAR{NOBS CLAS0 W} INTO ID0;
       READ ALL WHERE(W=S & CLAS0=I) VAR VARCONT INTO B;
       L0=J(ENE[S,I],G,0);
       DO K=1 TO G;
         IF PE[S,K] <= 1E-04 THEN PE[S,K] = 1E-04;
         L0[,K]=LOG(ALFA[K])+LOG(PE[S,K])-0.5*
                VECDIAG((B-REPEAT(MEDI[K,],ENE[S,I],1))*SINV*
                (B-REPEAT(MEDI[K,],ENE[S,I],1))`);
       END;
       L1=-(P/2)*1.83788-0.5*LOG(DETS)+L0;
       L2=EXP(L1);
       L3=LOG(L2[,+]);
       L[(CONT3+1:CONT3+ENE[S,I]),(1:G+1)] = L1||L3;
       TAO1=EXP(L1-(REPEAT(L3,1,G)));
       ID[(CONT3+1:CONT3+ENE[S,I]),(1:3)] = ID0;
       TAO[(CONT3+1:CONT3+ENE[S,I]),(1:G)] = TAO1;
       B1[(CONT3+1:CONT3+ENE[S,I]),(1:P)] = B;
       CONT3=CONT3+ENE[S,I];
     END;
  END;
END;
GROUP=TAO[,<:>]; MAXP=TAO[,];
C=ID||B1||TAO||GROUP||MAXP;
C1=J(N,P+G+5,0);
DO I=1 TO N;
  T1=TAO[I,];
  IF (T1[,] < 0.75) THEN C1[I,]=C[I,];
END;
LOGLTOT=SUM(L[,G+1]);
RESET NONAME;
PRINT LOGLTOT [FORMAT=20.5];
RESET NAME;
FINISH DOS;

START TRES;        /* MAXIMUM LIKELIHOOD ESTIMATORS (MAXIMIZATION) */
DO I=1 TO G;
  ALFA[I]=SUM(TAO[,I])/N;
END;
MED=J(M*G,P,0); TOT=J(G,P,0); MEDI=J(G,P,0); DIV=J(G,1,0);
CONT=0; CONT3=0;
```

```
DO S=1 TO M;
  DO I=1 TO G;
    TT=TAO[(CONT3+1:CONT3+ENES[S]),I];
    BB= B1[(CONT3+1:CONT3+ENES[S]),];
    PE[S,I]=SUM(TT)/(N*ALFA[I]);
    IF PE[S,I] > 0 THEN DO;
      CONT=CONT+1;
      MED[CONT,]=TT`*BB/(N*PE[S,I]*ALFA[I]);
    END;
  END;
  CONT3=CONT3+ENES[S];
END;
CONT=0;
DO S=1 TO M;
  DO I=1 TO G;
    IF PE[S,I] > 0 THEN DO;
      CONT=CONT+1;
      TOT[I,]=TOT[I,] + MED[CONT,] * N * PE[S,I] * ALFA[I];
      DIV[I]=DIV[I] + N * PE[S,I] * ALFA[I];
    END;
  END;
END;
DO I=1 TO G;
MEDI[I,] = TOT[I,] * (1/DIV[I]);
END;
SC=J(P,P,0);CONT3=0;
DO S=1 TO M;
  DO I=1 TO G;
    TT=TAO[(CONT3+1:CONT3+ENES[S]),I];
    BB= B1[(CONT3+1:CONT3+ENES[S]),];
    IF PE[S,I] > 0 THEN DO;
      SC=SC+(TT#(BB-REPEAT(MEDI[I,],NROW(BB),1)))`*
              (BB-REPEAT(MEDI[I,],NROW(BB),1));
    END;
  END;
  CONT3=CONT3+ENES[S];
END;
V=SC/N;
SINV=INV(V);
DETS=DET(V);
FINISH TRES;

START EM;                        /* LOOP UNTIL LOG-LIKELIHOOD CONVERGE */
DO WHILE ((ABS(LOGLTOT-LOGLTOT0)/ABS(LOGLTOT0)) > CRIT);
  ITERA=ITERA+1;
  LOGLTOT0=LOGLTOT;
/*L8*/ IF ITERA = 50 THEN CRIT=10;
/*L9*/ ELSE CRIT = 1E-8;
  RUN TRES;
  RUN DOS;
END;
FINISH EM;

******                   RUNNING                       ****** ;

RUN CERO;
RUN UNO;
```

```
RUN DOS;
LOGLTOT0 = LOGLTOT - 1;
CRIT=0;
ITERA=1;
RUN EM;
TAOM=TAO[,]; CALI=TAOM[:];
FINGROUP={FINGROUP}; MAXPN={MAXPROB};
NAMES=VARCLAS`||VARCONT`||VARTAO`||FINGROUP||MAXPN;
CREATE FINAL FROM C [COLNAME=NAMES];
APPEND FROM C;
CREATE BAJAS FROM C1 [COLNAME=NAMES];
APPEND FROM C1;
PRINT "FINAL RESULTS:" ;
RESET NONAME;
PRINT LOGLTOT "        FINAL LOG-LIKELIHOOD";
RESET NAME;
PRINT PE [FORMAT=7.5];
PRINT V [FORMAT=9.4];
PRINT DETS;
PRINT MEDI [FORMAT=9.4];
PRINT ITERA "        NUMBER OF ITERATIONS";
PRINT "    AVERAGE OF THE MAXIMA OF THE PROBALITIES:";
RESET NONAME;
PRINT CALI [FORMAT=8.4];
RESET NAME;
QUIT;

DATA T1;SET FINAL;
INIGROUP=CLAS0; DROP CLAS0;
TITLE2 'INITIAL AND FINAL CLASSIFICATIONS';
PROC FREQ; TABLES INIGROUP*FINGROUP / NOCOL NOPERCENT;
PROC FREQ; TABLES W*FINGROUP / NOPERCENT;
PROC PRINT; VAR NOBS W INIGROUP FINGROUP MAXPROB;
DATA T2;SET BAJAS;
INIGROUP=CLAS0;DROP CLAS0 MAXPROB;
IF INIGROUP EQ 0 THEN DELETE;
TITLE2 'OBSERVATIONS CLASSIFIED WITH LEAST THAN 75% OF PROBABILITY';
PROC PRINT;
RUN;
TITLE '            ';
RUN;

PROC SORT DATA=DATA0 OUT=T1;BY NOBS;
PROC SORT DATA=FINAL OUT=T2;BY NOBS;
DATA FIN;MERGE T1 T2;BY NOBS;
PROC SORT DATA=FIN OUT=C1; BY FINGROUP;
PROC MEANS NOPRINT DATA=C1;BY FINGROUP;
/*L10*/     VAR ad sd ph eh grw;
/*L11*/ OUTPUT OUT=C2 MEAN= ad sd ph eh grw;
DATA MED;SET C2;SIZE=_FREQ_;DROP _FREQ_ _TYPE_;
PROC PRINT;
/*L12/ proc candisc data=fin mah;var ad sd ph eh grw; class fingroup;
RUN;
```

Results

1. Frequency analysis showing the formation of each value of the *W* variable:

```
The FREQ Procedure
```

```
Cumulative   Cumulative
                                                          Per-  Cum.   Per-
Q1 Q2 Q3 Q4 Q5 Q6 Q7 Q8 Q9 Q10 Q11 Q12 Q13 Q14 Q15 Freq. cent  Freq.  cent

0  0  0  0  1  0  0  0  1  0   1   0   1   0   1   1  6.67   1    6.67
0  0  0  0  1  0  1  0  1  0   1   0   1   1   1   1  6.67   2   13.33
0  0  0  1  0  0  0  0  1  0   1   0   0   1   0   1  6.67   3    0.00
0  0  0  1  0  0  1  0  1  0   0   1   1   0   0   1  6.67   4    6.67
0  0  1  0  0  0  0  1  0  1   0   1   0   0   1   1  6.67   5   33.33
0  0  1  0  1  0  1  0  1  0   1   0   1   1   1   1  6.67   6   40.00
0  1  0  0  0  0  1  0  1  0   1   1   0   1   1   1  6.67   7   46.67
0  1  0  1  1  0  0  1  0  1   0   1   1   0   1   1  6.67   8   53.33
0  1  1  0  0  1  0  1  0  0   1   0   1   1   1   1  6.67   9   60.00
1  0  1  1  0  1  0  1  0  0   1   0   1   0   1   1  6.67  10   66.67
1  0  1  1  1  1  1  0  0  1   0   1   0   0   0   2 13.33  12   80.00
1  0  1  1  1  1  0  1  0  0   1   0   0   0   0   1  6.67  13   86.67
1  1  0  1  0  0  0  0  0  1   0   0   1   0   1   1  6.67  14   93.33
1  1  1  0  0  1  0  1  0  0   1   0   1   1   0   1  6.67  15  100.00
```

2. Names of the categorical (VARCLAS) and continuous (VARCONT) variables; number of levels of the *W* variable (M), number of groups (G), and number of continuous variables (P); numbers of observations by cell (ENE), total number of observations (N), number of observations by group (ENEI); and number of observations by multinomial cell (ENES):

VARCLAS	VARCONT	M	G	P	VARTAO
NOBS	ad	14	5	5	GROUP01
CLAS0	sd				GROUP02
W	ph				GROUP03
	eh				GROUP04
	grw				GROUP05

ENE					N	ENEI	ENES
0	0	0	0	1	15	2	1
0	0	0	0	1		4	1
0	0	1	0	0		2	1
0	0	1	0	0		4	1
0	1	0	0	0		3	1
0	0	0	0	1			1
0	1	0	0	0			1
0	1	0	0	0			1

1	0	0	0	0	1
0	0	0	1	0	1
0	0	0	2	0	2
0	0	0	1	0	1
0	1	0	0	0	1
1	0	0	0	0	1

3. Description of the initial grouping: variance-covariance matrix (V), det(V) = DETS, vectors of means by group (MEDI), proportion of observations by group (ALFA), and proportion of observations by cell (PE):

V

6.8643	7.5542	-0.7639	-1.2623	13.8824
7.5542	10.7997	0.3649	0.6961	-110.3075
-0.7639	0.3649	42.2084	26.2318	88.4037
-1.2623	0.6961	26.2318	36.8485	157.3504
13.8824	-110.3075	88.4037	157.3504	11672.031

DETS

44954302

MEDI

87.5500	93.1400	108.1650	46.3600	227.8700
87.2275	90.0825	106.1675	38.1950	440.5850
88.9500	91.9300	107.0000	38.6500	394.4850
89.9400	91.4400	120.9150	46.4725	542.7450
87.7700	89.4767	107.7800	44.1033	406.7600

ALFA

0.13333
0.26667
0.13333
0.26667
0.20000

PE

0.00010	0.00010	0.00010	0.00010	0.33297
0.00010	0.00010	0.00010	0.00010	0.33297
0.00010	0.00010	0.49940	0.00010	0.00010
0.00010	0.00010	0.49940	0.00010	0.00010
0.00010	0.24975	0.00010	0.00010	0.00010
0.00010	0.00010	0.00010	0.00010	0.33297
0.00010	0.24975	0.00010	0.00010	0.00010
0.00010	0.24975	0.00010	0.00010	0.00010
0.49940	0.00010	0.00010	0.00010	0.00010
0.00010	0.00010	0.00010	0.24963	0.00010
0.00010	0.00010	0.00010	0.49963	0.00010
0.00010	0.00010	0.00010	0.24963	0.00010

```
0.00010 0.24975 0.00010 0.00010 0.00010
0.49940 0.00010 0.00010 0.00010 0.00010
```

4. Convergence of the log-likelihood:

```
LOG LIKELIHOOD :

            -277.82415
            -277.79638
            -277.69873
            -276.90376
            -271.46840
            -262.19260
            -258.50714
            -258.50701
            -258.50701
```

5. Final results: Description of the resulting groups:

```
-258.507              FINAL LOG-LIKELIHOOD

                            V

    4.0629     5.5976     0.5649    -0.2552     0.3318
    5.5976     9.1068     4.0598     2.7156   -81.5348
    0.5649     4.0598    37.4202    24.9494   -78.6771
   -0.2552     2.7156    24.9494    36.7351    89.2941
    0.3318   -81.5348   -78.6771    89.2941  8410.8310

    DETS

3426193.4
                          MEDI

   86.3166    91.4333   106.0000    44.8765   239.6610
   86.7680    89.4840   107.6680    38.8621   452.1069
   92.9700    96.7700   100.3300    35.7700   290.7808
   89.9400    91.4400   120.9150    46.4725   542.7451
   89.7300    90.2050   110.8350    45.2000   478.5200

    ITERA

        9          NUMBER OF ITERATIONS

    AVERAGE OF THE MAXIMUM PROBABILITIES:

    1.0000

INITIAL AND FINAL CLASSIFICATIONS

The FREQ Procedure
```

6. Two-way tables of the observed changes produced by the MLM and the distribution of the *W* variable into the final groups:

```
Table of INIGROUP by FINGROUP

INIGROUP      FINGROUP

Frequency|
Row Pct  |        1|        2|        3|        4|        5|  Total
---------+--------+--------+--------+--------+--------+
       1 |      2 |      0 |      0 |      0 |      0 |      2
         | 100.00 |   0.00 |   0.00 |   0.00 |   0.00 |
---------+--------+--------+--------+--------+--------+
       2 |      0 |      4 |      0 |      0 |      0 |      4
         |   0.00 | 100.00 |   0.00 |   0.00 |   0.00 |
---------+--------+--------+--------+--------+--------+
       3 |      0 |      1 |      1 |      0 |      0 |      2
         |   0.00 |  50.00 |  50.00 |   0.00 |   0.00 |
---------+--------+--------+--------+--------+--------+
       4 |      0 |      0 |      0 |      4 |      0 |      4
         |   0.00 |   0.00 |   0.00 | 100.00 |   0.00 |
---------+--------+--------+--------+--------+--------+
       5 |      1 |      0 |      0 |      0 |      2 |      3
         |  33.33 |   0.00 |   0.00 |   0.00 |  66.67 |
---------+--------+--------+--------+--------+--------+
Total           3        5        1        4        2       15

Table of W by FINGROUP

W             FINGROUP

Frequency|
Row Pct  |
Col Pct  |        1|        2|        3|        4|        5|  Total
---------+--------+--------+--------+--------+--------+
       1 |      0 |      0 |      0 |      0 |      1 |      1
         |   0.00 |   0.00 |   0.00 |   0.00 | 100.00 |
         |   0.00 |   0.00 |   0.00 |   0.00 |  50.00 |
---------+--------+--------+--------+--------+--------+
       2 |      0 |      0 |      0 |      0 |      1 |      1
         |   0.00 |   0.00 |   0.00 |   0.00 | 100.00 |
         |   0.00 |   0.00 |   0.00 |   0.00 |  50.00 |
---------+--------+--------+--------+--------+--------+
       3 |      0 |      1 |      0 |      0 |      0 |      1
         |   0.00 | 100.00 |   0.00 |   0.00 |   0.00 |
         |   0.00 |  20.00 |   0.00 |   0.00 |   0.00 |
---------+--------+--------+--------+--------+--------+
       4 |      0 |      0 |      1 |      0 |      0 |      1
         |   0.00 |   0.00 | 100.00 |   0.00 |   0.00 |
         |   0.00 |   0.00 | 100.00 |   0.00 |   0.00 |
---------+--------+--------+--------+--------+--------+
       5 |      0 |      1 |      0 |      0 |      0 |      1
         |   0.00 | 100.00 |   0.00 |   0.00 |   0.00 |
         |   0.00 |  20.00 |   0.00 |   0.00 |   0.00 |
---------+--------+--------+--------+--------+--------+
       6 |      1 |      0 |      0 |      0 |      0 |      1
```

	100.00	0.00	0.00	0.00	0.00		
	33.33	0.00	0.00	0.00	0.00		
7	0	1	0	0	0		1
	0.00	100.00	0.00	0.00	0.00		
	0.00	20.00	0.00	0.00	0.00		
8	0	1	0	0	0		1
	0.00	100.00	0.00	0.00	0.00		
	0.00	20.00	0.00	0.00	0.00		
9	1	0	0	0	0		1
	100.00	0.00	0.00	0.00	0.00		
	33.33	0.00	0.00	0.00	0.00		
10	0	0	0	1	0		1
	0.00	0.00	0.00	100.00	0.00		
	0.00	0.00	0.00	25.00	0.00		
11	0	0	0	2	0		2
	0.00	0.00	0.00	100.00	0.00		
	0.00	0.00	0.00	50.00	0.00		
12	0	0	0	1	0		1
	0.00	0.00	0.00	100.00	0.00		
	0.00	0.00	0.00	25.00	0.00		
13	0	1	0	0	0		1
	0.00	100.00	0.00	0.00	0.00		
	0.00	20.00	0.00	0.00	0.00		
14	1	0	0	0	0		1
	100.00	0.00	0.00	0.00	0.00		
	33.33	0.00	0.00	0.00	0.00		
Total	3	5	1	4	2		15

7. Initial (INIGROUP) and final (FINGROUP) classification by observation (NOBS), and probability of membership of each observation into the final group:

```
INITIAL AND FINAL CLASSIFICATIONS

Obs     NOBS     W     INIGROUP     FINGROUP     MAXPROB

1        10      1        5            5         1.00000
2        11      2        5            5         1.00000
3         8      3        3            2         1.00000
4         4      4        3            3         1.00000
5         2      5        2            2         1.00000
6         7      6        5            1         1.00000
7        14      7        2            2         1.00000
```

8	13	8	2	2	1.00000
9	1	9	1	1	1.00000
10	5	10	4	4	1.00000
11	6	11	4	4	1.00000
12	9	11	4	4	1.00000
13	15	12	4	4	1.00000
14	12	13	2	2	0.99996
15	3	14	1	1	1.00000

8. Description of the observations classified in a group with membership probability less than or equal to 0.75. *In this example, there were no observations classified with 0.75 or less probability.*

9. Means of the continuous variables by FINGROUP:

Obs	FINGROUP	ad	sd	ph	eh	grw	SIZE
1	1	86.3167	91.4333	106.000	44.8767	239.660	3
2	2	86.7680	89.4840	107.668	38.8620	452.106	5
3	3	92.9700	96.7700	100.330	35.7700	290.780	1
4	4	89.9400	91.4400	120.915	46.4725	542.745	4
5	5	89.7300	90.2050	110.835	45.2000	478.520	2

10. Canonical analysis:

The CANDISC Procedure

Observations	15	DF Total	14	
Variables	5	DF Within Classes	10	
Classes	5	DF Between Classes	4	

Class Level Information

FINGROUP	Variable Name	Frequency	Weight	Proportion
1	_1	3	3.0000	0.200000
2	_2	5	5.0000	0.333333
3	_3	1	1.0000	0.066667
4	_4	4	4.0000	0.266667
5	_5	2	2.0000	0.133333

11. Mahalanobis distances between groups:

Pairwise Squared Distances Between Groups

$$D^2(i|j) = \left(\overline{X}_i - \overline{X}_j\right)' COV^{-1}\left(\overline{X}_i - \overline{X}_j\right)$$

Squared Distance to FINGROUP

From FINGROUP	1	2	3	4	5
1	0	9.69890	31.89182	32.03905	53.71280
2	9.69890	0	47.17016	25.33743	64.65959
3	31.89182	47.17016	0	21.83177	11.25790
4	32.03905	25.33743	21.83177	0	16.44569
5	53.71280	64.65959	11.25790	16.44569	0

12. Canonical analysis:

Eigenvalues of Inv(E)*H = CanRsq/(1-CanRsq)			Test of H0: The canonical correlations in the current row and all that follow are zero		

| Eigenvalue | | Cumulative | | Likelihood | | | | Pr > F |

Eigen- value	Differ- ence	Propor- tion	Cumula- tive	Likelihood Ratio	Approx. F Value	Num DF	Den DF	Pr > F
13.0060	9.4873	0.7566	0.7566	0.00896176	3.28	20	20.85	0.0048
3.5187	3.0719	0.2047	0.9613	0.12551810	1.87	12	18.812	0.1089
0.4468	0.2281	0.0260	0.9873	0.56717249	0.87	6	16	0.5349
0.2187		0.0127	1.0000	0.82057834	0.98	2	9	0.4107

13. Correlations between the canonical and the original variables:

Pooled Within Canonical Structure

Variable	Can1	Can2	Can3	Can4
ad	0.237833	0.038504	-0.573316	0.675213
sd	0.066067	-0.143236	-0.342918	0.869577
ph	0.130221	0.426431	0.662903	0.598061
eh	0.077913	0.038632	0.779939	0.275478
grw	0.130604	0.581677	-0.090215	-0.512354

14. Class means on canonical variables:

FINGROUP	Can1	Can2	Can3	Can4
1	-2.195407354	-2.087516850	0.619110701	0.208971868
2	-2.871781847	0.763304861	-0.330874168	-0.251208708
3	2.982907431	-2.512395225	-1.568231850	0.543463597

| 4 | 2.006569437 | 1.682283872 | 0.234528031 | 0.361097995 |
| 5 | 4.967973060 | -0.885357009 | 0.213579232 | -0.679363820 |

Chapter 16

Mixed Linear Model Approaches for Quantitative Genetic Models

Jixiang Wu
Jun Zhu
Johnie N. Jenkins

Purpose

Computer software for estimating variance and covariance components, correlations, and predicting genetic effects.

Software Description

We describe a suite of genetic software that employs mixed linear model approaches. The various components relate to three categories, viz, genetic models for diallel crosses (Table 16.1), seed traits (Table 16.2), and developmental traits (Table 16.3). It can also be used to analyze regional agronomic trials.

This software has several features:

1. Handles complicated genetic models for agronomic traits, seed traits, and developmental traits
2. Analyzes unbalanced data
3. Utilizes jackknifing techniques to test the significance of each genetic parameter
4. Provides some important references containing results
5. Fast computation

System Requirements

Hardware: PC 486 or above; 16MB RAM, or more; 10 MB or more
available hard disk space
Operating System: Microsoft Windows 95/98, Microsoft Windows
NT 3.5 or above.

Installing

The software suite is available upon request to the author or from the
Web site: <http://msa.ars.usda.gov/ms/msstate/csrl/jenkins.htm>.
All related files are compressed into RUNWIN32.EXE, a self-extrac-
ting file. To install the software,
1. Copy RUNWIN32.ZIP to a subdirectory (e.g. *c:\WIN32*) on hard
disk using Windows Explorer or Windows NT Explorer.
2. Double click RUNWIN32.ZIP and all files will be extracted into the
current subdirectory WinZip 6.3 extraction software.

Tasks Performed by the Software

This software performs analyses on agronomic traits for diallel cross
models; seed models; developmental models; and regional trials.

A. Programs for Diallel Cross Models

Table 16.1 shows programs for diallel crosses.
This software can be used to estimate genetic variance components and
genetic covariance components and to predict genetic effects and heterosis
for AD, ADM, and ADAA models.

B. Programs for Seed Models

Table 16.2 includes programs for seed models. These programs can also
be used to estimate genetic variance components, genetic covariance components, and to predict genetic effects of diploid seed and triploid seed
models.

TABLE 16.1. Diallel Crosses

Genetic Models	Jackknife by cell	Jackknife by block
AD[1]	GENAD	GENAD
	GENVAR1C	GENVAR1R
	GENCOV1C	GENCOV1R
	GENHET1C	GENHET1R
ADM[2]	GENADM	GENADM
	GENVAR1C	GENVAR1R
ADAA[3]	GENADAA	GENADAA
	GENVAR1C	GENVAR1R
	GENCOV1C	GENCOV1R

[1]Additive-dominance models
[2]Additive-dominance maternal models
[3]Additive-dominance additive × additive epistasis models

TABLE 16.2. Seed Models

Genetic models	Jackknife by cell	Jackknife by block
Diploid	GENDIPLD	GENDIPLD
	GENVAR0C	GENVAR0R
	GENCOV0C	GENCOV0R
	GENHET0C	GENHET0R
Triploid	GENTRIPL	GENTRIPL
	GENVAR0C	GENVAR0R
	GENCOV0C	GENCOV0R
	GENHET0C	GENHET0R

C. Programs for Developmental Genetic Models

Table 16.3 includes programs for developmental traits. These programs can be used to create conditional data files, to estimate conditional genetic variance components, and to predict conditional genetic effects for AD, ADM, ADAA for diploid or triploid seed models. Some programs have appeared in Tables 16.1 and 16.2.

TABLE 16.3. Developmental Traits

Genetic Models	Jackknife by cells	Jackknife by block
AD	GENAD	GENAD
	GENCOND1	GENCOND1
	GENVAR1C	GENVAR1R
ADM	GENADM	GENADM
	GENCOND1	GENCOND1
	GENVAR1C	GENVAR1R
ADAA	GENADAA	GENADAA
	GENCOND1	GENCOND1
	GENVAR1C	GENVAR1R
Diploid	GENDIPLD	GENDIPLD
	GENCOND0	GENCOND0
	GANVAR0C	GENVAR0R
Triploid	GENTRIPL	GENTRIPL
	GENCOND0	GENCOND0
	GENVAR0C	GENVAR0R

Use of the Software Package

A. Diallel Model Analysis

Step 1. Build a data file. The arrangement of data is shown in file dial1.txt. on the Web page. The first five columns in this file represent environment (e.g., year or location), female, male, generation, and block. In column 1, enter environment number (1...e); in column 2 and 3, enter female and male number, respectively (1...p); and in column 5 enter block number (1...b). The data identifiers should be consecutive positive integers, each beginning with 1. The generation codes for column 4 are 0 for parents, 1 for F_1, and 2 for F_2. Enter data in columns 6 to n.

Step 2. Create an information matrix based on genetic models. Run GENAD for AD model, GENADM for ADM model, and GENADE for ADAA model. For example, when GENAD is chosen, prompts will automatically appear on screen as follows:

Input name of your data file: (e.g., dial 1.txt)
Do you have block effects within location (or environment)? Y/N

When running GENADM, you will see an extra prompt:

Do you analyze triploid endosperm? Y/N

Two files will be automatically created, e.g., *dial1.dat, dial1.mat,* where, *dial1.mat* contains matrix information, and *dial1.dat* contains the data of traits to be analyzed.

Step 3. Estimate the variance components and predict genetic effects. For example, when GENVAR1C or GENVAR1R is selected on the screen you will see the following prompts:

Input name of your data file: (input the data file name as given in Step 2).
What kind of parents did you use? Input 1 for inbred or 0 for outbred.
Choose prediction method. Do you want to use LUP or AUP? Input L for LUP and O for AUP.
Input coefficients for each parent: 1 for first group, –1 for second group, 0 for others.
Input sampling number for the jackknife procedure if running GENVAR1C.

The results are automatically stored in a file named, for example, *dial1.var.*

Step 4. Estimate covariance components and correlation coefficients. After finishing step 3, you can run GENCOV1C or GENCOV1R, and you will see the following prompts:

Input name of your data file: (the name given in Step 2);
What kind of parents did you use? Input 1 for inbred or 0 for outbred;
Input sampling number for the jackknife procedure: if running GENCOV1C.

The results are automatically stored in a file named, for example, *dial1.cor.*

Step 5. Predict heterosis. After running step 2, you can run GENHET1C or GENHET1R. Follow the prompts that automatically appear on the screen after you have chosen to run either model:

Input name of your data file: (input the name used in Step 2);
What kind of parents did you use? Input 1 for inbred or 0 for outbred;

Input sampling number for the jackknife procedure: if running GENHET1C.

The results are automatically stored in a file named, for example, *dial1.pre.* Note: During the process, other temporary files such as *matrix.var, matrix.uq2, matrix.uq3, matrix.uq4, matrix.uq5,* or *matdjc.var* will be created. The user should delete these files after finishing all analyses.

B. Seed Model Analysis

Step 1. Build a data file. The arrangement of data is shown in file *ctseed.txt.* The first five columns in this data file represent environment (e.g., year, location), female, male, generation, and block. In column 1, enter environment number (1 . . . e), in column 2, female number, in column 3, male number (1 . . . p), and in column 5, block number (1. . . b). The data identifiers should be consecutive integers, each beginning with 1. The generation codes for column 4 are 0 for parent, 1 for F_1, 2 for F_2, 3 for $BC1 = (F_1 \times P_1)$, 4 for $BC2 = (F_1 \times P_2)$, 5 for $RBC1 = (P_1 \times F_1)$, and 6 for $RBC2 = (P_2 \times P_2)$. Enter data in columns 6 to *n.*

Step 2. Construct an information matrix based on genetic models; GENDIPLD for diploid seed model and GENTRIPL for triploid seed model. When GENDIPLD is run, the following prompts appear on the screen:

Input name of your data file: (for example, enter ctseed.txt)
Do you have block effect within location? Y/N

Note: Two files will be automatically created, *ctseed.dat, ctseed.mat,* where *ctseed.mat* contains matrix information, and *ctseed.dat* contains data on traits to be analyzed.

Step 3. Estimate variance components and predict genetic effects. For example, for GENVAR0C or GENVAR0R, the following prompts will appear on the screen:

Input name of your data file: (name given in Step 2)
Choose prediction method. Do you want to use LUP or AUP? For LUP, input L, for AUP input O.
Input coefficients for each parent: 1 for first group, –1 for second group, and 0 for others;

Input sampling number for the jackknife procedure: if running GENVAR0C.

The results are automatically stored in the *ctseed.var* file.

Step 4. Estimate covariance components and correlation coefficients. For example, run GENCOV0C or GENCOV0R. When running this program, on-screen prompts include:

Input name of your data file: (name given in Step 2)
Input sampling number for the jackknife procedure: if running GENCOV0C.

The results are automatically stored in the *ctseed.cov* file. Note: The user should delete temporary files created during the execution of the program, after finishing all analyses.

C. Developmental Genetic Model Analysis

Step 1. Construct the file. The file format is the same as for the diallel and seed models.

Step 2. Convert traits to conditional traits.

(a) Construct information matrix based on genetic models.

For example, for AD model run GENAD and follow the on-screen prompts.

Input name of your data file: filename.txt
Do you have blocks within location? Y/N

(b) Run GENCOND1 or GENCOND0, where GENCOND1 is for AD, ADM, and ADAA models; and GENCOND0 is for diploid and triploid seed models.

Step 3. Now run steps 2 through 5 from A (diallel models) or B (seed models), the only difference being the change in the name of input file from *filename.txt* to *filename.doc* (the latter is a conditional data file).

D. Crop Regional Trial Analysis

Software included: GENTEST, GENETESTM, and GENTESTW. These programs can be used to estimate variance components, compare the significance of differences among varieties, and to evaluate the stability of each variety.

Step 1. Build the data file. The arrangement of data is shown in the file *msbean.txt*. The four columns represent variety, year, location, and replication. In the first column, enter variety number (check variety should be the highest number; this is important if you choose to transform data relative to the check). In the second column, enter year number. In the third column, location number; and in the fourth column enter replication number. The data identifiers should be consecutive positive integers beginning with 1.

Step 2. Construct an information matrix based on chosen genetic models.

> For example, run GENTEST and follow these on-screen prompts:
> Input name of your data file:

Step 3. Estimate stability for a single trait.

> For example, run GENTESTM and follow the on-screen prompts:
> Input name of your data file: (from Step 1).
> Do you want to transform data relative to check genotype? Y/N
> How many linear contrasts do you want?
> Input coefficients for each variety: 1 for first group, −1 for second
>   group, 0 for others.

The results are automatically stored in the *region.var* file. The results include variance components, linear contrasts among different genotypes, and stability of each genotype for each trait.

Step 4. Estimate stability for multiple traits.

> For example, run GENTESTW and follow on-screen prompts:
> Input name of your data file: (from Step 1).
> Input weight or values for each trait (sum of these weights = 1.0).
> How many linear contrasts do you want?
> Input coefficients for each variety: 1 for first group, −1 for second
>   group, 0 for others.

The results are automatically stored in the *region.cov* file. These results include variance and covariance components and stability of each genotype for multiple traits.

The following references may help the reader to understand the use of software packages and Internet sites.

Atchley, W.R. and Zhu, J. (1997). Developmental quantitative genetics, conditional epigenetic variability and growth in mice. *Genetics* 147:765-776.

Cockerham, C.C. (1980). Random and fixed effects in plant genetics. *Theoretical and Applied Genetics* 56:119-131.

Cockerham, C.C. and Weir, B.S. (1977). Quadratic analysis of reciprocal crosses. *Biometrics* 33:187-203.

Eisen, E.J., Bohren, B.B., and McKean, H.E. (1966). Sex-linked and maternal effects in the diallel cross. *Australian Journal of Biological Science* 19:1061-1071.

Fisher, R.A. (1925). *Statistical Methods for Research Workers,* First Edition. Oliver & Boyd, Edinburgh and London.

Griffing, B. (1956). Concept of general and specific combining ability in relation to diallel crossing systems. *Australian Journal of Biological Science* 9:463-493.

Hallauer, A.R. and Miranda, J.B. (1981). *Quantitative Genetics in Maize Breeding.* Iowa State University Press, Ames, Iowa.

Hartley, H.D. and Rao, J.N.K. (1967). Maximum-likelihood estimation for the mixed analysis of variance model. *Biometrika* 54:93-108.

Henderson, C.R. (1963). Selection index and expected genetic advance. In Hanson, W.D. and Robinson, H.F. (Eds.), *Statistical Genetics and Plant Breeding* (pp. 141-163). Washington, DC: National Academy of Science, National Research Council.

Miller, R.G. (1974). The jackknife: A review. *Biometrika* 61:1-15.

Patterson, H.D. and Thompson, R. (1971). Recovery of inter-block information when block sizes are unequal. *Biometrika* 58:545-554.

Rao, C.R. (1971). Estimation of variance and covariance components MINQUE theory. *Journal of Multivariate Analysis* 1:257-275.

Rao, C.R. and Kleffe, J. (1980). Estimation of variance components. In Krishnaiah, P.R. (Ed.), *Handbook of Statistics,* Vol. 1 (pp. 1-40). North-Holland, New York.

Searle, S.R., Casella, G., and McCulloch, C.E. (1992). *Variance Components.* John Wiley and Sons, New York.

Shi, C.H., Zhu, J., Zeng, R.C., and Chen, G.L. (1997). Genetic and heterosis analysis for cooking quality traits of indica rice in different environments. *Theoretical and Applied Genetics* 95:294-300.

Wu, J.X., Zhu, J., Ji, D.F., and Xu, F.H. (1995). Genetic analysis for heterosis of fiber traits in Upland cotton (Chinese). *Acta Gossypii Sinica* 7(4):217-222.

Yan, J.Q., Zhu, J., He, C.X., Benmoussa, M., and Wu, P. (1998). Molecular dissection of developmental behavior of plant height in rice (*Oryza sativa* L.). *Genetics* 150:1257-1265.

Yan, X.F., Xu, S.Y., Xu, Y.H., and Zhu, J. (1998). Genetic investigation of contributions of embryo and endosperm genes to malt kolbach index, alpha-amylase activity and wort nitrogen content in barley. *Theoretical and Applied Genetics* 96(5):709-715.

Zeng, Z.-B. (1994). Precision mapping of quantitative trait loci. *Genetics* 136:1457-1468.

Zhu, J. (1989). Estimation of genetic variance components in the general mixed model. Doctoral dissertation, North Carolina State University, Raleigh, NC.

Zhu, J. (1993a). Methods of predicting genotype value and heterosis for offspring of hybrids (Chinese). *Journal of Biomathematics* 8(1):32-44.

Zhu, J. (1993b). Mixed model approaches for estimating covariances between two traits with unequal design matrices (Chinese). *Journal of Biomathematics* 8(3):24-30.

Zhu, J. (1994). General genetic models and new analysis methods for quantitative traits (Chinese). *Journal of Zhejiang Agricultural University* 20(6):551-559.

Zhu, J. (1996). Analysis methods for seed models with genotype × environment interactions (Chinese). *Acta Genetica Sinica* 23(1):56-68.

Zhu, J. (1998). Mixed model approaches of mapping genes for complex quantitative traits. In Wang, L.Z. and Dai, J.R. (Eds.), *Proceedings of Genetics and Crop Breeding of China* (pp. 19-20). Chinese Agricultural Science and Technology Publication House, Beijing.

Zhu, J., Wang, G.J., and Zhang, R.C. (1997). Genetic analysis on gene effects and GE interaction effects for kernel nutrient quality traits of Upland cotton (Chinese). *Journal of Biomathematics* 12(2):111-120.

Zhu, J. and Weir, B.S. (1994a). Analysis of cytoplasmic and maternal effects: I. A genetic model for diploid plant seeds and animals. *Theoretical and Applied Genetics* 89:153-159.

Zhu, J. and Weir, B.S. (1994b). Analysis of cytoplasmic and maternal effects: II. Genetic models for triploid endosperm. *Theoretical and Applied Genetics* 89:160-166.

Zhu, J. and Weir, B.S. (1996). Diallel analysis for sex-linked and maternal effects. *Theoretical and Applied Genetics* 92(1):1-9.

Chapter 17

Best Linear Unbiased Prediction (BLUP) for Genotype Performance

Mónica Balzarini
Scott Milligan

Importance

Whenever the examined genotypes (varieties, clones, cultivars, etc.) in an experiment can be regarded as random samples from larger sets or populations, genotype effects should also be considered random in the model. Commonly, random genotype effects are assumed to broaden inferences, i.e., to allow inference about a reference population. Moreover, pairwise comparisons between specific genotypes that have been examined will also be feasible. In addition to estimating genotype variance components, which are of intrinsic interest with random genotype effects, genotypes can be compared by calculating best linear unbiased predictors (BLUPs) of genotype effects. In this context, "best" means minimum mean squared prediction error (MSPE). The smaller the MSPE, the greater the relationship between the dependent and independent variables.

BLUPs are used to predict random effects in mixed models. The predictable function, $\mu + G_i$, $i = 1,...,g$, allows inference about the performance of the ith genotype from a trial involving g randomly selected varieties (broad inference). Thus, the BLUPs of $\mu + G_i$ in a mixed model have a similar role as the genotype mean in a fixed-effect model. BLUPs are called *shrinkage estimators* because they are obtained by regression toward the overall mean based on the variance components of the model effects. A simple version of BLUP to estimate genetic performance of the ith genotype, in a model containing unrelated genotypes as random effects, is

$$\text{BLUP } (\mu + G_i) = \overline{Y}.. + F^G \left(\overline{Y}_i. - \overline{Y}.. \right)$$

where F^G is the shrinkage factor that $=$

$$\frac{\sigma_G^2}{\sigma_G^2 + \sigma_e^2 / n},$$

where σ_G^2 is genetic variance or variance among genotypes, σ_e^2 is the error variance or variance within genotype, and n is the number of observations per genotype. Thus, for this simple model, the estimator moves the genotype mean toward the overall mean depending on the magnitude of a trait's heritability. A large heritability value implies little shrinkage or more reliability of genotype means, and in such a case, genotype BLUPs resemble genotype means. Therefore, the smaller the heritability value, the larger the shrinkage of extreme genotype means toward μ with a reduction in the risk of misinterpretation.

The BLUP for the predictable function, $\mu + G_i + E_j + G \times E_{ij}$, $i = 1,...,g$ and $j = 1,...,l$, in a model also involving environment and genotype-by-environment interaction (GE) random effects, assuming independent and equal variance random effects, is

$$\text{BLUP}\left(\mu + G_i + E_j + G \times E_{ij}\right) = \overline{Y}... + F^G\left(\overline{Y}_i.. - \overline{Y}...\right)$$
$$+ F^{GE}\left(\overline{Y}... - \overline{Y}i.. - \overline{Y}.j. \_ \overline{Y}ij.\right)$$

where F^G, F^E, and F^{GE} are the following shrinkage factors:

$$F^G = \frac{\sigma_G^2 + \dfrac{\sigma_{GE}^2}{l}}{\sigma_G^2 + \sigma_{GE}^2 + \dfrac{\sigma_e^2}{nl}} \quad F^E = \frac{\sigma_E^2 + \dfrac{\sigma_{GE}^2}{g}}{\sigma_E^2 + \sigma_{GE}^2 + \dfrac{\sigma_e^2}{ng}} \quad F^{GE} = \frac{\sigma_{GE}^2}{\sigma_{GE}^2 + \dfrac{\sigma_e^2}{n}}$$

where σ_{GE}^2 represents the *GE* variance component, σ_E^2 is the variance component associated with environment effects, and g and l are the number of environments and genotypes in the experiment, respectively.

A general form of BLUP of random effects can be obtained by expressing them as a solution of the mixed model equation (extended normal equations). In matrix form, a normal mixed model is

$$y = X\beta + Zu + e$$

where y is an n vector of observable random variables (data), X and Z are known design matrices, β is a p vector of fixed effects, and u (random effects) and e (error term) are unobservable normal random m and n vectors with covariance matrices G and R, respectively. The variance-covariance matrix is $V = ZGZ'+R$. Estimates of β and u can be written as follows:

$$\hat{\beta} = (X'\hat{V}^{-1}X)^- X'\hat{V}^{-1}y$$

$$\hat{u} = \hat{G}Z'\hat{V}^{-1}(y - X\hat{\beta})$$

Since variance components are usually unknown, estimated covariance matrices are used in place of true matrices. Thus, the vector $\hat{u}$ is better referred to as an empirical BLUP (EBLUP) to indicate that it is approximated from the data. The vector $\hat{u}$ can be interpreted as a weighted deviation of genotype means from the overall mean after adjusting for fixed effects. Provided that G is nonsingular, the estimated covariance matrix of $\hat{\beta}$ and $\hat{u}$ can be written as follows:

$$\hat{C} = \begin{bmatrix} \hat{C}_{11} & \hat{C}_{21}' \\ \hat{C}_{21} & \hat{C}_{22} \end{bmatrix}, \text{ where}$$

$$\hat{C}_{11} = (X'\hat{V}^{-1}X)^-$$

$$\hat{C}_{21} = -\hat{G}Z'\hat{V}^{-1}X\hat{C}_{11}$$

$$\hat{C}_{22} = (Z'\hat{R}^{-1}Z + \hat{G}^{-1})^{-1} - \hat{C}_{21}X'\hat{V}^{-1}Z\hat{G}$$

The elements of $\hat{C}_{22}$ provide the estimated prediction error variances that allow comparisons of BLUPs. A predictable function $K'\beta + M'u$ is estimated by $K'\hat{\beta} + M'\hat{u}$. Thus, the BLUP of $\mu + G_i$ is $\hat{\mu} + \hat{G}_i$. As a linear combination of fixed and random model effects, its prediction error variance can be easily derived from $\hat{C}$. These estimates address the objective of assessing and comparing genotype performance in a mixed model context. The square root of the prediction error variance (a standard error analog) can be used to approximate Prediction Intervals for the pairwise BLUP differences.

If the genotypes were genetically related, a matrix of genetic relationships, A, may be used to adjust the matrix G and, in turn, the BLUPs. These relationships may be computed from pedigree or molecular-based analyses. The covariance matrix for the genetic effects is commonly written as $G = \sigma_G^2 A$, where elements in A are used to represent genetic relatedness between any two genotypes, and it is expressed as a proportion of genetic variance. Note that $A = I$ represents the special case of unrelated genotypes. Several covariance structures for G and R may be reasonable. Likewise, following the aforementioned procedure, different BLUP versions can be obtained.

A SAS macro was developed to obtain BLUPs of genotype effects, prediction errors, BLUPs of the predictable functions $\mu + G_i$, and pairwise, approximated prediction intervals for the differences between the BLUPs of two genotypes. The macro uses restricted maximum likelihood (REML) variance component estimates obtained from PROC MIXED of SAS (SAS, 1997). It calls for an IML module to obtain and compare the BLUPs derived with or without (Model = GGR) (model = GI), assuming a given genetic relationship among the examined genotypes.

EXAMPLE

Two sample data sets, one with a large genotypic variance and the other with a small genotypic variance, are provided to show how a simple (assuming genotypes are not related) version of BLUP for genotypes (broad inference) works. Each data set contains four genotypes and four replicates per genotype.

```
data AltaH2;          data BajaH2;
input genotype y;     input genotype y;
datalines;            datalines;
1 28.41               1 18.41
1 27.35               1 17.35
1 27.25               1 17.25
1 28.42               1 18.42
2 19.90               2 19.90
2 18.93               2 18.93
2 19.52               2 19.52
2 19.61               2 19.61
3 19.65               3 19.45
3 18.79               3 19.65
3 18.86               3 21.86
3 19.45               3 20.79
4 21.80               4 21.80
4 18.61               4 18.61
4 22.37               4 22.52
4 22.52               4 17.37
```

Assuming the macro file is being retrieved from a floppy disk, write:

```
%include 'a:BLUP_GP.sas' ;
%BLUP_GP(Model=GI,Workds=AltaH2,Ldata=,S2_G=,S2_A=,S2_E=,
    Class=,Fixed=, C_matrix=,PredRes=);
```

The macro arguments that can be modified by the user are:

```
Model :enter GI  for  Independent or unrelated Genotypes
          enter GR for  Related Genotypes
Workds:work data set name
Ldata :name for the data set containing a Genetic
          relationship coefficient matrix. Required if model=GR
S2_G  :enter a known Genetic VarianceValue (otherwise it is estimated)
S2_A  :enter a known Additive Variance Value (optional)
S2_E  : enter a known Error Variance Value (otherwise it is estimated)
Class  : classification variables excluding genotype (optional)
Fixed  : model fixed effects separated by blanks (optional)
C_Matrix :YES to print mixed model parameter covariance matrix
PredRes  : YES to print Predicted and Residual Values
```

If the model GI (unrelated genotypes) is fitted to both data sets, the BLUPs of genotype effects and the BLUPs of $\mu + G_i$ shown in the output will be as follows:

AltaH2 data set		BajaH2 data set	
BLUP (G_i)	BLUP $(m + G_i)$	BLUP (G_i)	BLUP $(m + G_i)$
5.799	27.764 A	−0.992	18.473 B
−2.435	19.529 C	0.015	19.480 A
−2.733	19.231 C	0.600	20.065 A
−0.630	21.335 B	0.376	19.841 A

The BLUP differences among the ten possible pairwise comparisons of genotype versus a 95 percent prediction interval column containing either 1 or 0 for each contrast are shown. The zeros indicate that the prediction interval for the BLUP difference contains a zero value, which can be interpreted as not different in genotypic performance. For the AltaH2 data set, all BLUP differences, except those between genotypes 2 and 3, are significant. For the BajaH2 data set, only the difference between genotype 1 and 3 is significant. Here we show the traditional genotype means and least significant difference (LSD) for both sample data sets to compare BLUP performance against ordinary mean performance:

AltaH2 data set		BajaH2 data set	
Genotype Mean	LSD signif-icance	Genotype Mean	LSD signif-icance
27.858	A	17.856	B
19.490	C	19.490	AB
19.187	C	20.438	A
21.325	B	20.075	A

Note: For a large trait heritability, BLUPs closely resemble means, but a small genotype variance component shrinks genotype performances toward the overall mean, yielding different significances in the BajaH2 data set. In a real data set (large number of genotypes and probably different genotype sample sizes), shrinkage might provide significant improvement over genotype means.

The data GR is a sample data set to enter the covariance coefficient matrix for a set of four genotypes. The columns parm (equal to 1) and row (from 1 to g) should be in the data set to conform with SAS PROC MIXED requirements for using the covariance structure type = LIN(1). This structure indicates that the covariance parameters in G are a linear combination of variance components. This variance component may be the additive variance (entered as macro argument) to obtain BLUPs with shrinkage factors based on the narrow-sense heritability (matrix coefficients should be $2*r_{ij}$, where r_{ij} is the coancestry coefficient between genotypes i and j). The variance component might also be a genetic variance, entered as a macro argument or estimated from the data. The sample data GR shows a covariance coefficient (0.5) between genotypes 2 and 3.

```
Data GR;
input parm   row col1-col4;
datalines;
1 1   1   0   0   0
1 2   0   1   0.5  0
1 3   0   0.5  1   0
1 4   0   0   0   1
```

The relationship coefficient matrix data set should be indicated in the macro argument LDATA, whenever model GGR is required. The output interpretation is the same as before, but BLUPs have been calculated assuming the given genetic relationship among the genotypes.

```
/*                                                              */
/*SAMPLE DATA SET FOR OBTAINING BLUPS FOR GENOTYPE PERFORMANCE  */
/*Data ALTAH2 shows a larger genotype variance than data BAJAH2 */
/*Data GR shows how to inputa  genetic relationship matrix      */
/*The following variable names should be the same in your data set:*/
/*genotype, y (for trait values),Parm (equals 1),row (from 1 to g) */
/*Coll to Colg to indicate the g columns of the relationship matrix*/

data AltaH2;
input genotype y bl;
datalines;
1 28.41 1
1 27.35 2
1 27.25 3
1 28.42 4
2 19.90 1
2 18.93 2
2 19.52 3
2 19.61 4
3 19.65 1
3 18.79 2
3 18.86 3
3 19.45 4
4 21.80 1
4 18.61 2
4 22.37 3
4 22.52 4
data BajaH2;
input genotype y;
datalines;
1 18.41
1 17.35
1 17.25
1 18.42
2 19.90
2 18.93
2 19.52
2 19.61
3 19.45
3 19.65
3 21.86
3 20.79
4 21.80
4 18.61
4 22.52
4 17.37
Data GR;
input parm  row coll-col4;
datalines;
1 1 1 0   0   0
1 2 0 1   0.5 0
1 3 0 0.5 1   0
1 4 0 0   0   1
;
```

```
/*                                                                      */
/*Macro BLUP_GP produces Genotype Performance BLUPs for models with */
/*only genotype effects as random. Two versions of genotype BLUPs   */
/*can be obtained (with or without relationship among genotypes)    */
/*The macro arguments are:                                          */
/*Model :enter GI for Independent or unrelated Genotypes            */
/*          enter GR for Related Genotypes                          */
/*Workds:work data set name                                         */
/*Ldata :name for the data set containing a Genetic                 */
/*         relationship coefficient matrix. Required if model=GR    */
/*S2_G  :enter a known Genetic Variance (otherwise it is estimated) */
/*S2_A  :enter a known Additive Variance (optional)                 */
/*S2_E  :enter a known Error Variance (otherwise it is estimated)   */
/*Class :classification variables excluding genotype (optional)     */
/*Fixed :model fixed effects separated by blanks (optional)         */
/*C_matrix:YES to print mixed model parameter covariance matrix     */
/*PredRes :YES to print Predicted and Residual Values               */
/*                                                                  */

options nodate nocenter;
%macro BLUP_GP(Model=GI,Workds=BajaH2,Ldata=GR,S2_G=,S2_A=,S2_E=,
               Class=,Fixed=, C_matrix=Yes,PredRes=);


proc mixed data=&workds noclprint noitprint covtest;
class genotype &class;
model y=&Fixed/pm p;

%if (&model=GI) %then %do;
 random Genotype/s ;
%end;

%if (&model=GGR) %then %do;
 random Genotype/ldata=&ldata type=lin(1) s ;
%end;

 make 'solutionR' noprint out=BLUP&model;
 make 'covparms'  out=V&model;
 make 'predmeans' noprint out=L&model;
 make 'predicted' noprint out=Pred&model;

/*                                                                  */
/*Calculate the Predictable Function Mu+Genotype effect             */
/*                                                                  */

Data BLUPs;
set BLUP&model;
Keep Genotype BLUP P_Error P_value;
BLUP=_EST_;
P_Error=_SEPRED_;
P_Value=_PT_;

%if (&S2_G<0) %then %do;
%if (&S2_A<0) %then %do;

proc print;
```

```
title1 'Genotype effect BLUPs, Prediction Error and P-Value for
    H0:BLUP=0';
run;
%end;
%end;
title1 '   ';

%if (&model=GGR) %then %do;
data AR;
set &ldata;
drop parm row;
%end;

data level;
set &workds;
keep genotype;

data VarComp;
set V&model;
if covparm='Residual' or Covparm='GENOTYPE' or covparm='LIN(1)';

data Y;
set &workds;
keep Y;

  proc iml;
  use Y;
  read all into Y;
  use L&model var{_resid_};
  read all into ya;
  use level;
  read all into level;
  use VarComp;
  read all into VC;

  Z=design(level);
  yp=inv(Z`*Z)*Z`*ya;

  S2e=VC(|2,1|);
  S2u=VC(|1,1|);

  %if (&S2_E>0) %then %do; S2e=&S2_E; print S2e;%end;

  %if (&S2_G>0) %then %do; S2u=&S2_G; print S2u;%end;

  %if (&S2_A>0) %then %do; S2u=&S2_A; print S2u;%end;

  %if (&model=GI) %then %do;
    AR=I(nrow(Z`));
    V=inv(Z`*Z)*Z`*(S2u*Z*AR*Z`+S2e*I(nrow(Z)))*Z*inv(Z`*Z)`;
  %end;

  %if (&model=GGR) %then %do;
    use AR;
    read all into AR;
    V=inv(Z`*Z)*Z`*(S2u*Z*AR*Z`+S2e*I(nrow(Z)))*Z*inv(Z`*Z)`;
  %end;
```

```
C=VC(|1,1|)*AR;
BLUPg=C*inv(V)*yp;

X=J(nrow(level),1);
G=S2u*AR;
R=S2e*I(nrow(Z));
V=Z*G*Z`+ R;
Beta=inv(X`*X)*X`*Y;

K=J(nrow(Z`),1)`;
M=I(nrow(Z`));

PF=K`*Beta+ M`*BLUPg;

/*                                                                  */
/*Calculate Prediction Error Variances                             */
/*                                                                  */

C11=Ginv(X`*inv(V)*X);
C21=-G*Z`*inv(V)*X*C11;
C22=inv(Z`*inv(R)*Z+inv(G))-C21*X`*inv(V)*Z*G;
COV_BU=(C11||C21`) //(C21||C22);

VarPF=K`*C11*K+M`*C22*M+2*M*C21*K;

ncon=Nrow(PF)*(Nrow(PF)-1)/2;
row=0;
LPF=shape(0,ncon,nrow(BLUPG));
Gi=shape(0,ncon,1);
Gj=shape(0,ncon,1);

do i=1 to nrow(BLUPg);
  do j=1 to nrow(BlUPg);
    If i<j then do;
                 row=row+1;
                 LPF[row,i]=1;
                 LPF[row,j]=-1;
                 Gi[row,1]=i;
                 Gj[row,1]=j;
                 end;
  end;
end;

Con_PF=LPF*PF;
Var_C=LPF*VarPF*LPF`;
PF_Diff=Con_PF;
Con_PF=Con_PF@j(1,nrow(LPF));
CValue=2*SQRT(Diag(Var_C));
LS_IC=Diag(Con_PF+CValue);
LI_IC=Diag(Con_PF-CValue);

Pred_Int=J(nrow(LPF),1);
  do i=1 to nrow(LPF);
     If LS_IC[i,i]>0 then do;If LI_IC[i,i]<0 then Pred_Int[i]=0;end;
  end;
```

```
/*                                                                    */
/*Printing Results                                                    */
/*                                                                    */
print 'Predictable Functions and Pairwise Differences';
print  BLUPg PF Gi GJ PF_Diff Pred_Int;
print 'Pred_Int=0 means no different BLUPs';

%if (&C_matrix=YES) %then %do;
  print COV_BU;
%end;

%if (&PredRes=YES) %then %do;
  proc print data=Pred&model;
%end;

%mend BLUP_GP;

%BLUP_GP;

run;
```

Chapter 18

Graphing GE and GGE Biplots

Juan Burgueño
José Crossa
Mateo Vargas

Purpose

- To analyze multienvironment trials and to study genotype × environment interaction (GEI) using linear/bilinear AMMI or SREG models.
- To graph biplots to describe GEI and to identify megaenvironments and cultivars with best performance.

Definitions

The linear/bilinear AMMI model is

$$\bar{y}_{ij.} = \mu + \tau_i + \delta_j + \sum_{k=1}^{t} \lambda_k \alpha_{ik} \gamma_{jk} + \bar{\varepsilon}_{ij}$$

and the linear/bilinear SREG model is

$$\bar{y}_{ij.} = \mu + \delta_j + \sum_{k=1}^{t} \lambda_k \alpha_{ik} \gamma_{jk} + \bar{\varepsilon}_{ij}$$

where $\bar{y}_{ij.}$ is mean of the ith cultivar in the jth environment; μ is the overall mean, τ_i is the genotypic effect; δ_j is the site effect; $\lambda_k (\lambda_1 \geq \lambda_2 \geq .. \geq \lambda_t)$ represents the singular values (scaling constants); $\alpha_{ik} = (\alpha_{1k}, ..., \alpha_{gk})$ and $\gamma_{jk} = (\gamma_{1k}, ..., \gamma_{ek})$ are the singular vectors for cultivars and environment, respectively, with $\Sigma_i \alpha_{ik}^2 = \Sigma_j \gamma_{jk}^2 = 1$ and $\Sigma_i \alpha_{ik} \alpha_{ik'} = \Sigma_j \gamma_{jk} \gamma_{jk'} = 0$ for $k \neq k'$; ε_{ij}:

residual error with NID $(0, \sigma^2 / r)$ (σ^2 is pooled error variance and r is number of replicates).

Originators

Gollob, H.F. (1968). A statistical model which combines features of factor analytic and analysis of variance. *Psychometrika* 33:73-115.
Mandel, J. (1961). Non-additivity in two-way analysis of variance. *Journal of the American Statistical Association* 56:878-888.

Biplot

Biplots graph scores of sites and genotypes of the first bilinear term against scores of sites and genotypes of the second bilinear term.

Originators

Gabriel, K. R. (1971). The biplot graphic display of matrices with application to principal component analysis. *Biometrika* 58:453-467.
Yan, W., Hunt, L.A., Sheng, Q, and Szlavnics, Z. (2000). Cultivar evaluation and mega-environment investigation based on the GGE biplot. *Crop Science* 40:597-605.

Software Available

Burgueño, J., Crossa, J., and Vargas, M. (2001). SAS PROGRAMS for graphing GE and GGE biplots. Centro Internacional de Mejoramiento de Maiz y Trigo (CIMMYT), INT, accessed online <http:www.cimmyt.org/biometrics>.

Key References

Burgueño, J., Crossa, J., and Vargas, M. (2001). SAS PROGRAMS for graphing GE and GGE biplots. Centro Internacional de Mejoramiento de Maiz y Trigo (CIMMYT), INT, accessed online <http:www.cimmyt.org/biometrics>.
Vargas, M. and Crossa, J. (2000). The AMMI analysis and the graph of the biplot in SAS. Centro Internacional de Mejoramiento de Maiz y Trigo (CIMMYT) INT. México, p. 42.
Yan, W., Hunt, L.A., Sheng, Q., and Szlavnics, Z. (2000). Cultivar evaluation and mega-environment investigation based on the GGE biplot. *Crop Science* 40:597-605.

Contact

Juan A. Burgueño. Centro Internacional de Mejoramiento de Maiz y Trigo (CIMMYT) INT. Apdo. 6-641 C.P. 06600 México, D.F., Mexico.

Data to Be Analyzed

Yield (kg·ha^{-1}), obtained from individual analysis by year:

year genotype	1990	1991	1992	1993	1994	1995
1	5991.67	6640.33	5518.67	6657.67	6701.67	5280.67
2	7160.67	6081.00	5638.00	6688.67	7280.33	5869.67
3	7793.00	6954.33	5399.67	7553.33	7196.00	6041.00
4	7715.00	7170.00	6536.00	6530.67	7610.67	6284.00
5	8082.67	7224.67	6229.00	8087.67	7092.33	5891.00
6	8179.00	7467.33	6680.00	7296.00	8510.33	6901.67
7	7780.00	7095.67	7175.00	8385.67	8579.33	6931.67
8	7864.00	7632.33	7075.00	8529.67	8591.67	7012.33

Case 1

Using the AMMI model for obtaining the GE biplot:

```
/* setup output */
OPTIONS PS = 5000 LS=78 NODATE;
FILENAME BIPLOT 'EXAMPLE1.CGM';
GOPTIONS DEVICE=CGMMWWC GSFNAME=BIPLOT GSFMODE=REPLACE;

/* read data file */
DATA RAW;
          INFILE 'c:\cimmyt\amii\EXAMPLE1.DAT';
          INPUT ENV $ GEN $ YIELD;
          YLD=YIELD/1000;

/* analysis linear-bilinear model */
PROC GLM DATA=RAW OUTSTAT=STATS ;
          CLASS ENV GEN;
          MODEL YLD = ENV GEN ENV*GEN/SS4;
DATA STATS2;
SET STATS ;
DROP _NAME_ _TYPE_ ;
IF _SOURCE_ = 'ERROR' THEN DELETE;

/* values obtained from previous analysis */
MSE=0.1580245;
DFE=94;
NREP=3;
SS=SS*NREP;
MS=SS/DF;
F=MS/MSE;
PROB=1-PROBF(F,DF,DFE);
PROC PRINT DATA=STATS2 NOOBS;
          VAR _SOURCE_ DF SS MS F PROB;
```

```
/* define AMMI model   */
PROC GLM DATA=RAW NOPRINT;
           CLASS ENV GEN;
           MODEL YLD = ENV GEN / SS4 ;
           OUTPUT OUT=OUTRES R=RESID;
PROC SORT DATA=OUTRES;
           BY GEN ENV;
PROC TRANSPOSE DATA=OUTRES OUT=OUTRES2;
           BY GEN;
           ID ENV;
           VAR RESID;
PROC IML;
USE OUTRES2;
READ ALL INTO RESID;
NGEN=NROW(RESID);
NENV=NCOL(RESID);
USE STATS2;
READ VAR {MSE} INTO MSEM;
READ VAR {DFE} INTO DFEM;
READ VAR {NREP} INTO NREP;
CALL SVD (U,L,V,RESID);
MINIMO=MIN(NGEN,NENV);
L=L[1:MINIMO,];
SS=(L##2)*NREP;
SUMA=SUM(SS);
PERCENT=((1/SUMA)#SS)*100;
MINIMO=MIN(NGEN,NENV);
PERCENTA=0;
           DO I = 1 TO MINIMO;
                     DF=(NGEN-1)+(NENV-1)-(2*I-1);
                     DFA=DFA//DF;
                     PORCEACU=PERCENT[I,];
                     PERCENTA=PERCENTA+PORCEACU;
                     PERCENAC=PERCENAC//PERCENTA;
           END;
DFE=J(MINIMO,1,DFEM);
MSE=J(MINIMO,1,MSEM);
SSDF=SS||PERCENT||PERCENAC||DFA||DFE||MSE;
L12=L##0.5;
SCOREG1=U[,1]#L12[1,];
SCOREG2=U[,2]#L12[2,];
SCOREG3=U[,3]#L12[3,];
SCOREE1=V[,1]#L12[1,];
SCOREE2=V[,2]#L12[2,];
SCOREE3=V[,3]#L12[3,];
SCOREG=SCOREG1||SCOREG2||SCOREG3;
SCOREE=SCOREE1||SCOREE2||SCOREE3;
SCORES=SCOREG//SCOREE;
CREATE SUMAS FROM SSDF;
APPEND FROM SSDF;
CLOSE SUMAS;
CREATE SCORES FROM SCORES;
APPEND FROM SCORES ;
CLOSE SCORES;

/* obtaining the biplot's polygon and its perpendiculars */
d1=scoreg[,1:2][cvexhull(scoreg[,1:2])[loc(cvexhull(scoreg[,1:2])>0),],];
d=d1//d1[1,];
xxx=J(nrow(d)-1,1,0);
```

```
yyy=J(nrow(d)-1,1,0);
ppp={0 1,1 0};
            do i=1 to nrow(d)-1 ;
                    dd=d[i:i+1,];
                            if dd[1,1]>dd[2,1] then ddd=ppp*dd;
                            else ddd=dd;
                    p=(ddd[2,2]-ddd[1,2])/(ddd[2,1]-ddd[1,1]) ;
                            if p<0 then ss=1 ;
                            else ss=-1 ;
                    r=tan((180-90-
abs(atan(p)*180/3.14156))*3.14156/180)*ss ;
                    aa=(ddd[1,2]+ddd[2,2])/2-p*(ddd[1,1]+ddd[2,1])/2;
                    xx=aa/(r-p) ;
                            if abs(r)<1 then xxx[i,]=1;
                            else xxx[i,]=1/abs(r);
                                    if xx<0 then xxx[i,]=-xxx[i,] ;
                                    else xxx[i,]=xxx[i,];
                    yyy[i,]=xxx[i,]*r;
            end;
kk=xxx||yyy;
xx1={V1 V2};
create pol from d[colNAME=xx1];
append from d ;
close pol;
xx2={V3 V4};
create perp from kk[colNAME=xx2];
append from kk ;
close perp;
data pol; set pol; TYPE="pol";
data perp; set perp; TYPE="per";
DATA SSAMMI;
SET SUMAS;
SSAMMI =COL1;
PERCENT =COL2;
PERCENAC=COL3;
DFAMMI =COL4;
DFE =COL5;
MSE =COL6;
DROP COL1 - COL6;
MSAMMI=SSAMMI/DFAMMI;
F_AMMI=MSAMMI/MSE;
PROBF=1-PROBF(F_AMMI,DFAMMI,DFE);
PROC PRINT DATA=SSAMMI NOOBS;
            VAR SSAMMI PERCENT PERCENAC DFAMMI MSAMMI F_AMMI PROBF;

/* prepare data for plotting */

PROC SORT DATA=RAW;
            BY GEN;
PROC MEANS DATA = RAW NOPRINT;
            BY GEN ;
            VAR YLD;
            OUTPUT OUT = MEDIAG MEAN=YLD;
DATA NAMEG;
            SET MEDIAG;
            TYPE = 'GEN';
            NAME = GEN;
            KEEP TYPE NAME YLD;
PROC SORT DATA=RAW;
            BY ENV;
```

```
PROC MEANS DATA = RAW NOPRINT;
            BY ENV ;
            VAR YLD;
            OUTPUT OUT = MEDIAE MEAN=YLD;
DATA NAMEE;
            SET MEDIAE;
            TYPE = 'ENV';
            NAME1 = 'S'||ENV;
            NAME = COMPRESS(NAME1);
            KEEP TYPE NAME YLD;
DATA NAMETYPE;
            SET NAMEG NAMEE;
DATA BIPLOT0 ;
            MERGE NAMETYPE SCORES;
            DIM1=COL1;
            DIM2=COL2;
            DIM3=COL3;
            DROP COL1-COL3;
data biplot ;
            set biplot0 pol perp;
PROC PRINT DATA=BIPLOT NOOBS;
            VAR TYPE NAME YLD DIM1 DIM2 DIM3;
Data labels;
            set biplot ;
            retain xsys '2' ysys '2' ;
            length function text $8 ;
            text = name ;
                    if type = 'GEN' then do;
                            color='black ';
                            size = 0.6;
                            style = 'hwcgm001';
                            x = dim1;
                            y = dim2;
                                if dim1 >=0
                                    then position='5';
                                    else position='5';
                            function = 'LABEL';
                            output;
                    end;
                    if type = 'ENV' then DO;
                            color='black ';
                            size = 0.6;
                            style = 'hwcgm001';
                            x = 0.0;
                            y = 0.0;
                            function='MOVE';
                            output;
                            x = dim1;
                            y = dim2;
                            function='DRAW' ;
                            output;
                                if dim1 >=0
                                    then position='5';
                                    else position='5';
                            function='LABEL';
                            output;
                            end;
                    if type = "per" then do;
                            color='red';
                            line=2;
```

```
                                 size = 0.6;
                                 style = 'hwcgm001';
                                 x=0.0;
                                 y=0.0;
                                 function='MOVE';
                                 output;
                                 x=v3;
                                 y=v4;
                                 function='DRAW';
                                 output;
                         end;

/* graphing the biplot */
Proc gplot data=biplot;
Plot dim2*dim1 v2*v1 / overlay Annotate=labels frame
            Vref=0.0 Href = 0.0
            cvref=black chref=black
            lvref=3 lhref=3
            vaxis=axis2 haxis=axis1
            vminor=1 hminor=1 nolegend;
            symbol1 v=none c=black h=0.7 ;
            symbol2 v=none c=blue i=j line=3 ;
            axis2
                     length = 6.0 in
                     order = (-1.0 to 1.0 by 0.2)
                     label=(f=hwcgm001 h=1.2 a=90 r=0 'Factor 2')
                     value=(h=0.8)
                     minor=none;
            axis1
                     length = 6.0 in
                     order = (-1.0 to 1.0 by 0.2)
                     label=(f=hwcgm001 h=1.2 'Factor 1')
                     value=(h=0.8)
                     minor=none;
run;
```

Output

```
The SAS System
General Linear Models Procedure
Class Level Information

Class           Levels            Values
ENV             6                 90 91 92 93 94 95
GEN             8                  1 2 3 4 5 6 7 8

Number of observations in data set = 48

General Linear Models Procedure

Dependent Variable:YLD
```

Source	DF	Sum of Squares	Mean Square	FValue	Pr>F
Model	47	36.27979794	0.77191059	.	.
Error	0	.	.		

```
Corrected Total    47   36.27979794
```

R-Square	C.V.	RootMSE	YLDMean
1.000000	0	0	7.053890

Source	DF	TypeIVSS	Mean Square	FValue	Pr>F
ENV	5	16.39991745	3.27998349	.	.
GEN	7	14.25289126	2.03612732	.	.
ENV*GEN	35	5.62698924	0.16077112	.	.

SOURCE	DF	SS	MS	F	PROB
ENV	5	49.1998	9.83995	62.2685	.0000000000
GEN	7	42.7587	6.10838	38.6547	.0000000000
ENV*GEN	35	16.8810	0.48231	3.0521	.0000096813

SSAMMI	PERCENT	PERCENAC	DFAMMI	MSAMMI	F_AMMI	PROBF

TYPE	NAME	YLD	DIM1	DIM2	DIM3
GEN	1	6.13178	0.00956	-0.40614	0.71637
GEN	2	6.45306	-0.20785	0.11243	-0.42252
GEN	3	6.82289	0.54723	0.38688	-0.17175
GEN	4	6.97439	-0.55039	0.48510	0.32201
GEN	5	7.10122	0.79807	0.22696	0.09611
GEN	6	7.50572	-0.50527	0.31346	-0.10796
GEN	7	7.65789	-0.12223	-0.61605	-0.38304
GEN	8	7.78417	0.03088	-0.50265	-0.04923
ENV	S90	7.57075	0.26094	0.88469	-0.35836
ENV	S91	7.03321	0.14700	0.24992	0.80484
ENV	S92	6.28142	-0.39629	-0.30402	0.23481
ENV	S93	7.46617	0.91858	-0.58360	-0.19014
ENV	S94	7.69529	-0.58882	-0.25741	-0.30403
ENV	S95	6.27650	-0.34140	0.01041	-0.18713

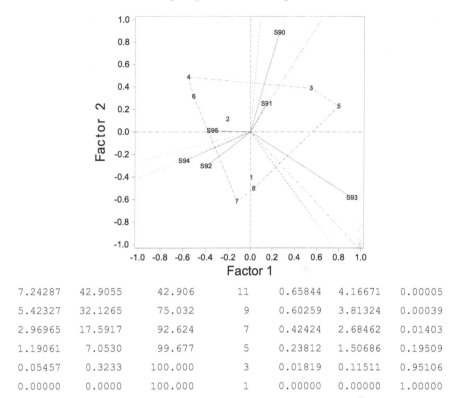

7.24287	42.9055	42.906	11	0.65844	4.16671	0.00005
5.42327	32.1265	75.032	9	0.60259	3.81324	0.00039
2.96965	17.5917	92.624	7	0.42424	2.68462	0.01403
1.19061	7.0530	99.677	5	0.23812	1.50686	0.19509
0.05457	0.3233	100.000	3	0.01819	0.11511	0.95106
0.00000	0.0000	100.000	1	0.00000	0.00000	1.00000

Case 2

Using the SREG model for obtaining the GGE biplot (required changes to the previous program are highlighted in bold):

```
AMMI Model                        SREG Model
PROC GLM DATA=RAW NOPRINT;        PROC GLM DATA=RAW NOPRINT;
CLASS ENV GEN;                       CLASS ENV;
MODEL YLD = ENV GEN/SS4;             MODEL YLD = ENV/SSR;
OUTPUT OUT=OUTRES R=RESID;           OUTPUT OUT=OUTRES R=RESID;
```

The range of the scores should be changed.

Previous codes   Modified codes

Axis2   Axis 2

length = 6.0 in order = (–1 to 1 by 10)

label=(f=hwcgm001 h=1.2 a=90 r=0
  'Factor 2')

value=(h=0.8)

minor=none;

axis 1

  length = 6.0 in order = (–1 to 1 by 10)

  label=(f=hwcgm001 h=1.2 'Factor 1')

  value=(h=0.8)

  minor=none;

length = 6.0 in order = (–1.2 to 1.2 by 10)

label=(f=hwcgm001 h=1.2 a=90 r=0
  'Factor2')

value=(h=0.8)

minor=none;

axis1

  length = 6.0 in order = (–1.2 to 1.2 by 10)

  label=(f=hwcgm001 h=1.2 'Factor 1')

  value=(h=**0.6**)

  minor=none;

Previous codes

if abs(r)<1 then xxx[i,]=**1**;

else xxx[i,]=**1**/abs(r);

Modified codes

if abs(r)<1 then xxx[i,]=**1.2**;

else xxx[i,]=**1.2**/abs(r);

Ouput

```
General Linear Models Procedure
Class Level Information

Class                      Levels    Values
ENV                        6         90 91 92 93 94 95
GEN                        8          1 2 3 4 5 6 7 8

Number of observations in data set = 48

General Linear Models Procedure

Dependent Variable: YLD

                        Sum of        Mean
Source       DF         Squares       Square       FValue    Pr>F
Model        47         36.27979794   0.77191059   .         .
Error        0          .             .
Corrected
 Total       47         36.27979794

R-Square     C.V.       RootMSE       YLDMean
1.000000     0          0             7.053890

Source       DF         TypeIVSS      MeanSquare   FValue    Pr>F
ENV          5          16.39991745   3.27998349   .         .
GEN          7          14.25289126   2.03612732   .         .
ENV*GEN      35          5.62698924   0.16077112   .         .
```

SOURCE	DF	SS	MS	F	PROB
ENV	5	49.1998	9.83995	62.2685	.0000000000
GEN	7	42.7587	6.10838	38.6547	.0000000000
ENV*GEN	35	16.8810	0.48231	3.0521	.0000096813

SSAMMI	PERCENT	PERCENAC	DFAMMI	MSAMMI	F_AMMI	PROBF
43.8940	73.5988	73.599	11	3.99037	25.2516	0.00000
7.1690	12.0206	85.619	9	0.79656	5.0407	0.00002
5.1750	8.6771	94.296	7	0.73929	4.6783	0.00016
2.2756	3.8156	98.112	5	0.45513	2.8801	0.01830
1.0970	1.8394	99.951	3	0.36567	2.3140	0.08092
0.0289	0.0485	100.000	1	0.02894	0.1832	0.66965

TYPE	NAME	YLD	DIM1	DIM2	DIM3
GEN	1	6.13178	-1.16627	-0.05995	-0.64719
GEN	2	6.45306	-0.71482	-0.24510	0.06686
GEN	3	6.82289	-0.31952	0.50436	0.40386
GEN	4	6.97439	-0.12234	-0.60031	0.39015
GEN	5	7.10122	0.02080	0.78280	0.28332
GEN	6	7.50572	0.56288	-0.48905	0.35601
GEN	7	7.65789	0.81001	-0.01615	-0.47082
GEN	8	7.78417	0.92925	0.12340	-0.38218

ENV	S90	7.57075	0.74276	0.23078	0.98338
ENV	S91	7.03321	0.52365	0.09414	0.12375
ENV	S92	6.28142	0.85870	-0.35358	-0.30693
ENV	S93	7.46617	0.84586	0.99020	-0.44343
ENV	S94	7.69529	0.94654	-0.52657	-0.19706
ENV	S95	6.27650	0.80649	-0.31772	0.03713

Chapter 19

Analysis for Regional Trials with Unbalanced Data

Jun Zhu

Purpose

To analyze unbalanced data of regional trials for comparing varieties by linear contrast tests.

Definitions

Statistical Model

Analysis of experimental data from regional trials is based on the following linear model, which regards the genotypic effects (G) as fixed and further partitions the random E ($E = Y + L + YL$) and GE interaction effects ($GE = GY + GL + GYL$) ($h = 1,2,...,g$; $i = 1,2,..., n_h$; $j = 1,2,..., n_{hi}$; $k = 1,2,..., r$):

$$y_{hijk} = G_h + Y_i + L_j + YL_{ij} + GY_{hi} + GL_{hj} + GYL_{hij} + B_{k(ij)} + e_{hijk}$$

where, Y = year effect, L = location effect, YL = year × location interaction effect, GY = genotype × year effect, GL = genotype × location effect, GYL = genotype × year × location interaction effect, B = block effect, and e = residual effect.

Analysis

Balanced data of regional trials can be easily analyzed using ANOVA methods. The experimental data of regional trials, however, are quite often unbalanced because of missing genotype records in specific locations and years. Mixed model approaches can then be applied for analyzing unbalanced data of a single trait and multiple traits from regional trials (Zhu, Lai, and Xu, 1993; Zhu, Xu, and Lai, 1993). The phenotypic data of trait y $(f = 1, 2, \ldots, t)$ can be expressed in a matrix form of a mixed linear model:

$$
\begin{aligned}
y_{(f)} &= Xb_{(f)} + U_Y e_{Y(f)} + U_L e_{L(f)} + U_{YL} e_{YL(f)} + U_{GY} e_{GY(f)} + U_{GL} e_{GL(f)} \\
&\quad + U_{GYL} e_{GYL(f)} + U_B e_{B(f)} + e_{\varepsilon(f)} \\
&= Xb_{(f)} + \sum_{u=1}^{8} U_u e_{u(f)}
\end{aligned}
$$

with variance matrix:

$$
\begin{aligned}
\mathrm{var}(y_{(f)} &= \sigma^2_{Y(f)} U_Y U_Y^T + \sigma^2_{L(f)} U_L U_L^T + \sigma^2_{YL(f)} U_{YL} U_{YL}^T + \sigma^2_{GY(f)} U_{GY} U_{GY}^T \\
&\quad + \sigma^2_{GL(f)} U_{GL} U_{GL}^T + \sigma^2_{GYL(f)} U_{GYL} U_{GYL}^T + \sigma^2_{B(f)} U_B U_B^T + \sigma^2_{\varepsilon(f)} I \\
&= \sum_{u=1}^{8} \sigma^2_{u(f)} U_u U_u^T = V_{(f)}
\end{aligned}
$$

The covariance between traits $y_{(f)}$ and $y_{(f)}$ is

$$
C_{(f, f')} = \sum_{u=1}^{8} \sigma_{u(f)/u(f')} U_u U_u^T .
$$

Both variance $(V_{(f)})$ and covariance $(C_{(ff')})$ matrices can be estimated by the MINQUE(1) method (Zhu, 1992; Zhu and Weir, 1996). Comparison of genotypes for trait f can be conducted by testing linear contrast among genotype effects $(\sum_{h=1}^{g} c_h G_{h(f)})$. The linear contrast can be estimated by

$$
C_{(f)} = c^T \hat{b} = c^T (X^T \hat{V}_{(f)}^{-1} X)^- X^T \hat{V}_{(f)}^{-1} y_{(f)}
$$

with sampling variance $\hat{\sigma}^2 (C_{(f)}) = c^T (X^T \hat{V}_{(f)}^{-1} X)^- c$.

If $|C_{(f)} / \hat{\sigma}(C_{(f)})| > z_{(a/2)}$, reject the null hypothesis $H_0 : \sum_{h=1}^{g} c_h G_{h(f)} = 0$ and accept the alternative hypothesis $H_1 : \sum_{h=1}^{g} c_h G_{h(f)} \neq 0$ at a significance level $= \alpha$.

To compare the weighted genotypic merits of t traits $(\sum_{f=1}^{t} w_f G_{h(f)})$, the weighted linear contrast can be estimated by

$$C_W = \sum_{h=1}^{g} c_h \sum_{f=1}^{t} w_f \hat{G}_{h(f)} = \sum_{f=1}^{t} w_f C_{(f)}$$

with sampling variance

$$\sigma^2(C_W) = \sum_{f=1}^{t} w_f^2 \sigma^2(C_{(f)}) + 2\sum_{f=1}^{t-1} \sum_{f'=f+1}^{t} w_f w_{f'} \sigma(C_{(f)}, C_{(f')})$$

where, $\hat{\sigma}(C_{(f)}, C_{(f')}) = c^T (X^T \hat{C}_{(f,f')}^{-1} X)^- c$ is the covariance between $C_{(f)}$ and $C_{(f)}$.

If $|C_W / \hat{\sigma}(C_W)| > z_{(\alpha/2)}$, reject the null hypothesis $H_0: \sum_{h=1}^{g} c_h \sum_{f=1}^{t} w_f G_{h(f)} = 0$ and accept the alternative hypothesis $H_1: \sum_{h=1}^{g} c_h \sum_{f=1}^{t} w_f G_{h(f)} \neq 0$ at a significant level = α.

Originators

Zhu, J. (1992). Mixed model approaches for estimating genetic variances and covariances. *Journal of Biomathematics* 7(1):1-11.

Zhu, J., Lai, M.G., and Xu, F.H. (1993). Analysis methods for unbalanced data from regional trial of crop variety: Analysis for multiple traits (Chinese). *Journal of Zhejiang Agricultural University* 19(3):241-247.

Zhu, J. and Weir, B.S. (1996). Diallel analysis for sex-linked and maternal effects. *Theoretical and Applied Genetics* 92(1):1-9.

Zhu, J., Xu, F.H., and Lai, M.G. (1993). Analysis methods for unbalanced data from regional trials of crop variety: Analysis for single trait (Chinese). *Journal of Zhejiang Agricultural University* 19(1):7-13.

Software Available

Zhu, J. (1997). GENTEST.EXE a computer software for constructing regional test models, GENTESTM.EXE for analyzing single traits of regional tests, and GENTESTW.EXE for analyzing multiple traits of regional tests. *Analysis Methods for Genetic Models* (pp. 285-292), Agricultural Publication House of China, Beijing (program free of charge). Contact Dr. Jun Zhu, Department of Agronomy, Zhejiang University, Hangzhou, China. E-mail: <jzhu@zju.edu.cn>.

EXAMPLE

Unbalanced data (COTTEST.TXT) to be analyzed (Variety = 3, Year = 2, Location = 8, Blk = 1):

Var	Year	Loca	Blk	Yield	Lint%
1	1	1	1	65.7	40.6
1	1	2	1	55.9	40.0
1	1	3	1	83.3	39.5
1	1	4	1	47.0	38.3
1	1	5	1	63.0	40.7
1	1	6	1	26.1	37.6
1	2	1	1	64.7	39.3
1	2	2	1	61.9	40.5
1	2	3	1	58.2	40.0
1	2	4	1	45.3	38.6
1	2	5	1	56.7	38.8
1	2	6	1	44.8	37.3
1	2	7	1	46.7	40.2
1	2	8	1	52.1	39.0
2	1	1	1	64.3	44.0
2	1	2	1	64.2	42.7
2	1	3	1	69.7	43.8
2	1	4	1	34.3	40.3
2	1	5	1	59.4	43.9
2	1	6	1	63.3	42.3
2	1	7	1	59.1	43.3
2	1	8	1	76.2	44.5
2	2	1	1	65.7	42.7
2	2	2	1	78.4	44.5
2	2	3	1	66.6	43.5
2	2	4	1	48.6	41.4
2	2	5	1	70.0	42.3
2	2	6	1	61.0	40.4
2	2	7	1	63.0	45.1
2	2	8	1	73.6	45.0
3	1	1	1	61.4	41.9
3	1	2	1	75.9	38.8
3	1	3	1	75.3	40.0
3	1	4	1	61.3	37.6
3	1	5	1	64.1	40.4
3	1	6	1	57.8	38.2
3	1	7	1	86.8	40.5
3	1	8	1	64.8	40.4
3	2	3	1	72.3	40.0

3	2	4	1	50.7	38.1
3	2	5	1	52.7	38.8
3	2	6	1	63.3	37.8
3	2	7	1	72.0	40.3
3	2	8	1	73.2	40.0

1. Use GENTEST.EXE for generating mating design matrix and data. Before running these programs, create a file for your analysis with four design columns, followed by trait columns. The four design columns are: variety, year, location, and block. There is a limitation (<100 traits) for the number of trait columns.
2. Run GENTESTM.EXE for analyzing each trait. Standard errors of estimates are calculated by jackknifing over locations for stability testing. Always run GENTESTM.EXE before analyzing multiple traits. This program will allow you to choose data transformation based on check variety. You will also need to input coefficients (1, 0, or −1) for conducting linear contrasts for different varieties. The results will be automatically stored in a file named COTTEST.VAR for analysis of single traits.
3. After you finish analysis for each trait, run GENTESTW.EXE for combining analysis of all traits studied. This program will allow you to choose weight coefficients for each trait. The results will be automatically stored in a file named COTTEST.VAR for analysis of multiple traits.

Output 1 for Single Trait Test

```
Traits =, 2
 Variance components =, 6
 File name is cottest.VAR
 Date and Time for Analysis: Thu Jun 22 20:36:15 2000

 Variance Components Estimated by MINQUE(1) with GENTESTW.EXE.
Contrast 1:  +  +  -
Contrast 2:  +  -
Contrast 3:  +  -  -

Analysis of trait Yield
Estimate of Var(Y) =, 0
Estimate of Var(L) =, 43.7906
Estimate of Var(YL) =, 0
Estimate of Var(GY) =, 1.79863
Estimate of Var(GL) =, 31.5667
Estimate of Var(e) =, 55.9106
```

```
Mean of Variety:,      Mean,    S.E.
Mean of Variety 1 =,   55.1,      3.86105
Mean of Variety 2 =,   63.5875,   3.71664
Mean of Variety 3 =,   66.5429,   3.86105


( 3),   V 1 ,   55.1000,      A
( 2),   V 2 ,   63.5875,   a   AB
( 1),   V 3 ,   66.5429,   a   B


Contrast,     C-value,       S.E.,       Standard Normal z-value
(1) This Linear Contrast Test Is for Varieties: (V1, V2) vs. (V3)
Contrast 1,  -14.398222,   7.480348,   1.924806
(2) This Linear Contrast Test Is for Varieties: (V1) vs. (V2)
Contrast 2,  -8.487495,    4.215852,   2.013234
(3) This Linear Contrast Test Is for Varieties: (V1) vs. (V2, V3)
Contrast 3,  -19.930346,   7.480348,   2.664361

Stability Analysis for Variety
Estimates and S.E. are obtained by Jackknifing over environments.

Stability Analysis for Variety 1:
 a = -25.4509, S.E. = 24.3941, 0.95 C.I. is < -73.2633 & 22.3616 >
 b = 1.32148, S.E. = 0.3862, 0.95 C.I. is < 0.564532 & 2.07844 >
 r = 0.83176, S.E. = 0.0856267, 0.95 C.I. is < 0.663932 & 0.999588 >

Stability Analysis for Variety 2:
 a = 8.36712, S.E. = 25.0719, 0.95 C.I. is < -40.7739 & 57.5081 >
 b = 0.879325, S.E. = 0.386812, 0.95 C.I. is < 0.121172 & 1.63748 >
 r = 0.718145, S.E. = 0.157704, 0.95 C.I. is < 0.409044 & 1.02725 >

Stability Analysis for Variety 3:
 a = 16.9362, S.E. = 13.7335, 0.95 C.I. is < -9.98134 & 43.8538 >
 b = 0.804752, S.E. = 0.227599, 0.95 C.I. is < 0.358659 & 1.25085 >
 r = 0.740325, S.E. = 0.104486, 0.95 C.I. is < 0.535531 & 0.945118 >

Stability in Order for Variety
Order by b ( 3),   V 3 ,   a = 16.9362 ,    b = 0.8048 ,   r = 0.7403
Order by b ( 2),   V 2 ,   a = 8.3671 ,     b = 0.8793 ,   r = 0.7181
Order by b ( 1),   V 1 ,   a = -25.4509 ,   b = 1.3215 ,   r = 0.8318

Analysis of trait Lint%
Estimate of Var(Y) =, 0
Estimate of Var(L) =, 0.812176
Estimate of Var(YL) =, 0.495318
Estimate of Var(GY) =, 0
Estimate of Var(GL) =, 0.175569
Estimate of Var(e) =, 0.244097

Mean of Variety:,      Mean,    S.E.
Mean of Variety 1 =, 39.3143,  0.428769
Mean of Variety 2 =, 43.1063,  0.411924
Mean of Variety 3 =, 39.4857,  0.428769


( 3),   V 1 ,   39.3143,   a   A
( 2),   V 3 ,   39.4857,   a   A
( 1),   V 2 ,   43.1063,
```

```
Contrast,  C-value,    S.E., Standard Normal z-value
(1) This Linear Contrast Test Is for Varieties: (V1, V2) vs. (V3)
Contrast 1,  3.449110,  0.552117,  6.247059
(2) This Linear Contrast Test Is for Varieties: (V1) vs. (V2)
Contrast 2,  -3.791962,  0.297600,  12.741823
(3) This Linear Contrast Test Is for Varieties: (V1) vs. (V2, V3)
Contrast 3,  -3.963402,  0.552117,  7.178548

Stability Analysis for Variety
Estimates and S.E. are obtained by Jackknifing over environments.

Stability Analysis for Variety 1:
 a = 7.89727,  S.E. = 4.23156, 0.95 C.I. is < -0.396587 & 16.1911 >
 b = 0.772582, S.E. = 0.103208, 0.95 C.I. is < 0.570295 & 0.974869 >
 r = 0.905533, S.E. = 0.0482835, 0.95 C.I. is < 0.810898 & 1.00017 >

Stability Analysis for Variety 2:
 a = 0.177973, S.E. = 4.68567, 0.95 C.I. is < -9.00594 & 9.36189 >
 b = 1.05051,  S.E. = 0.115322, 0.95 C.I. is < 0.824481 & 1.27654 >
 r = 0.917149, S.E. = 0.0463384, 0.95 C.I. is < 0.826326 & 1.00797 >

Stability Analysis for Variety 3:
 a = 2.71998,  S.E. = 4.52641, 0.95 C.I. is < -6.15178 & 11.5917 >
 b = 0.902994, S.E. = 0.113154, 0.95 C.I. is < 0.681213 & 1.12478 >
 r = 0.937269, S.E. = 0.0332955, 0.95 C.I. is < 0.87201 & 1.00253 >

Stability in Order for Variety
Order by b ( 3),  V 1 ,  a = 7.8973 ,  b = 0.7726 ,  r = 0.9055
Order by b ( 2),  V 3 ,  a = 2.7200 ,  b = 0.9030 ,  r = 0.9373
Order by b ( 1),  V 2 ,  a = 0.1780 ,  b = 1.0505 ,  r = 0.9171

Time Used (Hour) = 0.009722
```

Output 2 for Multiple Trait Test

```
Traits =, 2
 Variance components =, 6
 File name is cottest.COV
 Date and Time for Analysis: Thu Jun 22 20:38:33 2000

 Variance Components Estimated by MINQUE(1) with GENTESTW.EXE.

<W1>: 0.6, <W2>: 0.4,

Analysis for Public Users

Estimated Var for <Yield>
Estimate for Var(Y) =,  -7.04805
Estimate for Var(L) =,  98.8959
Estimate for Var(YL) =, -1.89563
Estimate for Var(GY) =,  4.06199
Estimate for Var(GL) =, 71.2897
Estimate for Var(e) =, 126.267
```

```
Estimated Cov for <Yield> & <Lint%>
Estimate for Cov (Y) =,  -0.0680642
Estimate for Cov (L) =,  29.2247
Estimate for Cov (YL) =, -2.47928
Estimate for Cov (GY) =,  1.12331
Estimate for Cov (GL) =,  1.57541
Estimate for Cov (e) =,   4.52009

Estimated Var for <Lint%>
Estimate for Var(Y) =,  -0.280077
Estimate for Var(L) =,   5.20919
Estimate for Var(YL) =, 3.1769
Estimate for Var(GY) =,-0.097269
Estimate for Var(GL) =, 1.12607
Estimate for Var(e) =,   1.56561

Analysis for multiple traits:
Combined Variety Mean:
Mean of Variety:,    Mean,   S.E.
Mean of Variety 1 =,  89.5086, 3.82078
Mean of Variety 2 =, 101.003,  3.68336
Mean of Variety 3 =, 100,      3.82078

 ( 3),  V 1 ,  89.5086,   a   A
 ( 2),  V 3 ,  100.0000,  a   A
 ( 1),  V 2 ,  101.0029,  a   A

Contrast,  C-value,   S.E.  , Standard Normal z-value
(1) This Linear Contrast Test Is for Varieties:
Cont. 1,  -9.488488,  49.575051,  0.191396
(2) This Linear Contrast Test Is for Varieties:
Cont. 2,  -11.494320,  15.667476,  0.733642
(3) This Linear Contrast Test Is for Varieties:
Cont. 3,  -21.985739,  49.575054,  0.443484

Stability Analysis for Variety
Estimates and S.E. are obtained by Jackknifing over environments.

Stability Analysis for Variety 1:
 a = -28.7855, S.E. = 34.6409, 0.95 C.I. is < -96.6817 & 39.1106 >
 b = 1.23013, S.E. = 0.351782, 0.95 C.I. is < 0.540642 & 1.91963 >
 r = 0.827958, S.E. = 0.0706079, 0.95 C.I. is < 0.689566 & 0.966349 >

Stability Analysis for Variety 2:
 a = 11.0869, S.E. = 35.9274, 0.95 C.I. is < -59.3307 & 81.5045 >
 b = 0.917704, S.E. = 0.35883, 0.95 C.I. is < 0.214397 & 1.62101 >
 r = 0.753344, S.E. = 0.1381, 0.95 C.I. is < 0.482668 & 1.02402 >

Stability Analysis for Variety 3:
 a = 21.659, S.E. = 19.4541, 0.95 C.I. is < -16.4711 & 59.7891 >
 b = 0.808956, S.E. = 0.204255, 0.95 C.I. is < 0.408615 & 1.2093 >
 r = 0.781107, S.E. = 0.0871521, 0.95 C.I. is < 0.610289 & 0.951926 >

Stability in Order for Variety
Order of b ( 3),  V 3 ,  a = 21.6590 ,  b = 0.8090 ,  r = 0.7811
Order of b ( 2),  V 2 ,  a = 11.0869 ,  b = 0.9177 ,  r = 0.7533
Order of b ( 1),  V 1 ,  a = -28.7855 ,  b = 1.2301 ,  r = 0.8280

Time Used (Hour) = 0.006667
```

Chapter 20

Conditional Mapping of QTL with Epistatic Effects and QTL-by-Environment Interaction Effects for Developmental Traits

Jun Zhu

Purpose

To map quantitative trait loci (QTL) for net effects due to gene expression from time $t - 1$ to t.

Definitions

Genetic Model

For multiple-environment data of doubled haploid (DH) or recombinant inbred line (RIL) populations, the conditional phenotypic value of the *j*th genetic entry in environment *h* at time *t*, given phenotypic value at time $t - 1$, can be expressed as the following conditional genetic model:

$$
\begin{aligned}
y_{hj(t|t-1)} = {} & \mu_{(t|t-1)} + a_{1(t|t-1)} x_{A_{1j}} + a_{2(t|t-1)} x_{A_{2j}} + aa_{(t|t-1)} x_{AA_j} \\
& + u_{E_{hj}} e_{E_{h(t|t-1)}} + u_{A_1 E_{hj}} e_{A_1 E_{h(t|t-1)}} + u_{A_2 E_{hj}} e_{A_2 E_{h(t|t-1)}} + u_{AAE_{hj}} e_{AAE_{h(t|t-1)}} \\
& + \sum_f u_{M_{fj}} e_{M_{f(t|t-1)}} + \sum_l u_{MM_{lj}} e_{MM_{l(t|t-1)}} + \sum_p u_{ME_{hpj}} e_{ME_{hp(t|t-1)}} \\
& + \sum_q u_{MME_{hqj}} e_{MME_{hq(t|t-1)}} + \varepsilon_{hj(t|t-1)}
\end{aligned}
$$

where $\mu_{(t|t-1)}$ is the conditional population mean; $a_{1(t|t-1)}$ and $a_{2(t|t-1)}$ are the conditional additive effects of loci Q_1 and Q_2, respectively; $aa_{(t|t-1)}$ is the conditional additive × additive epistatic effect of loci Q_1 and Q_2; $x_{A_{1j}}$, $x_{A_{2j}}$, and x_{AA_j} are coefficients of these conditional genetic main effects; $e_{E_{h(t|t-1)}}$ is the conditional random effect of environment h with coefficient $u_{E_{hj}}$; $e_{A_1E_{h(t|t-1)}}$ (or $e_{A_2E_{h(t|t-1)}}$) is the conditional additive × environment interaction effect with coefficient $u_{A_1E_{hj}}$ (or $u_{A_2E_{hj}}$) for Q_1 (or Q_2); $e_{AAE_{h(t|t-1)}}$ is the conditional epistasis × environment interaction effect with coefficient $u_{AAE_{hj}}$; e_{M_f} is the conditional marker main effect with coefficient u_{M_j}; $e_{MM_{l(t|t-1)}}$ is the conditional marker × marker interaction effect with coefficient u_{MM_j}; $e_{ME_{hp(t|t-1)}}$ is the conditional marker × environment interaction effect with coefficient $u_{ME_{hpj}}$; $e_{MME_{hq(t|t-1)}}$ is the marker × marker × environment interaction effect with coefficient $u_{MME_{qhj}}$; and $\varepsilon_{hj(t|t-1)}$ is the conditional residual effect.

Mixed Linear Model

The conditional epistasis QTL model can be expressed in the matrix form as follows:

$$
\begin{aligned}
y_{(t|t-1)} &= Xb_{(t|t-1)} + U_E e_{E(t|t-1)} + U_{A_1E} e_{A_1E)t|t-1)} + U_{A_2E} e_{A_2E(t|t-1)} \\
&\quad + U_{AAE} e_{AAE(t|t-1)} + U_M e_{M(t|t-1)} + U_{MM} e_{MM(t|t-1)} \\
&\quad + U_{ME} e_{ME(t|t-1)} + U_{MME} e_{MME(t|t-1)} + e_{\varepsilon(t|t-1)} \\
&= Xb_{(t|t-1)} + \sum_{u=1}^{9} U_u e_{u(t|t-1)} \\
&\sim N(Xb_{(t|t-1)}, V_{(t|t-1)} = \sum_{u=1}^{9} \sigma^2_{u(t|t-1)} U_u R_u U_u^T)
\end{aligned}
$$

where $y_{(t|t-1)}$ is the conditional phenotype vector; $b_{(t|t-1)}$ is the conditional fixed parameter vector for conditional population mean and QTL effects; X is the known incidence matrix of the fixed parameters; $e_{1(t|t-1)} = e_{E(t|t-1)} \sim N(0, \sigma^2_{E(t|t-1)}I)$ is the vector of conditional environment effects; $e_{2(t|t-1)} = e_{A_1E(t|t-1)} \sim N(0, \sigma^2_{A_1E(t|t-1)}I)$ is the vector of conditional $A_1 \times E$ interaction effects; $e_{3(t|t-1)} = e_{A_2E(t|t-1)} \sim N(0, \sigma^2_{A_2E(t|t-1)}I)$ is the vector of conditional $A_2 \times E$ interaction effects; $O, e_4(t/t-1) = e_{AAE}(t/t-1) \sim N(0, \sigma^2_{AAE(t|t-1}R_{AAE})$ is the vector of conditional $AA \times E$ interaction effects;

$e_{5(t|t-1)} = e_{M(t|t-1)} \sim N(0, \sigma^2_{M(t|t-1)} R_M)$ is the vector of conditional marker main effects; $e_{6(t|t-1)} = e_{MM(t|t-1)} \sim N(0, \alpha^2_{MM(t|t-1)} R_{MM})$ is the vector of conditional interaction marker main effects; $e_{7(t|t-1)} = e_{ME(t|t-1)} \sim N(0, \sigma^2_{MM(t|t-1)} R_{ME})$ is the vector of conditional $M \times E$ interaction effects; $e_{8(t|t-1)} = e_{MME(t|t-1)} \sim N(0, \sigma^2_{MME(t|t-1)} R_{MME})$ is the vector of conditional $MM \times E$ interaction effects; $e_{9(t|t-1)} = e_{\varepsilon(t|t-1)} \sim N(0, \sigma^2_{\varepsilon(t-1)} I)$ is the vector of conditional residual effects; $U_u (u=1, 2, ..., 8)$ is the known incidence matrix of the conditional random effects, and $U_9 = I$.

Analysis Methodology

With observed phenotypic data at time $t-1$ ($y_{(t-1)}$) and time t ($y_{(t)}$), conditional phenotypic data $y_{(t|t-1)}$ can be obtained via mixed model approaches (Zhu, 1995). Then a mixed-model-based composite interval mapping (MCIM) can be used for mapping QTLs with conditional epistatic effects and QTL × environment interaction effects (Zhu, 1998; Zhu and Weir, 1998; Wang et al., 1999). The likelihood function (L) for the parameters of conditional fixed effects $b_{(t|t-1)}$ and conditional variance components $[\sigma^2_{u(t|t-1)}]$ is

$$L(b_{(t|t-1)}, V_{(t|t-1)}) = 2\pi)^{-\frac{n}{2}} \left| V_{(t|t-1)} \right|^{-\frac{1}{2}}$$
$$\times \exp\left[-\tfrac{1}{2}(y_{(t|t-1)} - Xb_{(t|t-1)})^T V^{-1}_{(t|t-1)}(y_{(t|t-1)} - Xb_{(t|t-1)}) \right]$$

with the log of the likelihood function (l)

$$l(b_{(t|t-1)}, V_{(t|t-1)}) = -\tfrac{n}{2}\ln(2\pi) - \tfrac{1}{2}\ln\left| V_{(t|-1)} \right| - \tfrac{1}{2}(y_{(t|t-1)}$$
$$- Xb_{(t|t-1)})^T V^{-1}_{(t|t-1)}(y_{(t|t-1)} - Xb_{(t|t-1)}).$$

For searching QTL, null hypothesis for genetic parameters (conditional QTL main effects and *QE* interaction effects) can be tested by the likelihood ratio statistic (*LR*):

$$LR = 2l_1(\hat{b}_{(t|t-1)1}, V_{(t|t-1)1}) - 2l_0(\hat{b}_{(t|t-1)0}, V_{(t|t-1)0}).$$

The maximum likelihood estimates of QTL effects in $b_{(t|t-1)}$ can be obtained by

$$\hat{b}_{(t|t-1)} = \left(X^T V^{-1}_{(t|t-1)} X \right)^{-1} X^T V^{-1}_{(t|t-1)} y_{(t|t-1)}$$

with variance-covariance matrix

$$\mathrm{var}(\hat{b}_{(t|t-1)}) = X^T V^{-1}_{(t|t-1)} X)^{-1}.$$

Conditional QE interaction effects (conditional additive × environment interaction $e_{A_iE(t|t-1)}$ and $e_{A_jE(t|t-1)}$, conditional epistasis × environment interaction $e_{AA_{ij}E(t|t-1)}$) can be obtained by the best linear unbiased prediction (BLUP) method:

$$\hat{e}_{u(t|t-1)} = \sigma^2_{u(t|t-1)} U^T_u Q_{(t|t-1)} y_{(t|t-1)}$$

with variance-covariance matrix

$$\mathrm{var}(\hat{e}_{u(t|t-1)} = \sigma^4_{u(t|t-1)} U^T_u Q_{(t|t-1)} U_u$$

where $Q_{(t|t-1)} = V^{-1}_{(t|t-1)} - V^{-1}_{(t|t-1)} X (X^T V^{-1}_{(t|t-1)} X)^{-1} X^T V^{-1}_{(t|t-1)}.$

Originators

Wang, D., Zhu, J., Li, Z.K., and Paterson, A.H. (1999). Mapping QTLs with epistatic effects and QTL × environment interactions by mixed linear model approaches. *Theoretical and Applied Genetics* 99:1255-1264.

Zhu, J. (1995). Analysis of conditional effects and variance components in developmental genetics. *Genetics* 141(4):1633-1639.

Zhu, J. (1998). Mixed model approaches of mapping genes for complex quantitative traits. In Wang, L.Z. and Dai J.R. (Eds.), *Proceedings of Genetics and Crop Breeding of China* (pp.19-20). Chinese Agricultural Science and Technology Publication House, Beijing.

Zhu, J. and Weir, B.S. (1998). Mixed model approaches for genetic analysis of quantitative traits. In Chen, L.S., Ruan, S.G., and Zhu, J. (Eds.), *Advanced Topics in Bio-*

mathematics: Proceedings of International Conference on Mathematical Biology (pp. 321-330).World Scientific Publishing Co., Singapore.

Software Available

Wang, D., Zhu, J., Li, Z.K., and Paterson, A.H. (1999). QTLMapper Version 1.0: A computer software for mapping quantitative trait loci (QTLs) with additive effects, epistatic effects and QTL × environment interactions. *User Manual for QTLMapper Version 1.0* (program free of charge). Contact Dr. Jun Zhu, Department of Agronomy, Zhejiang University, Hangzhou, China. E-mail: <jzhu@zju.edu.cn>.

Zhu, J. (1997). GENCOND1.EXE, a computer software for calculating conditional phenotypic data. *Analysis Methods for Genetic Models* (pp. 278-285), Agricultural Publication House of China, Beijing (program free of charge).

Chapter 21

Mapping QTL with Epistatic Effects and QTL-by-Environment Interaction Effects

Jun Zhu

Purpose

To map quantitative trait loci (QTL) with additive, epistatic, and QTL-by-environment interaction effects for doubled haploid (DH) or recombinant inbred line (RIL) populations.

Definitions

Genetic Model

If multiple-environment data of DH or RIL populations are used for mapping QTL, the phenotypic value of the jth genetic entry in environment h can be expressed as shown in the genetic model

$$
\begin{aligned}
y_{hj} = {} & \mu + a_1 x_{A_{1j}} + a_2 x_{A_{2j}} + aa x_{AA_j} \\
& + u_{E_{hj}} e_{E_h} + u_{A_1 E_{hj}} e_{A_1 E_h} + u_{A_2 E_{hj}} e_{A_2 E_h} + u_{AAE_{hj}} e_{AAE_h} \\
& + \sum_f u_{M_{fj}} e_{M_f} + \sum_l u_{MM_{lj}} e_{MM_l} + \sum_p u_{ME_{hpj}} e_{ME_{hp}} + \sum_q u_{MME_{hqj}} e_{MME_{hq}} + \varepsilon_{hj}
\end{aligned}
$$

where μ is the population mean; a_1 and a_2 are the additive effects of loci Q_1 and Q_2, respectively; aa is the additive $\times$ additive epistatic effect of loci Q_1 and Q_2; $x_{A_{1j}}$, $x_{A_{2j}}$, and x_{AA_j} are coefficients of these genetic main effects; e_{E_h} is the random effect of environment h with coefficient $u_{E_{hj}}$; $e_{A_1 E_h}$ (or $e_{A_2 E_h}$) is the additive $\times$ environment interaction effect with coefficient $u_{A_1 E_{hj}}$

(or $u_{A_2E_{hj}}$) for Q_1 (or Q_2); e_{AAE_h} is the epistasis × environment interaction effect with coefficient $u_{AAE_{hj}}$; e_{M_f} is the marker main effect with coefficient u_{M_f}; e_{MM_l} is the marker × marker interaction effect with coefficient u_{MM_l}; $e_{ME_{hp}}$ is the marker × environment interaction effect with coefficient $u_{ME_{hpj}}$; $e_{MME_{hq}}$ is the marker × marker × environment interaction effect with coefficient $u_{MME_{qhj}}$; and ε_{hj} is the residual effect.

Mixed Linear Model

The epistatic QTL model can be expressed in matrix form as

$$
\begin{aligned}
y &= Xb + U_E e_E + U_{A_1E} e_{A_1E} + U_{A_2E} e_{A_2E} + U_{AAE} e_{AAE} \\
&\quad + U_M e_M + U_{MM} e_{MM} + U_{ME} e_{ME} + U_{MME} e_{MME} + e_\varepsilon \\
&= Xb + \sum_{u=1}^{9} U_u e_u \\
&\sim N(Xb, V = \sum_{u=1}^{9} \sigma_u^2 U_u R_u U_u^T)
\end{aligned}
$$

where y is the phenotype vector; b is the fixed parameter vector for population mean and QTL effects; X is the known incidence matrix of the fixed parameters; $e_1 = e_E \sim N(0, \sigma_E^2 I)$ is the vector of environment effects; $e_2 = e_{A_1E} \sim N(0, \sigma_{A_1E}^2 I)$ is the vector of A1 × E interaction effects; $e_3 = e_{A_2E} \sim N(0, \sigma_{A_2E}^2 I)$ is the vector of A2 × E interaction effects; $e_4 = e_{AAE} \sim N(0, \sigma_{AAE}^2 R_{AAE})$ is the vector of AA × E interaction effects; $e_5 = e_M \sim N(0, \sigma_M^2 R_M)$ is the vector of marker main effects; $e_6 = e_{MM} \sim N(0, \sigma_{MM}^2 R_{MM})$ is the vector of interaction marker main effects; $e_7 = e_{ME} \sim N(0, \sigma_{ME}^2 R_{ME})$ is the vector of M × E interaction effects; $e_8 = e_{MME} \sim N(0, \sigma_{MME}^2 R_{MME})$ is the vector of MM × E interaction effects; $e_9 = e_\varepsilon \sim N(0, \sigma_\varepsilon^2 I)$ is the vector of residual effects; $U_u (u = 1, 2, \ldots, 8)$ is the known incidence matrix of the random effects, and $U_9 = I$.

Analysis Methodology

An approach of mixed-model-based composite interval mapping (MCIM) can be constructed for handling epistatic effects and QTL × environment interaction effects. The likelihood function (L) for the parameters of fixed effects b and variance components $[\sigma_u^2]$ is

$$L(b,V) = (2\pi)^{-\frac{n}{2}} |V|^{-\frac{1}{2}} \exp\left[-\tfrac{1}{2}(y-Xb)^T V^{-1}(y-Xb)\right]$$

with the log of the likelihood function (l)

$$l(b,V) = -\tfrac{n}{2}\ln(2\pi) - \tfrac{1}{2}\ln|V| - \tfrac{1}{2}(y-Xb)^T V^{-1}(y-Xb).$$

For searching QTL, the null hypothesis for genetic parameters (QTL main effects and Q × E interaction effects) can be tested by the likelihood ratio statistic (LR):

$$LR = 2l_1(\hat{b}_1, v_1) - 2l_0(\hat{b}_0, v_0).$$

The maximum likelihood estimates of QTL effects in b can be obtained by

$$\hat{b} = (X^T V^{-1} X)^{-1} X^T V^{-1} y$$

with variance-covariance matrix

$$\text{var}(\hat{b}) = (X^T V^{-1} X)^{-1}.$$

Q × E interaction effects (additive × environment interaction e_{A_iE} and e_{A_jE}, epistasis × environment interaction $e_{AA_{ij}E}$) can be obtained by the best linear unbiased prediction (BLUP) method:

$$\hat{e}_u = \sigma_u^2 U_u^T Qy$$

with variance-covariance matrix

$$\text{var}(\hat{e}_u) = \sigma_u^4 U_u^T Q U_u$$

where $Q = V^{-1} - V^{-1} X (X^T V^{-1} X)^{-1} X^T V^{-1}$.

Originators

Wang, D., Zhu, J., Li, Z.K., and Paterson, A.H. (1999). Mapping QTLs with epistatic effects and QTL × environment interactions by mixed linear model approaches. *Theoretical and Applied Genetics* 99:1255-1264.

Zhu, J. (1998). Mixed model approaches of mapping genes for complex quantitative traits. In Wang, L.Z. and Dai, J.R. (Eds.), *Proceedings of Genetics and Crop Breeding in China* (pp. 19-20). Chinese Agricultural Science and Technology Publication House, Beijing.

Zhu, J. and Weir, B.S. (1998). Mixed model approaches for genetic analysis of quantitative traits. In Chen, L.S., Ruan, S.G., and Zhu, J. (Eds.), *Advanced Topics in Biomathematics: Proceedings of International Conference on Mathematical Biology* (pp. 321-330). World Scientific Publishing Co., Singapore.

Software Available

Wang, D., Zhu, J., Li, Z.K., and Paterson, A.H. (1999). *User Manual for QTLMapper Version 1.0: A Computer Software for Mapping Quantitative Trait Loci (QTLs) with Additive Effects, Epistatic Effects and QTL × Environment Interactions* (program free of charge). Contact Dr. Jun Zhu, Department of Agronomy, Zhejiang University, Hangzhou, China. E-mail: <jzhu@zju.edu.cn>.

EXAMPLE

Data of DH population with ninety-six lines and fifty-four markers on three chromosomes (provided by Drs. N. Huang and P. Wu). Data analysis method is described in detail in the user manual for QTLMapper Version 1.0 (Wang et al., 1999).

Data file (ckge.map) for map information:

```
_Chromosomes 3
_MarkerNumbers 18 15 21
_DistanceUnit cM

*MapBegin*
Marker#      ch1           ch2           ch3
1             0             0             0
2            19.236        12.9949        7.7618
3            16.2488        5.3402       13.2518
4             4.8552       22.2875        6.9239
5             4.8047       27.7327        9.8037
```

```
6          15.3881         6.3438          2.7929
7          15.5969        29.4517         17.5239
8          15.0048        10.2825         41.7545
9           3.8375         8.9339         37.3036
10          3.2747        12.824          15.8394
11         34.4392         8.4598         18.7639
12          2.5322         5.1683          2.5121
13         23.7979        10.1262          5.0168
14          8.2644         5.2896         28.9405
15         13.3483        13.2089          1.9109
16         33.5319                        22.7256
17          2.5622                        15.2455
18          9.2129                        32.48
19                                         7.1483
20                                         9.4924
21                                        18.718
```

MapEnd

Data file for marker and trait information:

```
_Population DH
_Genotypes  96
_Observations 192
_Environments  yes
_Replications  no
_TraitNumber  5
_TotalMarker  54
_MarkerCode  P1=1  P2=2  F1=3  F1P1=4  F1P2=5
```

MarkerBegin

Ind	M1	M2	M3	M4	M5	M6	M7	M8	M9	M10	M11	M12	M13	M14	M15	M16	M17	M18	M19	M20	M21	M22	M23	M24	M25	M26	M27
1	2	1	1	1	1	1	1	2	2	2	2	2	2	2	1	1	1	2	2	2	2	1	1	2	2	2	
2	1	1	1	1	1	1	1	2	2	2	2	1	1	1	1	1	1	2	2	2	2	2	1	1	1		
3	1	1	2	2	2	2	2	2	2	1	1	1	1	1	1	1	1	2	2	2	1	1	1	1	1		
4	2	2	2	2	2	2	2	2	1	.	1	1	1	2	2	2	2	2	2	1	1	2	2	2			
5	2	2	1	.	1	1	1	2	2	2	.	.	2	.	1	1	1	2	2	2	.	2	2	2	2		
6	1	1	2	2	2	2	2	2	1	1	.	1	1	2	2	2	2	1	1	1	1	1	2	2			
7	1	1	2	.	2	2	2	2	2	2	2	2	1	1	1	1	2	2	2	2	2	1	2	2			
8	1	2	2	2	2	2	2	2	1	1	1	1	1	1	1	1	2	2	2	2	2	2	2	2			
9	2	2	2	.	2	2	2	2	1	1	1	1	1	1	1	2	2	2	2	1	1	1	1				
10	1	2	2	.	2	2	2	2	2	2	2	2	2	2	1	1	1	1	1	2	2	2					
11	2	2	2	.	2	2	2	2	1	1	.	1	2	2	2	2	2	2	1	1	1	1					
12	1	1	1	1	1	1	1	1	1	1	1	2	2	1	1	1	1	2	2	2	2	2	1				
13	2	2	.	2	2	2	2	2	2	2	2	.	1	1	1	1	2	2	2	2	2	2	2	2			
14	2	1	1	1	1	1	2	2	2	2	2	2	2	2	2	2	1	1	2	1	1	1	2	2			
15	2	2	2	2	2	2	2	2	2	2	1	1	2	2	2	2	2	2	2	2	2	2	2				
16	2	2	2	2	2	2	2	2	2	2	2	2	2	2	1	1	1	1	2	2	2	2	2				
17	2	2	2	2	2	2	2	2	2	2	2	2	2	2	2	1	2	2	2	2	2	2	2				
18	1	1	2	2	2	2	2	2	2	1	1	1	1	1	1	2	2	2	2	1	1	1	1				
19	1	1	.	1	1	1	2	2	2	1	.	.	2	2	1	1	1	1	.	2	.	2	1	1			
20	1	1	2	2	2	2	2	2	2	2	2	2	1	2	2	2	2	1	1	1	2	2					
21	2	2	2	2	2	1	2	2	2	2	2	2	2	1	1	1	2	1	1	1	1	2	2				
22	1	2	2	2	2	2	1	1	1	1	1	.	1	1	1	1	1	2	2	2	2	2	1	1			
23	1	1	1	.	1	2	2	2	2	1	1	1	1	1	2	2	2	2	2	2	2	2	2	2			
24	2	2	2	.	.	2	2	2	1	2	1	1	1	1	2	1	2	1	1	.	2	2	2	2	.		
25	2	2	2	2	2	2	2	2	2	1	1	.	1	1	1	1	1	1	1	1	1	1	2	2	2		

```
26  1 1 1 1 1 2 2 2 2 2 1 1 1 1 1 2 2 2 2 1 1 1 2 2 2 2 2
27  1 1 1 . 2 2 2 2 2 2 2 2 2 2 . 2 2 2 1 2 2 2 2 2 2 2 2
28  1 2 1 1 1 1 1 1 1 1 1 1 1 1 . 1 1 1 1 2 2 2 1 1 1 1 1
29  1 1 2 2 2 2 2 2 2 2 1 1 2 2 . 2 2 2 1 1 1 1 2 2 2 2 2
30  1 1 2 1 2 1 1 1 1 1 1 1 2 2 2 2 2 1 1 1 1 1 1 1 . 1
31  2 2 2 2 2 2 2 2 2 2 2 2 2 2 2 2 2 2 2 2 . 2 . 2 1 2 2
32  2 2 . 1 2 . 1 2 2 2 2 2 2 2 2 2 2 . 1 1 1 . 1 1 . 1
33  2 2 2 2 2 2 2 2 2 2 1 1 1 1 2 2 2 2 2 2 2 2 1 1 1 . 1
34  . 2 2 . 2 2 2 2 2 2 2 2 2 2 2 1 1 1 1 1 . 1 1 2 2 2 2
35  . 2 2 2 2 2 2 2 2 2 1 1 1 . 1 1 2 1 1 . 2 2 1 2 . 1
36  1 1 . 1 1 1 1 1 1 1 2 . 2 2 1 2 2 . 2 2 2 . 2 1 . 1
37  1 2 2 2 2 2 2 2 2 1 1 2 2 2 1 1 1 1 1 1 2 2 2 2 2 2
38  1 1 1 1 1 1 2 2 2 2 1 1 1 1 1 2 2 2 2 2 2 2 2 2 2 2 2
39  2 2 2 2 2 2 2 2 2 2 2 2 2 2 . 2 2 1 2 2 2 1 1 2 2 2 2
40  1 1 1 1 1 1 . 2 2 2 2 2 2 2 2 2 2 2 2 2 1 1 1 1 1 1 1
41  1 1 1 1 1 1 1 1 2 2 2 2 2 2 2 2 2 2 2 1 1 1 1 1 1 1
42  2 2 2 . 2 2 2 2 2 1 1 2 2 2 1 1 1 2 1 1 1 1 2 2 2 2
43  2 2 2 2 2 2 2 2 2 1 1 1 1 1 2 2 2 2 2 2 1 1 1 1 1
44  2 2 2 2 2 2 2 2 2 2 2 2 2 1 1 1 2 2 2 2 1 1 1 . 1
45  1 1 1 1 1 2 2 2 2 2 2 1 1 1 1 1 1 1 1 1 2 2 2 2
46  2 2 2 2 2 1 1 1 1 1 1 1 1 1 1 1 1 1 . 1 2 2 2
47  . 1 . 2 . 2 2 2 1 2 . 2 2 2 2 2 2 2 2 2 2 1 1 . . 1
48  2 2 2 2 2 2 1 1 1 1 1 1 2 2 2 2 2 2 2 2 2 2 2 2 1
49  1 2 2 2 2 2 2 2 2 2 2 2 2 2 1 1 1 2 2 2 1 1 1 1 1 1
50  1 2 2 2 2 2 2 2 2 2 1 1 1 1 1 1 1 1 2 2 1 1 1 2 2 2
51  1 2 2 2 2 2 2 2 2 2 2 1 1 1 1 1 1 1 2 2 2 2 2 2 2
52  1 1 1 1 1 1 2 2 2 2 2 2 2 2 2 2 1 1 1 1 . 1 2 2 2
53  2 2 2 2 2 2 2 2 2 2 2 2 2 2 1 1 1 2 1 1 . 1 2 2 2
54  1 1 1 1 1 2 2 2 2 2 2 2 2 1 2 1 2 2 2 2 2 1 1 1
55  1 1 1 2 1 2 2 2 2 2 2 2 2 2 2 2 2 2 2 . 1 1 1 1
56  1 1 1 2 1 2 2 2 2 2 2 2 2 2 2 2 2 2 2 1 1 1 1 1
57  1 1 1 1 1 1 1 1 1 1 2 2 2 2 2 . 1 1 1 2 2 2 2 2 2 1
58  2 2 2 2 2 2 2 2 2 2 2 2 2 2 2 2 2 2 2 2 2 2 2 2 2
59  2 2 2 2 2 2 2 2 2 1 1 1 1 1 2 2 2 2 2 2 2 2 2 2 2
60  2 1 1 1 1 1 1 1 1 1 1 1 1 1 1 2 2 2 2 2 2 2 2 2 1
61  2 2 2 2 2 2 2 2 2 2 2 2 1 1 1 2 2 2 2 2 2 2 2 2
62  2 2 2 . 2 2 2 2 2 2 2 2 2 2 1 1 1 1 . 2 2 2 2
63  1 1 1 1 1 1 1 2 2 2 2 2 2 2 2 2 2 2 2 1 2 2 2 2 2
64  2 2 2 2 2 2 2 2 2 . . 1 1 1 . 2 2 2 2 2 . 1 1 1 1 1
65  1 1 1 1 1 2 2 2 2 2 2 1 1 1 1 1 1 1 1 1 2 2 2 2 2
66  2 2 2 2 . . 2 2 2 2 2 2 2 2 2 2 2 2 2 . 1 1 1 1 1
67  2 2 2 2 2 2 2 2 2 1 1 1 1 1 2 2 2 1 1 1 1 2 2
68  1 1 1 1 1 2 2 2 2 2 2 2 2 1 1 1 1 2 2 2 1 1 1 2 2
69  1 1 1 . 2 2 2 2 2 2 . 2 2 2 2 2 2 1 2 2 2 . 2 2 2 2
70  2 2 2 2 2 2 2 2 2 2 2 2 2 2 2 2 2 1 1 2 2 2 2 2 2
71  1 2 2 . 2 2 2 2 2 . 2 2 2 2 2 2 2 2 2 2 1 1 1 2
72  1 1 1 . 1 1 1 1 1 2 2 1 . 1 2 2 1 . 1 2 . 2
73  1 1 1 . 1 2 2 2 2 . 1 . 1 1 2 2 1 2 2 2 1 1 1 . 1
74  2 2 2 . 2 2 2 2 2 . 1 1 1 1 1 1 1 2 2 2 1 1 1 1 1
75  1 1 1 1 1 1 2 2 2 . 1 1 1 1 1 1 1 2 2 2 2 2 1 1 1
76  1 1 1 1 1 1 1 2 2 2 2 1 1 1 1 1 1 2 2 1 1 1 1 1 1
77  2 1 2 . 2 2 2 2 2 2 1 1 1 1 1 1 1 1 . 1 1 1 1
78  1 1 1 1 1 1 1 1 1 1 2 2 2 2 . 1 1 1 2 2 2 2 2 2 1
79  1 1 1 2 1 2 2 2 2 2 1 1 1 2 2 2 2 2 2 1 1 1 1 1 1
80  2 2 2 2 2 1 2 2 2 2 2 2 2 2 1 1 1 2 2 2 1 1 1 1 1
81  2 1 2 2 2 2 2 2 2 2 2 2 1 1 1 1 1 1 1 1 1 1 1 1 1
82  2 2 2 2 2 2 2 2 2 2 1 1 . 1 1 1 1 1 1 1 . 1 1 . 1
```

```
83  1 1 2 2 2 2 2 2 2 1 . . 1 1 2 2 2 1 2 2 1 1 1 2 2 2
84  1 1 1 1 1 2 2 2 2 2 1 1 1 1 1 1 1 1 1 1 1 1 1 1 2 2
85  2 2 2 2 2 2 2 2 2 2 2 2 2 2 2 1 1 1 1 2 2 2 2 2 1 1
86  2 2 1 1 1 1 1 2 2 2 2 2 2 2 2 2 1 1 1 2 2 . 2 2 2 2
87  2 2 2 2 2 2 2 2 2 2 2 1 1 . . 2 2 2 2 1 1 1 1 1 1
88  1 1 1 1 1 2 2 2 2 2 2 1 1 1 1 1 1 2 2 2 2 2 2 2 2 2
89  2 2 2 2 2 2 2 2 2 2 2 2 2 2 1 1 1 2 2 2 1 1 1 2 2 2
90  1 1 2 . 2 2 2 2 2 2 2 2 2 2 2 2 2 2 2 2 2 . 2 2 2 2
91  1 2 2 2 2 2 2 2 2 2 2 2 2 1 1 1 1 1 1 1 . 1 2 2 2
92  2 2 2 . 2 2 2 2 . 2 1 1 . 2 . 2 2 2 1 1 1 . 2 2 2 2
93  1 1 1 1 1 1 2 2 2 2 2 2 2 . . 2 2 1 1 1 1 1 1 1 1 1
94  1 1 1 1 1 1 1 2 2 2 2 2 2 2 2 2 2 1 1 1 1 1 2 2 2
95  1 1 1 1 1 2 2 2 2 2 1 1 1 1 1 1 1 1 1 1 1 2 2 2 2 2
96  1 1 1 1 1 1 1 2 2 . 2 2 1 1 1 1 1 2 2 . 2 2 2 2 1 1 1
```

```
     M28 M29 M30 M31 M32 M33 M34 M35 M36 M37 M38 M39 M40 M41 M42 M43 M44 M45 M46 M47 M48 M49 M50 M51 M52 M53 M54
      2   2   2   2   2   2   2   2   2   2   2   2   2   1   2   1   1   1   1   1   1   1   2   2   .   2       ;
      1   1   1   1   1   1   2   1   1   1   1   1   2   2   2   1   1   1   1   1   1   1   1   1   1   1       ;
      1   1   1   1   .   1   1   1   1   1   1   2   2   2   .   2   2   2   2   2   2   2   2   .   2           ;
      1   2   2   2   2   1   2   2   2   2   2   2   2   2   2   2   2   2   2   2   2   2   .   2               ;
      2   .   .   2   .   .   1   .   2   2   .   2   2   .   2   1   2   .   2   2   2   .   2   .   .   1       ;
      1   1   1   1   1   .   2   2   2   2   2   2   1   2   2   2   2   1   1   1   .   1   1   .   1           ;
      2   2   2   2   2   2   1   1   1   2   2   2   1   1   2   1   1   1   1   1   1   .   1   1   .   1       ;
      2   2   2   1   2   1   1   1   2   2   2   2   2   2   .   2   2   2   1   1   1   1   2   2   2   2       ;
      1   1   1   1   1   1   1   1   1   1   1   1   2   1   .   1   1   1   1   1   1   1   1   .   1           ;
      2   2   2   2   2   2   2   2   2   2   2   2   2   1   1   1   1   1   1   1   1   1   1   2   2   1       ;
      1   1   1   1   1   1   1   1   1   1   1   1   2   1   .   1   1   1   1   1   1   1   1   .   1           ;
      2   2   2   1   1   1   1   1   2   2   2   2   2   1   1   1   2   1   1   1   1   1   1   1               ;
      1   1   1   1   1   1   1   1   1   1   1   1   2   1   .   1   1   1   1   1   1   1   1   .   1           ;
      2   2   2   2   2   2   2   2   2   2   2   2   1   1   .   1   1   1   2   2   2   2   2   .   2           ;
      2   2   2   2   2   2   1   1   1   1   1   1   1   1   1   1   1   1   1   1   .   1   .   2               ;
      2   2   2   2   2   2   1   1   1   1   1   1   1   2   1   2   2   2   1   1   1   1   1   1   1           ;
      1   1   1   1   1   2   2   2   2   2   2   1   1   1   1   1   1   1   1   1   1   2   2   2   2           ;
      2   .   2   .   .   2   .   2   2   .   2   2   .   2   .   1   1   1   1   1   .   1   1   .   1           ;
      2   2   2   2   2   2   1   1   1   1   1   1   1   2   1   .   1   1   1   1   1   .   1   1   1           ;
      2   2   2   2   2   2   2   2   2   2   2   1   1   1   .   1   1   1   1   1   1   1   1   1   1           ;
      1   1   1   1   1   1   2   2   2   1   1   1   1   .   1   1   1   1   1   1   1   1   .   1               ;
      2   2   2   2   2   2   2   2   2   2   2   2   2   1   1   1   1   1   1   1   1   1   1   2               ;
      2   2   .   2   2   2   .   2   2   2   1   2   1   2   1   2   1   1   1   1   .   1   1   2   .           ;
      2   2   2   2   1   1   1   1   2   1   1   2   2   2   .   2   2   2   2   2   2   2   2   1               ;
      2   2   2   2   2   2   2   2   2   2   2   2   1   .   1   1   1   1   1   2   1   1   1   1   2           ;
      1   1   1   1   1   1   1   1   1   1   1   1   1   2   2   1   .   1   1   1   1   1   2   1   2   2   2   ;
      2   2   2   1   1   1   1   1   1   1   1   1   2   2   1   .   2   2   2   2   2   2   2   2   2           ;
      1   2   2   2   2   2   2   2   1   1   1   2   2   .   2   2   2   1   1   1   1   1   1   2               ;
      2   2   2   2   .   1   1   1   1   .   1   .   1   2   2   2   2   1   1   1   1   .   2               ;
      1   .   1   1   1   1   2   2   2   2   2   1   .   1   1   1   1   2   2   2   .   2   .   2           ;
      1   1   1   2   2   1   1   1   2   2   2   1   1   1   1   1   1   2   2   2   1   1   1   1           ;
      2   2   .   2   .   1   1   1   .   1   .   2   1   1   1   1   2   2   2   2   2   .   2               ;
      1   2   1   2   2   2   2   2   2   1   1   .   1   1   1   2   2   2   2   2   .   2               ;
      1   .   1   1   1   1   1   1   .   2   2   2   .   1   1   1   1   2   2   2   .   2   2   .   1       ;
      1   2   2   2   2   2   2   2   2   2   2   2   2   2   2   2   2   2   1   2   2   2   .   2           ;
      1   2   2   2   2   1   1   1   1   1   1   1   2   2   2   2   2   2   1   1   1   .   1               ;
      2   2   2   2   .   2   2   2   2   2   2   1   2   2   2   2   2   1   2   2   2   .   2               ;
      1   1   1   1   1   1   2   2   2   2   1   1   1   1   1   1   1   .   1   1   1   2   2   .   2       ;
      1   1   1   1   1   1   2   2   2   2   1   1   1   1   1   1   1   1   1   2   2   .   2               ;
```

```
2 2 2 2 2 2 1 2 2 2 2 2 2 2 2 2 2 2 . 1 1 1 2 2 . 2 ;
1 1 1 1 1 1 1 1 1 2 2 2 2 2 2 2 2 2 2 2 2 2 2 2 2 2 ;
1 1 1 1 1 1 1 . 1 1 2 2 2 1 1 1 1 1 1 1 1 1 1 1 1 1 ;
2 2 2 2 2 2 2 2 2 2 2 2 2 2 2 2 2 1 1 1 1 1 1 1 1 1 ;
2 2 . 1 1 1 2 2 2 2 2 2 1 1 1 1 1 1 1 1 . 1 1 1 . 2 ;
1 1 1 1 1 1 2 2 2 2 2 2 2 1 1 1 1 1 1 1 1 1 2 2 2 2 ;
1 1 1 1 1 1 2 2 2 2 2 2 1 1 1 1 1 1 1 1 1 1 1 2 1 ;
2 2 2 2 2 2 2 2 2 2 2 2 2 2 2 2 2 2 2 2 2 2 2 2 2 2 ;
1 1 1 1 1 1 2 2 2 2 2 2 2 2 2 2 2 2 2 2 2 2 2 2 2 2 ;
2 2 2 2 2 2 2 2 2 2 2 2 2 2 1 1 1 1 1 1 1 1 1 1 1 1 ;
2 2 2 2 2 2 2 2 2 2 2 2 2 2 2 2 2 2 2 2 2 2 2 2 2 2 ;
1 1 1 1 1 1 2 2 2 2 2 2 2 2 2 2 2 2 2 2 . 2 1 1 1 ;
1 . 1 1 1 . 2 2 2 2 2 2 1 2 2 2 2 . . 2 2 2 2 1 1 . 1 ;
1 1 1 1 1 1 2 2 2 2 2 2 1 2 2 2 2 2 . 2 2 2 2 1 1 1 1 ;
1 1 1 1 1 1 2 2 2 2 2 2 . 1 1 1 1 1 1 1 1 1 1 1 . 2 ;
2 2 2 1 2 1 2 2 2 2 2 2 . 1 1 2 2 2 2 2 2 1 1 1 1 ;
1 1 1 2 2 2 2 2 2 1 1 1 2 2 2 2 2 2 2 2 2 1 1 . 1 ;
1 1 1 1 1 1 1 2 2 2 2 2 2 1 1 1 1 1 1 1 1 1 1 1 1 1 ;
2 2 2 2 2 2 2 2 2 2 2 2 2 2 1 1 1 1 1 1 1 1 2 2 2 2 ;
2 1 1 1 1 1 1 1 1 2 2 2 2 2 2 1 1 1 1 2 2 2 2 2 . 1 ;
1 1 1 1 1 1 2 2 2 2 1 1 1 1 1 1 1 1 2 2 2 2 2 2 . 2 ;
1 1 . 2 . 1 2 2 2 2 2 2 2 2 . 2 1 2 2 1 1 1 1 1 . 1 ;
2 2 2 2 2 2 2 2 2 2 2 2 2 2 2 2 2 2 1 1 1 1 1 . 1 ;
1 1 1 1 1 1 2 2 2 2 2 2 2 2 2 2 2 2 2 . . . 1 . ;
2 2 2 2 2 2 1 1 1 1 1 1 1 1 1 1 1 1 1 1 1 . 1 2 . 2 ;
2 2 2 2 2 2 1 1 2 2 2 2 2 2 1 1 1 1 1 1 1 1 2 2 2 2 ;
2 . 2 2 . . 2 2 2 2 . 2 2 1 1 1 2 . . 2 2 2 2 2 . . 1 ;
2 2 2 2 2 2 1 1 2 2 2 1 1 1 1 2 2 2 2 2 2 2 2 2 2 2 ;
2 2 2 2 2 2 1 1 1 1 1 1 1 1 1 1 1 1 1 1 1 1 1 1 1 1 ;
2 2 2 2 2 . 1 1 1 1 1 1 1 1 2 2 1 1 1 . 2 2 2 1 1 . 1 ;
1 1 1 1 1 . 1 2 2 2 2 2 2 2 1 1 1 1 1 2 2 2 2 2 . 2 ;
2 2 2 2 2 2 2 2 1 1 1 1 1 2 1 1 1 1 1 1 1 1 1 1 . 1 ;
1 1 1 1 1 1 1 1 . 2 2 2 2 2 1 1 1 1 1 1 1 1 1 1 . 1 ;
1 1 1 1 1 1 2 2 2 2 2 2 2 2 2 2 2 2 2 1 2 1 1 1 1 ;
1 1 1 1 1 1 2 2 2 2 2 2 2 . 1 1 1 1 1 1 1 1 1 1 . 2 ;
1 1 1 1 1 1 2 2 2 2 2 2 2 1 1 1 1 1 1 1 1 1 1 1 1 1 ;
1 1 1 2 1 2 2 2 2 2 2 2 2 1 1 2 2 2 2 2 2 2 2 2 . 1 ;
1 1 1 1 1 1 1 2 2 2 2 2 2 2 2 2 2 2 . 2 1 1 1 . 1 ;
1 1 1 1 1 . 1 1 1 . 2 2 1 2 1 1 1 . 1 1 1 . 1 . 2 . 2 ;
2 2 2 2 2 . 2 2 2 2 2 2 2 2 1 1 1 . 1 1 1 2 1 1 1 2 2 ;
2 2 2 2 2 2 2 2 2 2 2 2 1 1 1 1 1 1 1 1 1 2 2 2 2 2 ;
1 1 1 1 1 1 2 2 2 2 2 2 2 1 1 1 1 1 1 1 2 1 1 1 1 2 ;
2 2 2 2 2 2 2 2 2 2 2 2 2 2 2 1 2 2 2 1 2 2 2 2 2 ;
1 1 1 1 1 1 2 2 2 2 2 2 1 2 2 1 1 1 1 . 1 2 1 1 1 1 ;
1 1 1 1 1 1 2 2 1 1 1 1 1 2 2 2 2 2 2 2 2 2 1 1 1 1 ;
2 2 2 2 2 2 2 2 2 2 2 2 2 2 2 2 2 2 2 2 . 2 2 1 ;
2 2 2 2 2 . 2 2 2 2 2 2 2 2 2 2 2 2 2 2 1 2 2 1 . 1 ;
2 2 2 2 2 . 2 2 2 2 2 2 2 2 2 . 2 . 1 1 1 1 1 . 1 ;
2 2 2 2 2 2 1 1 1 1 1 2 2 2 1 1 1 1 . 1 1 1 2 2 . 2 ;
1 1 1 1 1 . 1 . 2 2 2 2 2 2 2 2 . 2 . 1 2 1 1 1 . 1 ;
2 2 2 1 1 1 1 1 1 1 1 1 1 2 2 2 2 1 1 2 1 . 2 2 2 ;
2 2 2 2 2 2 1 1 1 1 1 1 1 1 2 2 2 2 2 2 1 2 2 1 . 1 ;
1 1 1 1 1 1 1 1 2 2 2 2 2 2 2 2 2 2 1 1 . 1 1 1 . 2 ;
```

MarkerEnd

TraitBegin

Env#	Ind#	SH5 ;	Env#	Ind#	SH5 ;	Env#	Ind#	SH5 ;	Env#	Ind#	SH5 ;
1	1	52.5 ;	1	49	77.6 ;	2	1	43.8 ;	2	49	66.7 ;
1	2	62.5 ;	1	50	50.4 ;	2	2	39.5 ;	2	50	50.1 ;
1	3	77.9 ;	1	51	60.0 ;	2	3	57.8 ;	2	51	56.7 ;
1	4	57.2 ;	1	52	68.6 ;	2	4	44.9 ;	2	52	56.4 ;
1	5	51.7 ;	1	53	58.0 ;	2	5	41.9 ;	2	53	53.3 ;
1	6	62.5 ;	1	54	67.2 ;	2	6	44.2 ;	2	54	58.4 ;
1	7	56.0 ;	1	55	65.2 ;	2	7	46.8 ;	2	55	60.1 ;
1	8	62.7 ;	1	56	66.7 ;	2	8	51.4 ;	2	56	57.2 ;
1	9	62.1 ;	1	57	67.1 ;	2	9	46.4 ;	2	57	53.1 ;
1	10	76.2 ;	1	58	59.6 ;	2	10	65.6 ;	2	58	54.9 ;
1	11	69.1 ;	1	59	67.5 ;	2	11	53.0 ;	2	59	56.0 ;
1	12	68.4 ;	1	60	67.7 ;	2	12	58.4 ;	2	60	52.6 ;
1	13	45.4 ;	1	61	60.3 ;	2	13	40.2 ;	2	61	52.0 ;
1	14	68.4 ;	1	62	70.9 ;	2	14	59.1 ;	2	62	57.0 ;
1	15	83.9 ;	1	63	78.8 ;	2	15	67.7 ;	2	63	65.0 ;
1	16	81.5 ;	1	64	70.9 ;	2	16	67.7 ;	2	64	59.0 ;
1	17	74.4 ;	1	65	52.2 ;	2	17	63.7 ;	2	65	46.3 ;
1	18	73.9 ;	1	66	70.7 ;	2	18	68.0 ;	2	66	55.3 ;
1	19	58.7 ;	1	67	66.3 ;	2	19	48.7 ;	2	67	58.4 ;
1	20	64.5 ;	1	68	55.0 ;	2	20	51.8 ;	2	68	48.9 ;
1	21	61.2 ;	1	69	75.3 ;	2	21	49.9 ;	2	69	59.6 ;
1	22	48.5 ;	1	70	75.5 ;	2	22	41.1 ;	2	70	56.8 ;
1	23	48.2 ;	1	71	57.5 ;	2	23	34.1 ;	2	71	43.4 ;
1	24	83.5 ;	1	72	49.7 ;	2	24	71.4 ;	2	72	42.5 ;
1	25	55.2 ;	1	73	75.5 ;	2	25	44.6 ;	2	73	66.5 ;
1	26	49.6 ;	1	74	52.5 ;	2	26	47.0 ;	2	74	40.1 ;
1	27	77.3 ;	1	75	64.6 ;	2	27	62.3 ;	2	75	57.3 ;
1	28	78.5 ;	1	76	57.2 ;	2	28	60.8 ;	2	76	52.8 ;
1	29	71.0 ;	1	77	52.1 ;	2	29	64.1 ;	2	77	43.4 ;
1	30	67.2 ;	1	78	53.9 ;	2	30	62.3 ;	2	78	46.8 ;
1	31	80.9 ;	1	79	72.0 ;	2	31	67.6 ;	2	79	63.1 ;
1	32	88.4 ;	1	80	70.0 ;	2	32	77.3 ;	2	80	57.8 ;
1	33	79.2 ;	1	81	52.3 ;	2	33	66.1 ;	2	81	43.2 ;
1	34	77.1 ;	1	82	55.5 ;	2	34	58.6 ;	2	82	43.3 ;
1	35	72.7 ;	1	83	63.7 ;	2	35	57.9 ;	2	83	48.3 ;
1	36	78.6 ;	1	84	62.1 ;	2	36	60.3 ;	2	84	57.5 ;
1	37	65.1 ;	1	85	61.0 ;	2	37	55.1 ;	2	85	50.2 ;
1	38	48.8 ;	1	86	71.8 ;	2	38	44.3 ;	2	86	59.9 ;
1	39	65.5 ;	1	87	76.6 ;	2	39	47.1 ;	2	87	58.6 ;
1	40	79.3 ;	1	88	55.4 ;	2	40	71.4 ;	2	88	44.1 ;
1	41	81.8 ;	1	89	60.9 ;	2	41	70.1 ;	2	89	43.6 ;
1	42	63.8 ;	1	90	58.9 ;	2	42	57.4 ;	2	90	44.7 ;
1	43	96.6 ;	1	91	44.4 ;	2	43	76.0 ;	2	91	37.0 ;
1	44	67.2 ;	1	92	73.6 ;	2	44	56.7 ;	2	92	56.3 ;
1	45	55.5 ;	1	93	73.8 ;	2	45	46.1 ;	2	93	63.6 ;
1	46	44.2 ;	1	94	82.4 ;	2	46	28.3 ;	2	94	69.8 ;
1	47	80.1 ;	1	95	53.7 ;	2	47	67.7 ;	2	95	41.8 ;
1	48	75.9 ;	1	96	64.2 ;	2	48	66.7 ;	2	96	55.4 ;

TraitEnd

How to use the software:

1. Run QTLMAPPER.EXE to analyze QTL positions and effects. First create two files: one is a map file (ckge.map) and the other is a marker and trait file (ckge.txt). Choose run from submenu and map epistatic QTL.
2. After finishing the general analysis, choose output submenu and screen putative additive-effect QTL or epistatic QTL. The results are presented in Output 1.
3. Run jackknife test in output submenu for detecting significant additive and epistatic effects. The results are presented in Output 2.

Output 1 for Contribution of QTL Effects

```
// Result file created by QTLMapper V 1.0
// Data file name:        D:\QTLSOURCE\ckge.txt
// Marker map file name:  D:\QTLSOURCE\ckge.map
// Environments:          yes
// Replications:          no
// Contents: relative contributions (H^2) for putative main-effect
     QTLs/epistatic QTLs
//    Calculations based on: D:\QTLSOURCE\ckge.jke
//    BGV control method:  A  (control marker main & interaction ef-
     fects)
# Date: 2000-07-04     Time: 14:05:56
```

Trait 1: SH5

Ch-Ini	Int.Namei	Sitei(M)	Ch-Inj	Int.Namej	Sitej(M)	H^2(Ai)	H^2(Aj)	H^2(AAij)	H^2(AEi)	H^2(AEj)	H^2(AAEij)
1-5	M5-M6	0.00	1-17	M17-M18	0.04	0.0000	0.0714	0.0000	0.0000	0.0001	0.0000
1-9	M9-M10	0.00	1-15	M15-M16	0.12	0.0000	0.2532	0.0000	0.0000	0.0001	0.0011
2-6	M24-M25	0.28	3-18	M51-M52	0.06	0.0000	0.0727	0.0000	0.0000	0.0001	0.0000
2-9	M27-M28	0.02	3-16	M49-M50	0.00	0.0625	0.0000	0.0000	0.0000	0.0001	0.0000

```
General contributions:
   Additive(A): H^2(A)=0.6131;   Epistasis: H^2(AA)=0.0000
   QE Interactions: H^2(AE)=0.0003; H^2(AAE)=0.0011

End
```

Output 2 for QTL A and AA Effects

```
// Result file created by QTLMapper V 1.0
// Data file name: D:\QTLSOURCE\ckge.txt
// Marker map file name: D:\QTLSOURCE\ckge.map
// Environments:  yes
// Replications:  no
// Contents: Jackknife test results for epistatic QTLs
// Jackknife based on: D:\QTLSOURCE\ckge.fle
// BGV control method: A (control marker main & interaction effects)
// Threshold probability: 0.005000

# Date: 2000-07-04 Time: 13:48:12

Trait 1: SH5
```

Ch-Ini	Int.Namei	Sitei(M)	Ch-Inj	Int.Namej	Sitej(M)	Ai	Probi	Aj	Probj	AAij	Probij
1-5	M5-M6	0.00	1-17	M17-M18	0.04	0.055	0.9410	-4.541	0.0000	1.044	0.1782
1-9	M9-M10	0.00	1-15	M15-M16	0.12	-1.458	0.3977	-8.553	0.0003	-4.077	0.0559
2-6	M24-M25	0.28	3-18	M51-M52	0.06	2.115	0.0270	-4.582	0.0000	-1.209	0.1388
2-9	M27-M28	0.02	3-16	M49-M50	0.00	4.250	0.0000	-0.887	0.2098	-0.175	0.8197

AEi1	Prob	AEi2	Prob	AEj1	Prob	AEj2	Prob	AAEij1	Prob	AAEij2	Prob
-0.122	0.5595	0.122	0.5590	-0.194	0.0532	0.194	0.0534	0.201	0.0558	-0.201	0.0558
0.081	0.7146	-0.080	0.7188	0.144	0.5623	-0.143	0.5645	0.393	0.0001	-0.394	0.0001
0.025	0.8673	-0.026	0.8604	-0.158	0.1465	0.159	0.1452	0.159	0.1237	-0.159	0.1233
0.137	0.2064	-0.137	0.2064	-0.130	0.2267	0.131	0.2254	-0.144	0.1296	0.144	0.1303

End

Chapter 22

Gene Segregation and Linkage Analysis

Jinsheng Liu
Todd C. Wehner
Sandra B. Donaghy

Purpose

To calculate single-gene goodness-of-fit testing to analyze gene linkage relationships, including calculations of chi-square, probability value, and two-locus-combined phases, for all gene pairs in segregation for the F_2, BC_{1P1}, and BC_{1P2} generations. Recombination frequency and standard error are calculated according to the linkage phase.

Genetic Analysis

Linkage is estimated using the chi-square method, a widely used standard for genetic data analysis (although it may produce inaccurate results in some cases). Recombination frequency (RF) and standard error (SE) are calculated according to phase (coupling or repulsion), using the following formulas (Sinnott and Dunn, 1939; Weir, 1994).

Definitions

F_2 (repulsion):

$$RF = p = \sqrt{\frac{-(bc+ad) + \sqrt{(bc+ad)^2 + ad(bc-ad)}}{(bc-ad)}}$$

F_2 (coupling):

$$RF = 1 - p$$
$$SE = \sqrt{(1-p^2)(2+p^2)/2n(1+2p^2)}$$

BC_1 (only coupling accepted):

$$RF = (b+c)/n$$
$$SE = \sqrt{RF(1-RF)/n}$$

where $a\,(A\_B\_)$, $b\,(A\_bb)$, $c\,(aaB\_)$ and $d\,(aabb)$ are genotype segregation ratios in F_2 or BC_1.

Originators

Sinnott, E.W. and Dunn, L.C. (1939). *Principles of Genetics*. McGraw-Hill, New York.
Weir, B.S. (1994). *Genetic Data Analysis: Methods for Discrete Population Data*. Sinauer, Sunderland, MA.

Software Available

Files can be found on the World Wide Web at <http://cuke.hort.ncsu.edu/cucurbit/ Wehner/software.html>. Or, send a 3.5" floppy disk to Todd C. Wehner, Department of Horticultural Science, North Carolina State University, Raleigh, NC 27695-7609.

Publication

Liu, J.S., Wehner, T.C., and Donaghy, S.B. (1997). SASGENE: A SAS computer program for genetic analysis of gene segregation and linkage. *Journal of Heredity* 88: 253-254.

Some References Using the Software

Wehner, T.C., Liu, J.S., Staub, J.E., and Fazio, G. (2003). Segregation and linkage of 14 loci in cucumber. *Journal of American Society of Horticulture Science*.

Contact

Dr. Todd Wehner, Department of Horticultural Science, North Carolina State University, Raleigh, NC 27695-7609, USA. E-mail: <todd_wehner@ncsu.edu>; Web site: <http:// cuke.hort.ncsu.edu>.

Revisions That Have Been Made

SASGene1.0 and 1.1 had an error in the formula for calculation of SE for RF in coupling. F_2 (coupling):

$$RF = 1 - p$$
$$SE = \sqrt{(1-p^2)(1+p^2)/2n(1+2p^2)}$$

SASGene1.2 has been corrected F_2 (coupling):

$$RF = 1 - p$$
$$SE = \sqrt{(1-p^2)(2+p^2)/2n(1+2p^2)}$$

EXAMPLE

Data to be analyzed:

Plot	Rep	Fam	Gen	Plnt	Bi	Rc	Dv	Sp	Ll	Df	F	B	D	U	Tu
1	1	28	1	1	B	N	N	N	L	N	M	W	D	N	W
2	1	28	1	2	B	N	N	N	L	N	M	W	D	N	W
3	1	28	1	3	B	N	N	N	L	N	M	W	D	N	W
4	1	28	1	4	B	N ·	N	N	L	N	M	W	D	N	W
5	1	28	1	5	B	N	N	L	N	M	W	D	N	W	
6	1	28	2	1	N	N	N	N	N	D	G	W	D	U	S
7	1	28	2	2	N	N	N	N	N	D	G	W	S	U	S
8	1	28	2	3	N	N	N	N	N	D	G	W	S	U	S
9	1	28	2	4	N	N	N	N	N	D	G	W	S	U	S
10	1	28	2	5	N	N	N	N	N	D	G	W	S	U	S
11	1	28	3	1	B	N	N	N	N	N	G	W	D	N	W
12	1	28	3	2	B	N	N	N	N	N	G	W	D	N	W
13	1	28	3	3	B	N	N	N	N	N	G	W	D	N	W
14	1	28	3	4	B	N	N	N	N	N	G	W	D	N	W
15	1	28	3	5	B	N	N	N	N	N	M	W	D	N	W
16	1	28	3	6	B	N	N	N	N	N	G	W	D	N	W
17	1	28	4	1	B	.	N	N	N	D	G	W	D	U	S
18	1	28	4	2	N	.	N	N	L	D	G	W	.	U	S
19	1	28	4	3	B	N	N	N	N	D	O	W	.	U	S
20	1	28	4	4	B	N	N	N	N	N	G	W	D	N	W
21	1	28	4	5	B	N	N	N	N	N	M	W	D	N	W
22	1	28	4	6	B	N	N	N	N	N	G	W	D	N	W
23	1	28	4	7	B	N	N	N	L	N	G	W	D	N	W
24	1	28	4	8	B	N	N	N	N	N	G	W	D	U	W
25	1	28	4	9	B	N	N	N	N	D	G	W	D	N	W
26	1	28	4	10	N	N	N	N	N	D	G	W	D	U	S
27	1	28	4	11	N	N	N	N	N	N	G	W	D	N	W

```
28   1   28   4   12   B   N   N   N   N   N   G   W   D   N   W
29   1   28   4   13   B   N   N   N   L   N   G   W   D   .   W
30   1   28   4   14   N   3   N   N   N   N   G   W   D   N   W
31   1   28   4   15   B   N   N   N   N   N   G   W   D   N   W
32   1   28   4   16   N   N   N   N   N   N   G   W   D   N   W
33   1   28   4   17   B   N   .   .   .   .   .   .   .   .   .
34   1   28   4   18   B   N   .   .   .   .   .   .   .   .   .
35   1   28   5   1    B   N   N   N   L   N   G   W   D   N   W
36   1   28   5   2    B   N   N   N   N   N   G   W   D   N   W
37   1   28   5   3    B   N   N   N   L   N   M   W   D   N   W
38   1   28   5   4    B   N   N   N   L   N   M   W   D   N   W
39   1   28   5   5    B   N   N   N   N   N   G   W   D   N   W
40   1   28   5   6    B   N   N   N   N   N   G   W   D   N   W
41   1   28   5   7    B   N   N   N   L   N   M   W   D   N   W
42   1   28   5   8    B   N   N   N   N   N   G   W   D   N   W
43   1   28   5   9    B   N   N   N   L   N   G   W   D   N   W
44   1   28   6   1    B   N   N   N   N   N   G   W   D   N   S
45   1   28   6   2    B   N   N   N   N   D   G   W   D   U   S
46   1   28   6   3    N   N   N   N   N   D   G   W   D   U   S
47   1   28   6   4    B   N   N   N   N   D   G   W   D   U   S
48   1   28   6   5    N   N   N   N   N   D   G   W   D   U   S
49   1   28   6   6    N   N   N   N   N   D   G   W   D   U   S
50   1   28   6   7    N   N   N   N   N   D   G   W   D   U   S
51   1   28   6   8    N   N   N   N   N   N   G   W   D   N   W
52   1   28   6   9    N   N   N   N   N   N   G   W   D   N   W
53   1   28   1   1    B   N   N   N   L   N   M   W   D   N   W
54   1   28   1   2    B   N   N   N   L   N   M   W   D   N   W
55   1   28   1   3    B   N   N   N   L   N   M   W   D   N   W
56   1   28   1   4    B   N   N   N   L   N   M   W   D   N   W
57   1   28   1   5    B   N   N   N   L   N   M   W   D   N   W
58   1   28   2   1    N   N   N   N   N   D   G   W   S   U   S
59   1   28   2   2    N   N   N   N   N   D   G   W   S   U   S
60   1   28   2   3    N   N   N   N   N   D   G   W   S   U   S
61   1   28   2   4    N   N   N   N   N   D   G   W   S   U   S
62   1   28   2   5    N   N   N   N   N   D   G   W   S   U   S
63   1   28   3   1    B   N   N   N   N   N   G   W   D   N   W
64   1   28   3   2    B   N   N   N   N   N   G   W   D   N   W
65   1   28   3   3    B   N   N   N   N   N   G   W   D   N   W
66   1   28   3   4    B   N   N   N   N   N   G   W   D   N   W
67   1   28   3   5    B   N   N   N   N   N   G   W   D   N   W
68   1   28   3   6    B   N   N   N   N   N   G   W   D   N   W
            .
            .
            .
469  5   30   1   1    N   R   N   S   N   N   M   B   D   U   W
470  5   30   1   2    N   R   N   S   N   N   M   B   D   U   W
471  5   30   1   3    N   R   N   S   N   N   M   B   D   N   W
472  5   30   1   4    N   R   N   S   N   N   G   B   D   N   W
473  5   30   1   5    N   R   N   S   N   N   G   B   D   U   W
474  5   30   2   1    N   N   N   N   N   N   G   W   S   U   S
475  5   30   2   2    N   N   N   N   N   D   G   W   S   U   W
476  5   30   2   3    N   N   N   N   N   D   G   W   S   U   S
477  5   30   2   4    N   N   N   N   N   N   G   W   S   U   S
478  5   30   2   5    N   N   N   N   N   D   G   W   S   U   S
479  5   30   3   1    B   N   N   N   N   D   M   B   D   N   W
480  5   30   3   2    B   N   N   N   N   N   G   B   D   N   W
481  5   30   3   3    B   N   N   N   N   D   G   B   D   N   W
```

```
482   5   30   3    4    B   N   N   N   N   D   G   B   D   N   W
483   5   30   3    5    B   N   N   N   N   N   G   B   D   N   W
484   5   30   3    6    B   N   N   N   N   D   G   B   D   N   W
485   5   30   4    1    B   N   N   N   N   N   G   B   D   N   W
486   5   30   4    2    B   N   N   N   N   N   G   W   D   U   S
487   5   30   4    3    N   R   N   S   N   N   G   W   D   U   W
488   5   30   4    4    B   N   N   N   N   D   G   B   S   U   W
489   5   30   4    5    B   N   N   N   N   D   G   B   S   U   W
490   5   30   4    6    B   N   N   N   N   N   G   B   D   N   W
491   5   30   4    7    B   N   N   N   N   N   G   W   D   U   S
492   5   30   4    8    B   N   N   N   N   D   G   B   D   N   W
493   5   30   4    9    N   N   N   N   N   N   G   W   D   N   S
494   5   30   4   10    N   N   N   N   N   D   G   B   D   N   S
495   5   30   4   11    B   N   N   N   N   N   G   B   D   N   W
496   5   30   4   12    N   N   N   N   N   D   G   B   D   U   S
497   5   30   4   13    B   N   N   N   N   D   G   W   S   U   W
498   5   30   4   14    B   N   N   N   N   D   G   B   D   N   W
499   5   30   4   15    N   N   N   N   N   N   G   B   D   N   W
500   5   30   4   16    N   N   N   N   N   N   G   W   D   U   S
501   5   30   4   17    N   N   N   N   N   N   G   B   D   N   W
502   5   30   4   18    N   N   N   N   N   N   G   B   D   N   W
503   5   30   5    1    B   N   N   N   N   D   G   B   D   N   W
504   5   30   5    2    N   R   N   S   N   N   G   B   D   U   W
505   5   30   5    3    B   N   N   N   N   D   N   B   D   U   W
506   5   30   5    4    N   R   N   S   N   N   M   .   .   .   .
507   5   30   5    5    B   N   N   N   N   N   G   B   D   N   W
508   5   30   5    6    B   N   N   N   N   D   G   B   D   N   W
509   5   30   5    7    B   N   N   N   N   N   G   B   D   N   W
510   5   30   5    8    B   R   N   S   N   N   M   B   D   N   W
511   5   30   5    9    N   R   N   S   N   N   M   .   .   .   .
512   5   30   6    1    B   N   N   N   N   D   G   B   D   N   W
513   5   30   6    2    N   N   N   N   N   D   G   B   D   U   S
514   5   30   6    3    B   N   N   N   N   D   G   B   D   U   W
515   5   30   6    4    B   N   N   N   N   D   G   W   D   U   S
516   5   30   6    5    N   N   N   N   N   D   G   B   D   N   W
517   5   30   6    6    B   N   N   N   N   D   G   W   D   N   W
518   5   30   6    7    N   N   N   N   N   D   G   W   D   U   W
519   5   30   6    8    N   N   N   N   N   D   M   W   D   U   S
520   5   30   6    9    B   N   N   N   N   D   G   W   D   N   S
              .
              .
              .
```

SAS Program (Five Files)

File 1: readme.txt

```
SASGENE 1.1
Program for Analysis of
Gene Segregation and Linkage
November 5, 1997
```

Instructions for Running SASGENE Macros

The SASGENE program for gene segregation and linkage analysis is written in SAS macro language. There are four SAS files. Three are macro files and one is an example. The first macro, SGENE, is for single-gene goodness-of-fit tests. The second macro, LINKAGE, is for analysis of gene linkage relationships. The third macro, CONVERT, is optional and converts gene values to "D" for dominant and "R" for recessive. STARTUP.SAS illustrates how to use the macros. The STARTUP.SAS file can easily be modified for other experiments of interest to the user.

The macros are written for version six and later versions of SAS. The amount of disk space required increases as the number of genes for the linkage analysis increases.

To use the macros, the user must create an input data file that will record data for the following fields: plot number, replication number, plant number, family number, generation number, and gene (or trait) names. Note that plot number, replication number, and plant number are used only for collecting data and are not used by the program for computing statistics. The user may specify any value for the family variable, but the macro requires values of 1, 2, 3, 4, 5, or 6 for the GNR (generation) variable (1 for P1, 2 for P2, 3 for F1, 4 for F2, 5 for BC1P1, 6 for BC1P2). Valid SAS variable names are used for the gene names. The genes (or traits) are variables (columns) and their values are observations (rows). Family and generation are identification variables. In the data file, the values of P1, P2, and F1 should not be omitted or the results may be incorrect.

The SGENE and LINKAGE macros require gene values to be coded as "D" for dominant, "R" for recessive, and "." or blank for a missing value. An optional macro, CONVERT, converts the original gene values to "D," "R," or missing. For each gene and family, the most frequent value for F1 is the dominant gene. Any other nonmissing values are treated as recessive, and any missing values are counted as missing.

An example of a SAS data set follows:

```
data orig;
    input PLOT REP FAMILY GNR BI $ RC $ DV $ SP $
    LL $ DF $ F $ B $ D $ U $ TU $;
    cards;
    1   1  20   1    N   R   N   S   N   N   M   B   D   N   W
    2   1  20   1    N   R   N   S   N   N   M   B   D   N   W
    3   1  20   1    N   R   N   S   N   N   M   B   D   N   W
    4   1  20   1    N   R   N   S   N   N   M   B   D   N   W
    .
    .
    .
run;
```

Either the macro code or a %INCLUDE (also known as %INC) statement is needed to define the macro to the SAS system. The user may include the macro into the program editor or use a %INC statement, such as %inc 'sgene.sas'. The %INC statement specifies the physical name of the external file where the macro is stored. The physical name is the name by which the host system recognizes the file. Depending on the host system and location of the file, the entire file name may need to be specified.

Examples:

```
%inc 'c:\mysas\sgene.sas';
%inc '~/sasmacro/sgene.sas';
```

The file, SGENE.SAS, contains the SAS macro, SGENE. File names, such as SGENE.SAS, usually carry the *sas* extension if the file is a SAS program or a SAS macro.

Once the macro is defined to SAS, the macro can be invoked. To invoke the macro, specify the %, the macro name (either SGENE, LINKAGE, or CONVERT), and the required parameters in parenthesis.

The SGENE macro has three parameters:

DS—name of the SAS data set to analyze
GENES—gene names from the SAS data set
P1—critical value for about half of the frequency of one parent to determine the expected segregation ratio (1:1 or 1:0) in BC1 generation

Example:

```
%sgene (ds=new,
        genes=BI RC DV SP LL DF F  B D U TU,
        p1=9);
```

The linkage macro has four parameters:

DS—name of the SAS data set to analyze
GENES—gene names from the SAS data set
P1, P2—critical value for about half of the frequency of the parents to determine if the phase is coupling or repulsion

Example:

```
%linkage (ds=new,
          genes=BI RC DV SP LL DF F  B D U TU,
          p1=9,
          p2=9);
```

The convert macro has three parameters:

    DS—name of the SAS data set to convert
    GENES—list of the desired gene names from the SAS data set
    DSOUT—name of the SAS data set after conversion

Example

```
%convert (ds=orig,
          genes=BI RC DV SP LL DF F  B D U TU,
          dsout=new);
```

Several additional files are stored in the same location as the introduction:

    STARTUP.SAS—example that illustrates how to use the macros
    ORIG.DAT—sample data for the startup.sas file
    CONVERT.SAS—file that contains the SAS macro convert
    SGENE.SAS—file that contains the SAS macro sgene
    LINKAGE.SAS—file that contains the SAS macro linkage

File 2: STARTUP.SAS

```
*********************************************************************
*                                                                  *
* SASGENE 1.1                                                      *
* Program for Analysis of                                          *
* Gene Segregation and Linkage                                     *
* November 5, 1997                                                 *
*                                                                  *
* Example of Invoking SASGENE macros                               *
*                                                                  *
*********************************************************************;

*********************************************************************
*                                                                  *
* Specify file names and include macros.                          *
*                                                                  *
* 1.  Specify the name of the file where the data are stored.      *
*     The name is enclosed in single quotes.                       *
```

```
*       example: filename in 'orig.dat' ;                           *
*                                                                   *
* 2.   Include the macros with the %INCLUDE (%inc) statement.       *
*      Specify the physical name of the external file where the macro *
*      is stored.  The physical name is enclosed in single quotes.  *
*      example: %inc 'convert.sas';                                 *
*                %inc 'sgene.sas';                                  *
*                %inc 'linkage.sas';                                *
*                                                                   *
*      Summary:                                                     *
*      The user only needs to change the information inside the     *
*      quotes on the FILENAME and %INCLUDE statements below.        *
*      The information inside the quotes specifies the name of the  *
*      external file where the data or macros are stored.  It may be *
*      necessary to specify the entire file name inside the quotes. *
*      example: %inc 'c:\sasmacro\convert.sas';                     *
*********************************************************************;

filename in 'example.dat';  /* name and location of data file     */
%inc 'convert.sas';    /* name and location of SAS macro CONVERT  */
%inc 'sgene.sas';      /* name and location of SAS macro SGENE    */
%inc 'linkage.sas';    /* name and location of SAS macro LINKAGE  */

*********************************************************************
*   include any desired titles and options                         *
*********************************************************************;
title  'Cucumber Gene Linkage Example';
options nodate pageno=1;
options linesize=80 pagesize=500;

*********************************************************************
*    Create SAS dataset                                            *
*       The user will need to modify the INPUT statement to specify *
*       the gene names from their experiment.  If list input is used, *
*       then missing values should be coded with a "."             *
*                                                                  *
*       Macros are expecting the following variable names:         *
*       family = family code                                       *
*       gnr    = generation code                                   *
*                                                                  *
*       Macros are expecting the following values for GNR variable: *
*       1 for P1                                                   *
*       2 for P2                                                   *
*       3 for F1                                                   *
*       4 for F2                                                   *
*       5 for BC1P1                                                *
*       6 for BC1P2                                                *
*                                                                  *
*       P1, P2 and F1 generations must be included                 *
*       for program to run (1 plant each is sufficient)            *
*********************************************************************;
data original;
    infile in missover pad; /* MISSOVER & PAD are options on INFILE */
    input plot rep family gnr plnt bi $ rc $ dv $ sp $ ll $ df $
        f $ b $ d $ u $ tu $ ;
    run;
```

```
********************************************************************
*    Invoke the CONVERT macro if the user needs to convert the gene    *
*    values to "D" or "R".  Otherwise delete the %convert statement.    *
*    The SGENE and LINKAGE macros are expecting the following gene      *
*    values:                                                            *
*     D for Dominant,                                                   *
*     R for Recessive,                                                  *
*     . or blank for missing value.                                     *
*                                                                       *
*    Specify the following parameters:                                  *
*        DS    - SAS dataset to convert                                 *
*        GENES - gene names from the SAS dataset                        *
*        DSOUT - output SAS dataset that has been converted             *
********************************************************************;
%convert(ds=original,
         genes=BI RC DV SP LL DF F B D U TU,
         dsout=new);

********************************************************************
*    Invoke the SGENE macro.                                            *
*    Modify the following parameters for your experiment:               *
*        DS    - SAS dataset to analyze (possibly the output dataset    *
*                from the CONVERT macro.                                *
*        GENES - gene names from the SAS dataset                        *
*        P1    - critical value for about half of the frequency of one  *
*                parent to determine the expected segregation ratio     *
*                (1:1 or 1:0) in BC1 generation.                        *
*                                                                       *
*                Indicates the number of plants of parent 1             *
*                that you feel must have the trait                      *
*                before you accept it as uniform                        *
*                (for example, 15 plants of P1 measured;                *
*                critical value set at 10,                              *
*                allowing 5 misclassifications)                         *
********************************************************************;
%sgene(ds=new,
       genes=BI RC DV SP LL DF F B D U TU,
       p1=9);

********************************************************************
*    Invoke the LINKAGE macro.                                          *
*    Modify the following parameters for your experiment:               *
*        DS    - SAS dataset to analyze (possibly the output dataset    *
*                from the CONVERT macro).                               *
*        GENES - gene names from the SAS dataset                        *
*        P1, P2- critical value for about half of the frequency of      *
*                the parents to determine if the phase is coupling or   *
*                repulsion.                                             *
*                                                                       *
*                Indicates the number of plants of parent 1             *
*                that you feel must have the trait                      *
*                before you accept it as uniform                        *
*                (for example, 15 plants of P1 measured;                *
*                critical value set at 10,                              *
*                allowing 5 misclassifications)                         *
********************************************************************;
%linkage(ds=new,
```

```
        genes=BI RC DV SP LL DF F B D U TU,
        p1=9,
        p2=9);
```

File 3: CONVERT.SAS

```
***********************************************************************
*                                                                     *
*       SASGENE 1.1                                                   *
*       Program for Analysis of                                      *
*       Gene Segregation and Linkage                                 *
*       November 5, 1997                                             *
*                                                                     *
**********************************************************************;

%macro convert
        (ds=_last_, /* SAS dataset to analyze(default:uses last one)*/
         genes=,    /* gene variable names                         */
         dsout=     /* name of new SAS dataset after conversion    */
         );

***********************************************************************
* Name:     CONVERT                                                  *
*                                                                     *
* Purpose:  Converts gene values to Dominant or Recessive           *
*                                                                     *
* Written:  09/14/95                                                 *
*                                                                     *
* Modified: 10/02/95                                                 *
*           03/05/97                                                 *
*                                                                     *
* Products: Base SAS                                                 *
*                                                                     *
* Example:  %convert(ds=save.orig,                                  *
*           genes=BI RC  DV SP LL DF F B D U TU SS NS,              *
*           dsout=new);                                             *
**********************************************************************;
proc format;
   value _gnrx
      1='P1'
      2='P2'
      3='F1'
      4='F2'
      5='BC1P1'
      6='BC1P2'
      ;
   run;
title2 'Gene Segregation and Linkage Analysis';
%local nogene word geneid i;

  /*  create nogenes macro variable                               */
  /*  nogenes is the number of genes listed in &genes             */
  %let nogenes=0;
  %if &genes ne %then %do;
      %let word=%scan(&genes,1);
```

```
        %do %while (&word ne );
            %let nogenes=%eval(&nogenes+1);
            %let word=%scan(&genes,&nogenes+1);
            %end;
        %end;
  /*  create geneid macro variable                              */
  /*  geneid is the names of the genes in quotes                */
  /*         used in array for identification in output         */
  %let word=%scan(&genes,1);
  %let geneid=%str(%'&word%');
  %do i=2 %to &nogenes;
      %let word=%scan(&genes,&i);
      %let geneid=%str(&geneid,%'&word%');
      %end;

proc sort data=&ds   out=_orig; by family; run;
data _generat; set _orig;
   length id 3;
   array y{*}    &genes;
   array yc{*} $ n1-n&nogenes (%unquote(&geneid));
   id=0;
   do _i_=1 to dim(y);
      id+1;
      code= y{_i_};
      gene=yc{_i_};
      output;
      end;
   keep family id gene gnr code;
   run;
proc sort data=_generat; by family id; run;

proc freq noprint;
   by family id gene;
   where code not=' ';
   tables code / out=_count;
   run;
proc means noprint; by family id ;
   var count;
   output out=_nocode n=n;
   run;
data _look; merge _count _nocode;  by family id;
   if n>2;
   run;
proc print label;
title3 'Observed frequencies for each gene locus and allele code';
title4 'These genes in this table have more than 2 codes:';
title5 '     some codes may have been misentered        ';
title6 'WARNING!!! Program will convert to 2 codes (D and R)
      ';
title7 '            Dominant will be assigned,
      ';
title8 '            other non-missing codes will be set to Recessive
      ';
   var family gene code count;
   label count='FREQUENCY';
   run;
```

```
/*  delete gene-family ids that do not make sense for analysis   */
/*  delete when the phenotype of P1 is the same as the           */
/*                  phenotype of P2                              */
title3 ' ';
proc freq       data=_generat noprint;
   by family id gene;
   tables code*gnr / out=_gnrcode(drop=percent) ;
   run;
data _gnrcode; set _gnrcode;
   if code=' ' then delete;
proc sort data=_gnrcode; by family id gene gnr descending count;
   run;
data _delete(keep=family id gene); set _gnrcode;
   by family id gene gnr;
   retain d1;
   if first.id then do;
      d1='           ';
      d2='           ';
      end;
   if first.gnr then do;
      if gnr=1 then d1=code;
         else if gnr=2 then do;
                        d2=code;
                        if d1=d2 then output _delete;
                        end;
      end;
   run;
proc print data=_delete(drop=id);
   title3 'These gene-family combinations will be deleted ';
   title4 'since the phenotype for P1 and P2 are the same ';
   title5 'and do not fit the assumptions of the analysis.';
   run;
data _generat _look; merge _generat _delete(in=yes);
   by family id gene;
   if yes then output _look;
      else output _generat;
   run;
proc freq       data=_look;
   by family id gene;
   tables code*gnr / missprint nocum nopercent norow nocol;
   label gnr='GENERATION';
   format gnr _gnrx.;
   run;

   /*   find the dominant gene by looking at generation 3 (F1)   */
title3 ' ';
proc freq noprint data=_generat;
   by family id gene;
   where gnr=3;
   tables code / out=_count;
   run;
proc sort; by family id count; run;

data _dom; set _count;
   by family id;
```

```
     array c  $ c1-c&nogenes;
     retain c1-c&nogenes;
     length c1-c&nogenes  $8;
     if first.family  then do;
        do _i_=1 to &nogenes;
           c{_i_}=' ';
           end;
        end;

     if last.id then c{id}=code;
     if last.family then output;
     keep family c1-c&nogenes;
     run;
data &dsout; merge _orig _dom;
     by family;
     array genes{*}     &genes;
     array dom{*}   $ c1-c&nogenes;

     do _i_=1 to dim(genes);
        if dom{_i_}=' ' then genes{_i_}=' ';/*useless data- no domi-
     nant*/
           else do;
              if genes{_i_}=dom{_i_}    then genes{_i_}='D';
                 else if genes{_i_}=' ' then genes{_i_}=' ';
                 else genes{_i_}='R';
              end;
        end;
     drop c1-c&nogenes _i_;
     run;
data _check; merge _orig _dom;
     by family;
     array genes      &genes;
     array dom $ c1-c&nogenes;
     array yc{*} $ n1-n&nogenes (%unquote(&geneid));
     id=0;
     do _i_=1 to dim(genes);
        id+1;
        gene=yc{_i_};
        old_code=genes{_i_};
        if dom{_i_}=' ' then new_code=' '; /*useless data- no dominant
     */
           else do;
              if genes{_i_}=dom{_i_}    then new_code='D';
                 else if genes{_i_}=' ' then new_code=' ';
                 else new_code='R';
              end;
           output;
        end;
     drop c1-c&nogenes n1-n&nogenes &genes;
     run;

title4 "Conversion to 'D' or 'R' for each gene and family";
proc freq;
     tables id*gene*family*new_code*old_code/list nopercent nocum
        nofreq;
```

```
    run;
proc datasets library=work memtype=data nolist;
   delete _check _count _dom _generat _look _nocode _orig
          _delete _gnrcode;
   quit;
%mend convert;
```

File 4: SGENE.SAS

```
****************************************************************************
*                                                                        *
*      SASGENE 1.1                                                        *
*      Program for Analysis of                                           *
*      Gene Segregation and Linkage                                      *
*      November 5, 1997                                                  *
*                                                                        *
****************************************************************************;

%macro sgene
        (ds=_last_,  /* SAS dataset to analyze(default:uses last one)*/
         genes=,     /* gene variable names                         */
         p1=         /* freq of parent(P1) to determine Dom. or Rec. */
         );

****************************************************************************
* Name:      SGENE                                                        *
*                                                                        *
* Purpose:   Single Locus Goodness of Fit Test                           *
*                                                                        *
* Written:   06/22/95                                                     *
*                                                                        *
* Modified:  10/03/95                                                     *
*            03/05/97                                                     *
*                                                                        *
* Example:   %sgene(ds=dst,                                               *
*            genes=BI RC DV SP LL DF F B  D  U  TU ,                      *
*            p1=9);                                                       *
****************************************************************************;

%local nogene word geneid i;
title2 'Gene Segregation and Linkage Analysis';
title3 'Single Locus Goodness of Fit Test';
title4 'Probability >.05 is accepted as Single Locus';
options missing=' ';
proc format;
   picture _prob
      low-0.05   ='9.999*'
      0.05<-<0.06='9.999 '
      0.06-high  ='9.99  '
      .          ='      '
      ;
   value _gnrx
     1='P1'
     2='P2'
     3='F1'
```

```
        4='F2'
        5='BC1P1'
        6='BC1P2'
        ;
        run;

   /*   create nogenes macro variable                              */
   /*   nogenes is the number of genes listed in &genes            */
   %let nogenes=0;
   %if &genes ne %then %do;
       %let word=%scan(&genes,1);
       %do %while (&word ne );
           %let nogenes=%eval(&nogenes+1);
           %let word=%scan(&genes,&nogenes+1);
           %end;
       %end;

   /*   create geneid macro variable                               */
   /*   geneid is the names of the genes in quotes                 */
   /*            used in array for identification in output        */
   %let word=%scan(&genes,1);
   %let geneid=%str(%'&word%');
   %do i=2 %to &nogenes;
       %let word=%scan(&genes,&i);
       %let geneid=%str(&geneid,%'&word%');
       %end;

data _gent(keep=id family gnr a gene aa bb ee)
     _look(keep=obs family gnr &genes);
   set &ds;
   length id aa bb ee 3
          obs 4
          a $ 1;
   array y{*}     &genes;
   array yc{*} $  n1-n&nogenes  (%unquote(&geneid)) ;

 /* create an obs. for each gene                                   */
 /* a     will be the response variable for phenotype of individual */
 /*       of each gene                                             */
 /* gene will be the character id of each gene name                */
 /* id   will be the numeric   id of the gene -used for sorting    */

   obs+1;
   id=0;
   do _i_=1 to dim(y);
        id+1;
        a=y{_i_}; gene=yc{_i_};
        a=upcase(a);
 /* ensure all values are in upper case                 */
 /* if the phenotype is dominant,  then aa=1             */
 /* if the phenotype is recessive, then bb=1             */
        aa=0; bb=0; ee=0;
        if      a ='D'  then aa=1;
        else if a ='R'  then bb=1;
        else if a =' '  then ee=1;
        else do;
```

```
            put '******* ERROR ******* '
                'Invalid value for gene ' gene
                ' (' gene '='a ')  at obs=' obs;
            output _look;
            end;
        output _gent;
        end;
    run;
    /* print any invalid data values for gene to notify user   */
title4 'Invalid data value for at least one gene'
       ' (value is not D, R, or missing)';
proc print data=_look;
    id obs;

    run;
proc datasets library=work nolist;
    delete _look;
    run;

title4 'Probability >.05 is accepted as Single Locus';
    /* compute the sums for number of dominant and recessive    */
    /* individuals in 6 generations                             */
proc means data= _gent noprint nway;
     class id family gnr;
     id gene;
     var aa bb ee;
     output out =_sum sum=d r missing;
     run;

    /* compute chi square and probability     */
data _chisq; set _sum;
    by id family;
    retain g1 omit;
    if first.family then do;
                        g1='   ';
                        omit='no ';
                        end;
    g1text='   ';

/* determine if the genotype of recurrent parent is dominant or    */
/* recessive;  this information is needed to choose                 */
/* expected 1:1 or 1:0 for chisq in BC1 and BC2                     */
if gnr =1 then do;
    t=sum(d,r);
    if t<=0 then omit='yes';
    if 0<d<&p1 then g1='REC';
        else if d>=&p1 then g1='DOM';
        else g1=' ';
    end;

if omit='yes' then delete;

if gnr >3 then do;
    t=d+r;
    chisq=0; df=0;
```

```
/* expected is 3:1 for chisq in F2 */
if gnr=4 then do;
      g1text='3:1';
      /*chisq for 3:1        */
      chisq=(d-t*0.75)**2/ (t*0.75) + (r-t*0.25)**2/(t*0.25);
      end;

   /* choose expected 1:1 or 1:0 for chisq in BC1 and BC2 */
   /* according to dominant or recessive recurrent parent */
   else if gnr =5 then do;
       if g1='DOM' then do;
           g1text='1:0';
           chisq=((d-t)**2)/ t ;
           end;
       else if g1='REC' then do;
           g1text='1:1';
           chisq=(d-t*0.5)**2/(t*0.5) +
                 (r-t*0.5)**2/(t*0.5);
           end;
       end;
   else if gnr =6 then do;
       if g1='DOM' then do;
           g1text='1:1';
           chisq=(d-t*0.5)**2/(t*0.5) + (r-t*0.5)**2/(t*0.5);
           end;
       if g1='REC' then do;
           g1text='1:0';
           chisq=((d-t)**2)/ t ;
           end;
       end;
   df=1;
   prob=probchi(chisq,df);
   prob=1-prob;
   end;
 drop omit;
 drop id _type_ t g1;
 run;

proc datasets library=work nolist;
   delete _gent _sum;
   run;

proc print noobs label uniform data=_chisq;
   by notsorted gene family;
   pageby gene;
   format chisq 8.2  prob _prob.  gnr _gnrx.;
   label gnr='GENERATION';
   label d='DOMINANT';
   label r='RECESSIVE';
   label g1text='EXPECTED';
   label _freq_='N';
   run;
%mend sgene;
```

File 5: LINKAGE.SAS

```
*********************************************************************
*                                                                   *
*       SASGENE 1.2                                                  *
*       Program for Analysis of                                     *
*       Gene Segregation and Linkage                                *
*       March 2, 1999                                               *
*                                                                   *
*                                                                   *
*       linkage.sas of SASGENE 1.2 differs from SASGENE 1.1         *
*       because there was an error in the calculation of            *
*       the SE for the F2 (coupling) as follows:                    *
*       in SASGENE 1.1 the formula was:                             *
*                   se=( (1-p*p)*(1+p*p)/(2*t*(1+2*p*p)) )**0.5;     *
*       in SASGENE 1.2 the formula is now:                          *
*                   se=( (1-p*p)*(2+p*p)/(2*t*(1+2*p*p)) )**0.5;     *
*                                                                   *
*********************************************************************;

%macro linkage
        (ds=_last_, /* SAS dataset to analyze(default:uses last one) */
         genes=,     /* gene variable names                          */
         p1=,        /* freq of P1 to determine Coupling or Repulsion */
         p2=         /* freq of P2 to determine Coupling or Repulsion */
         );

*********************************************************************
* Name:    LINKAGE                                                  *
*                                                                   *
* Purpose: Linkage Analysis for                                     *
*          Recombination Frequency Data in F2, BC1P1 & BC2P2 Pop.   *
*                                                                   *
* Written: 06/22/95                                                 *
*                                                                   *
* Modified: 10/03/95                                                *
*           03/05/97                                                *
*                                                                   *
* Example: %linkage(ds=dst,                                         *
*          genes=BI RC DV SP LL DF F B  D  U  TU ,                  *
*          p1=9                                                     *
*          p2=9                                                     *
*          );                                                       *
*                                                                   *
* Note:    The number of genes listed affects the amount of        *
*          time the program takes to execute.  The resources for    *
*          your platform will determine the number of genes you     *
*          can use.  Increasing the number of genes increases       *
*          the work space that is needed.                           *
*********************************************************************;

%local nogenes word geneid i;
title2 'Gene Segregation and Linkage Analysis';
title3 'Recombination Frequency Data in F2, BC1P1 & BC1P2 Population';
title4 'Prob with * indicates gene pair might be linked';
options missing=' ';
```

```
proc format;
   picture _prob
      low-0.05   ='9.999*'
      0.05<-<0.06='9.999 '
      0.06-high  ='9.99  '
      .          ='       '
      ;
   value _gnrx
      4='F2'
      5='BC1P1'
      6='BC1P2';
   run;
/*   create nogenes macro variable                                */
/*   nogenes is the number of genes listed in &genes              */
%let nogenes=0;
%if &genes ne %then %do;
      %let word=%scan(&genes,1);
      %do %while (&word ne );
          %let nogenes=%eval(&nogenes+1);
          %let word=%scan(&genes,&nogenes+1);
          %end;
      %end;

/*   create geneid macro variable                                 */
/*   geneid is the names of the genes in quotes                   */
/*           used in array for identification of output           */
%let word=%scan(&genes,1);
%let geneid=%str(%'&word%');
%do i=2 %to &nogenes;
      %let word=%scan(&genes,&i);
      %let geneid=%str(&geneid,%'&word%');
      %end;

data _gent;
   set &ds;
   length id aa bb cc dd ee  3
          m n              $ 1;
   array  y{*}     &genes;
   array yc{*} $  n1-n&nogenes  (%unquote(&geneid)) ;

/* create an obs. for each pair of genes                          */
/* m    will be the response variable for gene1                   */
/* n    will be the response variable for gene2                   */
/* id   will be the numeric id of the (i,j)th combination of gene pair*/
/* gene1         character id of the i-th part of (i,j) pair       */
/* gene2         character id of the j-th part of (i,j) pair       */

 obs+1;
 id=0;
   do _i_=1 to dim(y)-1;
      do _j_=_i_+1 to dim(y);
         id+1;
         m=y{_i_};   n=y{_j_};
         m=upcase(m);
         n=upcase(n);
         gene1=yc{_i_}; gene2=yc{_j_};
         aa=0; bb=0; cc=0; dd=0; ee=0;
```

```
         if      m ='D' and n ='D' then aa=1;
        else if m ='D' and n ='R' then bb=1;
        else if m ='R' and n ='D' then cc=1;
        else if m ='R' and n ='R' then dd=1;
        else if m =' ' or  n =' ' then ee=1;
        else put '*******ERROR******* '
                'Invalid data value on obs=' obs ' for '
                  yc{_i_}'=' m ' or '  yc{_j_}'=' n ;
      output;
      end;
    end;
  keep id family gnr m n gene1 gene2 aa bb cc dd ee;
  run;

 /* compute the sums of dominant and recessive individuals  */
 /* a=AABB b=AAbb c=aaBB d=aabb    */
proc means data= _gent noprint nway;
  class id family gnr;
  id gene1 gene2;
  var aa bb cc dd ee;
  output out =_sum(drop=_type_ ) sum=a b c d missing ;
  run;


data _P12; set _sum;
  by id family;
  retain phase
         p1dd p1dr p1rd p1rr
         p2dd p2dr p2rd p2rr
         omit;
  if first.family then do;
     p1dd=.; p1dr=.; p1rd=.; p1rr=.;
     p2dd=.; p2dr=.; p2rd=.; p2rr=.;
     phase=' ';
     omit='no ';
     end;

 if gnr=1 then do;
     t=sum(a,b,c,d);
     if t<=0 then omit='yes';
     if a=>&p1 then p1dd=1;
     if b=>&p1 then p1dr=1;
     if c=>&p1 then p1rd=1;
     if d=>&p1 then p1rr=1;
     end;
   else if gnr=2 then do;
     if a=>&p2 then p2dd=1;
     if b=>&p2 then p2dr=1;
     if c=>&p2 then p2rd=1;
     if d=>&p2 then p2rr=1;

 /* determine if phase= "C"(coupling),        */
 /*                     "R"(repulsion), or     */
 /*                     " "(useless phase).    */

     if      p1dd=1 and p2rr=1 then phase='C';
        else if p1rr=1 and p2dd=1 then phase='C';
```

```
            else if p1dr=1 and p2rd=1 then phase='R';
            else if p1rd=1 and p2dr=1 then phase='R';
       end;

   if omit='yes' then delete;

/* compute the chisq, probability, recombination frequencies (rf) */
/* and standard error (se).                                       */
if phase not=' ' and gnr>3 then do;
   t=sum(a,b,c,d);
   chisq=0; df=0;
   if gnr=4 then do;
       chisq=( a**2)/(t*9/16)+( b**2)/(t*3/16)+( c**2)/(t*3/16)
                  +( d**2) /(t*1/16) -t ;

       div=b*c-a*d;
       if div ne 0 then do;
          p=( (-(b*c+a*d)+((b*c+a*d)**2+a*d*(b*c-a*d))**0.5 )/div)
            **0.5;
          se=( (1-p*p)*(2+p*p)/(2*t*(1+2*p*p)) )**0.5;
          end;
       if phase='C' then rf=1-p;
          else if phase='R' then rf=p;
       end;

     else if 5<=gnr <=6 then do;
        chisq=  (a-t*0.25)**2/(t*0.25)+(b-t*0.25)**2/(t*0.25)
            + (c-t*0.25)**2/(t*0.25)+ (d-t*0.25)**2/(t*0.25);
        rf= (b+c)/t;
        se=(rf*(1-rf)/t)**0.5;
        end;
     df=3;
     prob=probchi(chisq,df);
     prob=1-prob;
     end;
drop omit;
drop p1dd p1dr p1rd p1rr p2dd p2dr p2rd p2rr    div p;
run;

/*  only print for good phases and generations 4, 5, and 6  */
data _gnr4to6;
   set _p12;
   if phase=' ' then delete;
   if gnr >3;
   run;
proc sort; by gnr phase id family; run;

proc print noobs label split='*' uniform data=_gnr4to6;
   by gnr;
   pageby gnr;
   var gene1 gene2  family phase _freq_ a b c d missing chisq df prob
     rf se;
   format chisq 5.1    rf se 6.3    prob _prob. gnr _gnrx.;
   label gnr='GENERATION';
   label family='FAM';
   label _freq_='N';
   label missing='MISS*-ING';
```

```
    label se='STD *ERROR';
    run;
%mend linkage;
```

SAS Output: Single-Gene Goodness-of-Fit

```
                   Cucumber Gene Linkage Example
                  Single Locus Goodness of Fit Test
             Probability >.05 is accepted as Single Locus
                       GENE=SS FAMILY=44
GENERATION    N   DOMINANT RECESSIVE MISSING EXPECTED CHISQ DF PROB
P1           45      45        0        0
P2           45       1       40        4
F1           54      49        5        0
F2          162     103       55        4      3:1    8.11  1  0.004*
BC1P1        81      78        3        0      1:0    0.11  1  0.73
BC1P2        81      38       42        1      1:1    0.20  1  0.65
```

SAS Output: Linkage Analysis

```
                      Cucumber Gene Linkage Example
            Recombination Frequency (RF) Data in F2, & BC1 Population
               Prob with * indicates gene pair might be linked
                            GENERATION=F2
                                        MISS-         STD
GENE1 GENE2 FAM PHASE  N   A   B   C   D  ING  CHISQ DF PROB    RF    ERROR
  U    SS   30   C   162  69  27  24  36   6   75.8  3  0.000* 0.323  0.036
  U    SS   44   C   162  77  27  26  28   4   35.5  3  0.000* 0.350  0.038
  U    NS   28   C   162  89  16  18  21  16   24.3  3  0.000* 0.265  0.034
  U    NS   30   C   162  74  22  33  27   6   35.0  3  0.000* 0.364  0.038
  RC   NS   30   R   162  83  34  24  15   6    4.8  3  0.18   0.559  0.042
```

Chapter 23

Mapping Functions

M. Humberto Reyes-Valdés

Purpose

To map genes and markers and to predict recombination frequencies.

Definitions

Mapping function: A mathematical function that relates map distances to recombination frequencies.

Genetic map distance (in morgans): The average number of crossovers per meiotic event between two loci. Genetic distance relates to physical distance, but they are not equivalent.

Morgan unit: A unit for expressing the distance between chromosome loci based on recombination. Haldane (1919) named it after T. H. Morgan.

Coincidence: Actual double recombinations/number expected with no interference. Each mapping function is based on an assumption about coincidence.

Originators

Carter, T.C. and Falconer, D.S. (1951). Stocks for detecting linkage in the mouse, and the theory of their design. *Journal of Genetics* 50:307-323.

Haldane, J.B.S. (1919). The combination of linkage values, and the calculation of distances between the loci of linked factors. *Journal of Genetics* 8:299-309.

Kosambi, D.D. (1944). The estimation of map distances from recombination values. *Annals of Eugenics* 12:172-175.

Pascoe, L. and Morton, N.E. (1987). The use of map functions in multipoint mapping. *American Journal of Human Genetics* 40:174-183.

Software Available

Reyes-Valdés, M.H. *GenMath* (a Mathematica application for genetics).

Key References Using the Formulas

Haley, C.S. and Knott, S.A. (1992). A simple regression method for mapping quantitative trait loci in line crosses using flanking markers. *Heredity* 69:315-324.
Reyes-Valdés, M.H. (2000). A model for marker-based selection in gene introgression breeding programs. *Crop Science* 40:91-98.

Contact

Dr. M. Humberto Reyes-Valdés, Universidad Autónoma Agraria Antonio Narro, Departamento de Fitomejoramiento, Buenavista, Saltillo, Coahuila, Mexico. C.P. 25315. E-mail: <mhreyes@uaaan.mx>.

EXAMPLES

Example 1

Convert the following recombination frequencies to map distances using the various mapping functions in Table 23.1:

$$[0.05, 0.10, 0.15, 0.20, 0.25]$$

TABLE 23.1. Mapping Functions

Author	Function	Coincidence
Haldane (1919)	$m = -\frac{1}{2}\log_e(1-2y)$	1
Kosambi (1944)	$m = \frac{1}{2}\tanh^{-1}(2y)$	$2y$
Pascoe and Morton (1987)	$m = -\frac{\log_e[(1-2y)^2/(1+2y+4y^2)]}{12} + \frac{\sqrt{3}\tan^{-1}[(1+4y)/\sqrt{3}]}{6} - 0.15115$	$(2y)^2$
Carter and Falconer (1951)	$m = \frac{1}{4}[\tanh^{-1}(2y) + \tan^{-1}(2y)]$	$(2y)^3$

Note: Where m = genetic map distance in morgans; y = recombination frequency; tan, $\tan^{-1}$ = tangent and inverse tangent, respectively; tanh, $\tanh^{-1}$ = hyperbolic tangent and inverse hyperbolic tangent, respectively.

The formulas in Table 23.1 can be used manually, but a computer program greatly facilitates calculations. Although the development of GenMath—a genetics package—is not fully complete, it can be used, at this time, for several applications, including mapping functions. The program runs in Mathematica software by Wolfram Research. With GenMath, proceed as follows:

1. Load GenMath in a Mathematica notebook. The prompts *In* and *Out* represent input and output, respectively.

```
In: <C:\genmath.m
```

2. To know the commands available in the package, type:

```
In: ?Global`Genmath`*
```

The output will be:

```
Out:
Abo        GenDis      Iden        PathAnalysis
Avef       HFun        IHFun       PMFun
CFun       Hw          IKFun       ReadConv
ChiTest    HwAbo       IPMFun      Ssd
Comp       Hwmean      KFun        Varc
Eftab      ICFun       Nsim
```

3. To know how a command works, e.g., HFun, type:

```
In: ?HFun
```

The output will be:

```
Out:
HFun[r] gives genetic distance in morgans for a given
   recombination  fraction r, based on Haldane mapping
   function
```

4. You can convert each value, one by one, as follows:

```
In: HFun[0.05]
Out: 0.0526803
```

5. Or you can convert the entire vector in one step.

```
In: Map[HFun,{0.05,0.1,0.15,0.2,0.25}]
Out: {0.0526803, 0.111572, 0.178337, 0.255413, 0.346574}
```

6. Since units in the output are in morgans, multiply by 100 to get centimorgans.

```
In: Map[HFun,{0.05,0.1,0.15,0.2,0.25}]100
Out: {5.26803, 11.1572, 17.8337, 25.5413, 34.6574}
```

7. To obtain the whole matrix of map distances in centimorgans with the use of the four functions, combine the commands: HFun (Haldane), KFun (Kosambi), PMFun (Pascoe and Morton), and CFun (Carter and Falconer).

```
In:
TableForm[{Map[HFun,{0.05,0.1,0.15,0.2,0.25}],
Map[KFun,{0.05,0.1,0.15,0.2,0.25}],
Map[PMFun,{0.05,0.1,0.15,0.2,0.25}],
Map[CFun,{0.05,0.1,0.15,0.2,0.25}]}100]
Out:
5.26803    11.1572    17.8337    25.5413    34.6574
5.01677    10.1366    15.476     21.1824    27.4653
5.00125    10.0201    15.1028    20.3322    25.8425
5.0001     10.0032    15.0244    20.1039    25.3238
```

Each row in this output corresponds to a given mapping function, in the same order depicted in Table 23.1. Notice that for low recombination frequencies (e.g., 0.05), map distances are similar with the use of all the mapping functions. However, as the recombination frequencies increase, map distances diverge. Thus, Haldane's mapping function is not a good choice for high recombination frequencies.

Example 2

Convert the following map distances, given in centimorgans, to recombination frequencies using the four mapping functions:

[10, 20, 30, 40, 50, 200]

One way to convert them is to find the analytical inverse of the mapping functions, i.e., to write y as a function of m, which may prove difficult when using the last two formulas in Table 23.1. With GenMath, use the commands for inverse mapping functions: IHFun, IKFun, IPMFun,

ICFun. To convert a single value, e.g., 10, with a given inverse mapping function, proceed as follows:

```
In: IPMFun[.1]
Out: 0.0998006
```

Notice that the map distance was divided by 100 to convert it to morgans before applying the command IPMFun.

To perform all the conversions and present them in table form, you can proceed as follows:

```
In:
TableForm[{Map[IHFun,{0.1,0.2,0.3,0.4,.5,2}],
Map[IKFun,{.1,.2,.3,.4,.5,2}],
Map[IPMFun,{.1,.2,.3,.4,.5,2}],
Map[ICFun,{.1,.2,.3,.4,.5,2}]}]//N
Out:
0.0906346   0.16484    0.225594   0.275336   0.31606    0.490842
0.0986877   0.189974   0.268525   0.332018   0.380797   0.499665
0.0998006   0.196886   0.285161   0.357854   0.41152    0.499987
0.0999684   0.198987   0.292644   0.372168   0.42959    0.5
```

Example 3

Plot recombination frequencies against map distances between 0 and 2 morgans using the functions of Haldane and Pascoe and Morton. With the use of GenMath, both plots can be combined as follows:

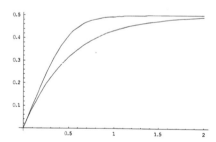

```
In:
Plot[{IHFun[x],IPMFun[x]},{x,0,2},PlotPoints->100]
Out:
```

The upper line corresponds to the function of Haldane, and the lower one to the function of Pascoe and Morton.

Final Remarks

All the formulas presented in this chapter assume coincidence = $(2y)^k$, where k is a constant that depends on a mapping function. The formulas include the most widely used mapping functions, but several other functions have also been developed. For an excellent account of this topic, the reader is referred to:

Crow, J.F. (1990). *Mapping functions. Genetics* 125:669-671.

Chapter 24

Bootstrap and Jackknife for Genetic Diversity Parameter Estimates

Julio Di Rienzo
Mónica Balzarini

Purpose

Bootstrap and jackknife are resampling (sample from a sample) techniques. Whenever distributional properties of parameter estimates cannot be analytically derived due to complex structures of sample data or statistics, bootstrap and jackknife procedures can provide empirical parameter estimates.

Definitions

Bootstrap for Mean and Standard Error

An original sample of size n should be available to obtain bootstrap samples. A bootstrap sample is a random sample of size n drawn with replacement from the original sample. The process is as follows:

1. Obtain a bootstrap sample and calculate the desired parameter estimator, say $\hat{\theta}$, from it.
2. Repeat step 1 K times. The bootstrap mean is the mean across all K runs of the estimator values,

$$\bar{\theta}^{B} = \frac{1}{K}\sum_{i=1}^{K}\hat{\theta}_{i},$$

and the bootstrap standard error of the estimator is

$$SE^B = \sqrt{\frac{1}{K-1} \sum_{i=1}^{K} (\hat{\theta}_i - \bar{\theta}^B)^2}.$$

Jackknife for Mean and Standard Error

An original sample of size n should be available to obtain jackknife samples. A jackknife sample is newly obtained from the original sample by leaving out one sample unit or object. The ith jackknife sample is the original data set with the ith object removed. The process is as follows:

1. Obtain a jackknife sample and calculate the desired parameter estimator, say $\hat{\theta}$, from it.
2. Repeat step 1 leaving out a different sample unit each time. For an original sample of size n, the total number of jackknife samples will be n. The jackknife mean is the mean across n estimator values,

$$\bar{\theta}^J = \frac{1}{n} \sum_{i=1}^{n} \hat{\theta}_i,$$

and the jackknife standard error of the estimator is

$$SE^J = \sqrt{\frac{n-1}{n} \sum_{i=1}^{a} (\hat{\theta}_i - \bar{\theta}^J)^2}.$$

The coefficient multiplying the sample variance is an inflation factor used to account for a smaller variation among jackknife samples than among bootstrap samples.

Application for Genetic Diversity Parameters

These computationally intensive techniques can be used to extract mean and standard errors of genetic diversity parameters from genomic data. With genotypes as sample units and several loci examined, the following multilocus statistics estimate genetic diversity:

1. Proportion of polymorphic loci (P) = the total number of polymorphic loci divided by the total number of loci examined. A locus is considered polymorphic if two or more alleles are detected.

2. Average number of alleles (Aa) = the total number of alleles counted in the sample divided by the total number of loci examined.
3. Effective number of alleles (Ae) = the reciprocal of the sum across all alleles.
4. Nei's expected heterozygosity (He),

$$He = \frac{1}{L} \sum_{j=1}^{L} \frac{2n}{2n-1} (1 - \sum_{i=1}^{a} x_i^2),$$

where n is the sample size, a is the number of alleles, x_i is the frequency of the ith allele at the jth locus, and L is the number of loci examined.
5. Nei's biased expected heterozygosity (BHe),

$$BHe = \frac{1}{L} \sum_{j=1}^{L} (1 - \sum_{i=1}^{a} x_i^2)$$

A computer program (Genetic_Diversity.exe) was develpoed to calculate bootstrap and jackknife means and standard errors of multilocus statistics. It may be obtained free of charge from <http://www.infostat.com.ar>Web page. It is a user-friendly program that allows reading of text files structured with genotypes and loci as row and column factors, respectively. By opening the program, a default file of five genotypes and thirteen loci is automatically loaded. The user may select either the bootstrap or jackknife procedure to calculate genetic diversity measures and their standard errors. If jackknife is chosen, an output sample size other than one (default value for the number of sample units left out each time) can be obtained. If bootstrap is chosen, the number of bootstrap replications, K, can be specified. The default value for K is 250. There is no maximum number of alleles that can be specified, but the limitation may be the number of different symbols available to identify them. The program does not distinguish between upper- and lowercase letters.

Jackknife and bootstrap routines can be easily adapted to other sample functions (not necessarily genetic diversity statistics). For questions, please contact Dr. Balzarini <mbalzari@agro.uncor.edu>.

Chapter 25

Software on Genetic Linkage and Mapping Available Through the Internet

Manjit S. Kang

Purpose

With the increasing use of Internet resources in all scientific fields, it becomes necessary to compile a meaningful list of software relative to mapping of markers and quantitative trait loci (QTL) that can be accessed through the Internet. Geneticists are now heavily engaged in mapping QTL for important plant and animal traits. Thus, they can benefit from such a list. The list assembled in this chapter, with modifications, is patterned after one that already exists at <http://linkage.rockefeller.edu/soft/> (attributed to Dr. Wentian Li of Rockefeller University). This list contains only those software programs that have a functional Web site. The Weizmann Institute of Science, Genome, and Bioinformatics also has a listing of some of the linkage and mapping-related software at the following Web site: <http://bioinformatics.weizmann.ac.il/repository/mapping_software.html>. Table 25.1 contains more than 100 such entries and a brief statement about the intended purpose of each software and its important features.

The reader should note that listed Web sites can change location without notice. No guarantee is made that they will remain operational. This list, or any other such list, should be regarded as informational in nature.

The reader is encouraged to check the Web sites listed in Table 25.1 to obtain additional information about software(s) of interest. Although many can be downloaded free of charge, others may require a fee.

TABLE 25.1. An Abbreviated Listing of Software on Genetic Linkage and Mapping

Name of Software	Features/Purpose	Web Site
ACT: Analysis of Complex Traits	Various modules can do the following: Calculate the proportion of genes which are identical by descent, shared in a nuclear family, assess increased allele sharing between all pairs of affected relatives, perform multivariate analysis of complex traits, estimate variance components using maximum likelihood and quasi-likelihood, and generate first-degree relationship coefficients for extended families.	http://www.epigenetic.org/Linkage/act.html
ALLASS: ALLele ASSociation	Nonparametric linkage and association mapping of disease genes. The ALLASS program implements a model for localizing disease genes by allelic association in a set of disease and normal haplotypes. The program can also model linkage disequilibrium between pairs of SNPs where SNP haplotypes are available.	http://cedar.genetics.soton.ac.uk/pub/PROGRAMS/ALLASS
ALP: Automated Linkage Preprocessor	ALP, a Microsoft Windows application, is designed to analyze microsatellite DNA fragments separated on an automated laser fluorescence sequencer (ALF, Pharmacia Biotech). ALP sizes DNA fragments, removes PCR stutter and other artifacts, if provided with pedigree data it performs genotyping checks to ensure a Mendelian inheritance pattern is followed, and formats data for Lathrop's linkage program package.	http://www.hgu.mrc.ac.uk/Softdata/ALP/
Analyze	Simplifies the performance of a large array of parametric and nonparametric tests for linkage and association on data entered in linkage format pedigree and parameter files.	ftp://ftp.ebi.ac.uk/pub/software/linkage_and_mapping/linkage_cpmc_columbia/analyze/
APM: Affected Pedigree-member Method	Linkage analysis.	http://watson.hgen.pitt.edu/register/docs/apm.html
Arlequin	An exploratory population genetics software environment able to handle large samples of molecular data (RFLPs, DNA sequences, microsatellites).	http://lgb.unige.ch/arlequin/
ASPEX: Affected Sibling Pairs EXclusion map	Performs multipoint exclusion mapping of affected sibling pair data for discrete traits. Allows genome-wide scan.	ftp://lahmed.stanford.edu/pub/aspex/

Name	Description	URL
Autoscan	Helps automate the tedious process of creating input files from genotype data of genome-wide scans.	http://www.genetics.ucla.edu/software/autoscan/
Beta	Performs nonparametric linkage analysis using allele sharing in sibling pairs.	http://cedar.genetics.soton.ac.uk/pub/PROGRAMS/BETA
BLOCK: BLOCKing Gibbs sampler for pedigree analysis	Perform general pedigree analysis on a general pedigree. Performs two-point linkage analysis on a general pedigree with an arbitrary number of alleles. Employs Markov chain Monte Carlo and Gibb's sampling.	http://www.cs.auc.dk/~claus/block.html
Borel (see also PANGAEA)	Programs for inference of genealogical relationships from genetic data, including siblingship inference.	ftp://ftp.u.washington.edu/pub/user-supported/pangaea/PANGAEA/BOREL/
CarthaGene	CarthaGene is a genetic/radiated hybrid mapping software. Uses multipoint maximum likelihood estimations of distances. Handles data made up of several distinct populations, which may each be either F2 backcross, recombinant inbred lines, F2 intercross, phase known outbreds, and/or radiated hybrids. Keeps best maps.	http://www.inra.fr/bia/T/CarthaGene/
CASPAR: Computerized Affected Sibling Pair Analyzer and Reporter	An exploratory program to study the genetics of complex (polygenic) diseases. Helps perform conditional linkage analyses, in which the population can be subdivided according to criteria at some loci and analyzed for linkage at other loci. Uses simulation to overcome the problems inherent in multiple testing.	http://www.ncbi.nlm.nih.gov/CBBresearch/Schaffer/caspar.html
Ceph2Map	Constructs linkage maps. Developed from CRI-MAP v2.4.	http://cedar.genetics.soton.ac.uk/pub/PROGRAMS/ceph2map
Clump	Utilizes the Monte Carlo method for assessing significance of a case-control association study with multiallelic markers. Useful for any 2 × N contingency table, especially when N is large.	http://www.mds.qmw.ac.uk/statgen/dcurtis/software.html
Combin	A software package developed for constructing ultradense linkage maps. Handles RFLP, SSR, and AFLP marker data.	http://www.dpw.wau.nl/pv/pub/combin/
COMDS: COMbineD Segregation and linkage analysis	Combined segregation and linkage analysis, incorporating severity and diathesis.	http://cedar.genetics.soton.ac.uk/pub/PROGRAMS/comds

TABLE 25.1 *(continued)*

Name of Software	Features/Purpose	Web Site
CoPE: Collaborative Pedigree drawing Environment	A JAVA program for drawing pedigrees and a standardized system for pedigree storage. Intended for epidemiologists and statisticians to share their familial data through networks.	http://www.infobiogen.fr/services/CoPE/
CRI-MAP	Allows automated construction of multilocus linkage maps.	http://compgen.rutgers.edu/multimap/crimap/index.html
CRI-MAP-PVM: CRI-MAP with Parallel Virtual Machine	A version of the CRI-MAP program for genetic likelihood computations that runs CRI-MAP's FLIPS and ALL functions in parallel on a distributed network of work stations.	http://compgen.rutgers.edu/multimap/crimappvm.html
Cyrillic	A program for drawing pedigrees and for linking their data to programs for calculating genetic risks, analyzing linkage to DNA markers, and aligning haplotypes.	http://www.cyrillicsoftware.com/
dGene	A simple dBASE III program for the management of pedigree and locus data. Permits easy extraction of genetic data for use with Mendel and Fisher.	http://www.biomath.medsch.ucla.edu/faculty/klange/software.html
DMap: Disequilibrium Map	Uses information from all disease locus-marker pairs while modeling the variability in disequilibrium values due to the evolutionary dynamics of the population.	http://lib.stat.cmu.edu/~bdevlin/
ECLIPSE (see also PANGAEA): Error Correcting Likelihoods In Pedigree Structure Estimation	It is a set of three programs (preproc, eclipse2, and eclipse3). Analyzes genetic-marker data for genotypic errors and pedigree errors.	http://stat.washington.edu/thompson/Genepi/pangea.shtml.
EH (EH+): Estimating Haplotype-frequencies	Estimates haplotype frequencies. Also provides log likelihood, chi-squares, and the degrees of freedom under H_0 (no allelic association) and H_1 (allelic association) hypotheses.	ftp://linkage.rockefeller.edu/software/eh/
ERPA: Extended Relative Pair Analysis	Performs nonparametric linkage analysis.	ftp://ftp.ebi.ac.uk/pub/software/linkage_and_mapping/statgen/dcurtis/

Name	Description	URL
ETDT: Extended Transmission/Disequilibrium Test	Performs TDT on markers with more than two alleles using a logistic regression analysis.	ftp://ftp.ebi.ac.uk/pub/software/linkage_and_mapping/statgen/dcurtis/
FASTLINK (see also LINKAGE): faster version of LINKAGE	Maps disease genes and their approximate locations.	http://www.ncbi.nlm.nih.gov/CBBresearch/Schaffer/fastlink.html
FAST-MAP: Fluorescent Allele-calling Software Toolkit: Microsatellite Automation Package	A pattern recognition program that facilitates fully automated genotyping of microsatellite markers.	http://www-2.cs.cmu.edu/~genome/FAST-MAP.html
FASTSLINK (see also SLINK): faster SLINK	Conditional simulation of genetic data on pedigrees.	http://watson.hgen.pitt.edu/register/soft_doc.html
FBAT: Family-Based Association Test	A program for implementing a broad class of family-based association tests that are adjusted for population admixture.	http://www.biostat.harvard.edu/~fbat/fbat.htm
Firstord	A method for preliminary ordering of loci based on two-point LOD scores.	http://www.mds.qmw.ac.uk/statgen/dcurtis/software.html
Fisher	A program for genetic analysis of biometric traits that are controlled by a combination of polygenic inheritance and complex environmental factors.	http://www.biomath.medsch.ucla.edu/faculty/klange/software.html
GAP: Genetic Analysis Package	A comprehensive package for the management and analysis of pedigree data. Provides powerful database management tools specifically designed for family data. Automatic pedigree drawing. Segregation and linkage analysis, based on traditional maximum likelihood methods and newer, more powerful, Monte Carlo methods that can model both genetic and environmental factors.	http://icarus2.hsc.usc.edu/epicenter/gap.html
GAS: Genetic Analysis System	An integrated computer program designed to automate and accelerate the acquisition and analysis of genomic data.	http://users.ox.ac.uk/~ayoung/gas.html
GASP: Genometric Analysis Simulation Program	Generates samples of family data based on user-specified genetic models. Verifies analysis algorithms relative to the underlying theory. Tests the statistical validity of newly developed methods of genetic segregation and linkage analysis and investigates the statistical properties of test statistics. Determines the power and robustness of these methods. Allows application of insights gained from these simulation experiments to ongoing collaborative genetic analyses.	http://www.nhgri.nih.gov/DIR/IDRB/GASP/

TABLE 25.1 *(continued)*

Name of Software	Features/Purpose	Web Site
GASSOC: Genetic ASSOCiation analysis software	Statistical methods for genetic associations using cases and their parents. Include TDT for multiple marker alleles.	http://www.mayo.edu/statgen/
Genehunter	Used for multipoint linkage analysis and nonparametric linkage analysis.	http://www.fhcrc.org/labs/kruglyak/Downloads/index.html
Genehunter-Imprinting	In German. Parametric (LOD score) analysis of traits conditioned by imprinted genes.	http://www.meb.uni-bonn.de/imbie/mitarbeiter/strauch/
GenoCheck	Identifies genotypes likely to be errors. Based on Fastlink.	http://www.crpc.rice.edu/softlib/geno.html
GenoDB: GENOtype DataBase	Manipulates large amounts of genotype data generated with fluorescently labeled dinucleotide markers.	http://osteoporosis.creighton.edu/
GGT: Graphical Geno-Typing package	Enables representation of molecular marker data by simple chromosome drawings in several ways. Commonly used marker file types that contain marker information serve as input for this program.	http://www.dpw.wau.nl/pv/PUB/ggt/
GRR: Graphical Representation of Relationships	Designed for detection of relationship specification errors in general pedigrees by use of genome scan marker data.	http://qtl.well.ox.ac.uk/GRR/
GSCAN: Genomic Software for Complex Analysis of Nucleotides	Linkage program based on a semiparametric method. Allows semiparametric two-point linkage, linkage disequilibrium, and combined linkage/linkage-disequilibrium analysis of general pedigree data for discrete traits, including pedigree consistency checks and pedigree drawing, gene-gene and gene-environment interaction incorporation, and Z-score computation.	http://cougar.fhcrc.org/~filq/html/main.htm
Haplo	Haplotyping with computation of conditional probabilities.	http://watson.hgen.pitt.edu/register/soft_doc.html
HAPPY: reconstructing HAPlotYpes	Two-stage analysis: ancestral haplotype reconstruction using dynamic programming followed by QTL testing by linear regression.	http://www.well.ox.ac.uk/~rmott/happy.html

272

Name	Description	URL
Hardy (see also PANGAEA)	Hardy contains program and documentation for Monte Carlo estimation of P values in sparse, two-dimensional contingency tables, or for Hardy Weinberg equilibrium.	http://www.stat.washington.edu/thompson/Genepi/Hardy.shtml
JoinMap	Software for the calculation of genetic linkage maps. Can handle many common types of mapping populations (BC1, F2, RILs, [doubled] haploids, full-sib family of outbreeders). Can combine (join) data derived from several sources into an integrated map.	http://www.plant.wageningen-ur.nl/default.asp?section=products&page=/products/mapping/joinmap/jmintro.htm
KINDRED	A program that stores and maintains data on families and members of families, and automatically draws pedigrees in a format suitable for presentation/publication.	http://icarus2.hsc.usc.edu/epicenter/kindred.html
LDB: Location DataBase	Gives locations for expressed sequences and polymorphic markers. Locations are obtained by integrating data of different types (genetic linkage maps, radiation hybrid maps, physical maps, cytogenetic data, and mouse homology) and constructing a single summary map. Integrates genetic linkage map and physical map.	ftp://cedar.genetics.soton.ac.uk/public_html/ldb.html
Linkage: general pedigrees (see also FASTLINK)	The core of the Linkage package is a series of programs for maximum likelihood estimation of recombination rates, calculation of LOD score tables, and analysis of genetic risks.	ftp://linkage.rockefeller.edu/software/linkage/
Linkbase	An easy and practical tool for connecting the genotype data produced by automatic sequencers (ABI Prism 377 [Perkin Elmer] and ALF [Pharmacia]) to linkage and sib-pair programs.	http://www.ktl.fi/molbio/software/linkbase/newintro.html
Loki (see also PANGAEA)	Analyzes quantitative traits observed on large pedigrees using Monte Carlo multipoint linkage and segregation analysis.	ftp://ftp.u.washington.edu/pub/user-supported/pangaea/PANGAEA/Loki/
Map/Map+/Map+H	Performs multiple pairwise linkage analysis and incorporates interference.	http://cedar.genetics.soton.ac.uk/pub/PROGRAMS/map+; http://cedar.genetics.soton.ac.uk/pub/PROGRAMS/map+h

TABLE 25.1 (continued)

Name of Software	Features/Purpose	Web Site
MAPL: MAPping and QTL analysis	Provides segregation ratio, linkage test, recombination value, grouping of markers, ordering of markers by metric multidimensional scaling, drawing maps, and graphical genotypes, as well as QTL analysis by interval mapping and ANOVA.	http://peach.ab.a.u-tokyo.ac.jp/~ukai/mapl98.html
Mapmaker/Exp	Constructs genetic linkage maps.	http://www-genome.wi.mit.edu/genome_software
Mapmaker/HOMOZ: HOMOZygosity mapping	Calculates multipoint LOD scores in pedigrees with inbreeding loops.	ftp://ftp-genome.wi.mit.edu/distribution/software/homoz/
Mapmaker/QTL	Helps map genes controlling quantitative traits.	ftp://ftp-genome.wi.mit.edu/distribution/software/newqtl/
Map Manager Classic	A program for Apple Macintosh personal computer that helps analyze the results of genetic mapping experiments using backcrosses, intercrosses, or recombinant inbred strains.	http://mapmgr.roswellpark.org/classic.html
Map Manager QT	A version of Map Manager with additional functions for analyzing quantitative traits. A graphic, interactive program to map quantitative trait loci by regression methods.	http://mapmgr.roswellpark.org/mmQT.html
Map Manager QTX	A version of Map Manager that combines the cross-platform design of Map Manager XP with enhanced QT analysis from Map Manager QT. Allows detection and localization of quantitative trait loci by fast regression-based single locus association, simple interval mapping, and composite interval mapping. Calculates empirical significance values by permutation tests. Allows a choice of Haldane, Kosambi, and Morgan mapping functions. Supports advanced backcross and advanced intercross designs.	http://mapmgr.roswellpark.org/mmQTX.html
MapPop	For selective mapping and bin mapping.	http://www.bio.unc.edu/faculty/vision/lab/mappop/

Name	Description	URL
Mapqtl	For calculation of QTL positions on genetic maps via interval mapping, composite interval mapping, or a nonparametric method.	http://www.plant.dlo.nl/default.asp?section=products&page=products/mapping/mapqtl/mqintro.htm
MCLEEPS: Monte Carlo Likelihood Estimation of Effective Population Size	For estimating effective population size from temporal changes in allele frequencies.	http://www.stat.washington.edu/thompson/Genepi/Mcleeps.shtml
MEGA2: Manipulation Environment for Genetic Analyses	A data-handling program for facilitating genetic linkage and association analyses.	http://watson.hgen.pitt.edu/mega2.html
Mendel	For genetic analysis of human pedigree data involving a small number of loci. Useful for segregation analysis, linkage calculations, genetic counseling, and allele frequency estimation.	http://www.biomath.medsch.ucla.edu/faculty/klange/software.html
MFLINK: Model Free Linkage analysis	Performs (nearly) model-free linkage analysis.	http://www.mds.qmw.ac.uk/statgen/dcurtis/software.html
MIM: Multipoint Identical-by-descent Method	For partitioning genetic variance of quantitative traits to specific chromosome regions using data on nuclear families.	ftp://morgan.med.utah.edu/pub/Mim/
Mld	A shuffling version of conditional tests for different combinations of allelic and genotypic disequilibrium on haploid and diploid data.	ftp://statgen.ncsu.edu/pub/zaykin/
MORGAN:MOnte CaRlo Genetic ANalysis (see also PANGAEA)	For segregation and linkage analysis, using Markov chain and Monte Carlo methods. Includes MCMC methods for multilocus gene identity by descent and homozygosity mapping.	http://www.stat.washington.edu/thompson/Genepi/Morgan.shtml
MultiMap	For automated construction of genetic maps. Developed for large-scale linkage mapping of markers genotyped in reference pedigrees. Adapted for automated construction of radiation hybrid maps.	http://compgen.rutgers.edu/multimap/index.shtml
NOPAR	Nonparametric linkage and association tests for quantitative traits.	http://cedar.genetics.soton.ac.uk/pub/PROGRAMS/nopar/

TABLE 25.1 (continued)

Name of Software	Features/Purpose	Web Site
PANGAEA: Pedigree Analysis for Genetics And Epidemiological Attributes	A nine-program package for genetic analyses including: Borel, Hardy, MORGAN, Pedpack, InSegT, Loki, MCLEEPS, Pedfiddler, and Eclipse.	http://www.stat.washington.edu/thompson/Genepi/pangaea.shtml
PAP: Pedigree Analysis Package	Computes the likelihood of specified parameter values; provides the probability of each genotype for pedigree members; simulates phenotypes for output into files; maximizes the likelihood over specified parameters (with or without standard errors); computes the standard errors of parameters for unknown estimates; simulates phenotypes and estimates parameter values; estimates expected LOD scores; and computes a grid of likelihood over one or two parameters. Also does TDT.	http://hasstedt.genetics.utah.edu/
PED: PEdigree Drawing software	A tool for fast and standardized drawing of pedigrees.	http://www.medgen.de/ped/index.htm
PedCheck	Detects marker-typing incompatibilities in pedigree data.	http://watson.hgen.pitt.edu/register/docs/pedcheck.html
Pedfiddler	A set of programs to manipulate pedigree graphs. Can be used as a stand-alone version of the graphics facilities found in Pedpack. Pedfiddler is not fully compatible with Pedpack because it is not intended for analysis but for graphical purposes only.	http://www.stat.washington.edu/thompson/Genepi/Pedfiddler.shtml
PedHunter	A software package that facilitates creation and verification of pedigrees within large genealogies.	http://www.ncbi.nlm.nih.gov/CBBresearch/Schaffer/pedhunter.html
Pedigree/Draw	For creation, editing, and drawing of pedigrees of human or non-human families. The package consists of applications, example files, and a user's guide.	http://www.sfbr.org/sfbr/public/software/pedraw/peddraw.html
PedJava	Uses browser technology to enter pedigrees into a database.	http://cooke.gsf.de/wjst/download.cfm

Program	Description	URL
PedPlot: PEDigree PLOTting program	Helps view a family structure by generating a plot from a pedfile/datafile pair.	http://www.chg.duke.edu/software/pedplot.html
PEDRAW/WPEDRAW: PEDigree DRAWing/Windows PEDigree DRAWing	A pedigree drawing program using LINKAGE or LINKSYS data files.	http://www.mds.qmw.ac.uk/statgen/dcurtis/software.html
PEDSYS: PEDigree database system	For management of genetic, pedigree, and demographic data. Designed principally for use with pedigree analysis of either human or nonhuman subjects.	http://www.sfbr.org/sfbr/public/software/pedsys/pedsys.html
Pointer	For complex segregation analysis with mixed models.	http://cedar.genetics.soton.ac.uk/pub/PROGRAMS/pointer
Progeny	Provides a method of tracking, managing, and viewing genetic data using the most advanced pedigree drawing and database technology.	http://www.progeny2000.com/
QTDT: Quantitative (trait) Transmission/Disequilibrium Test	Provides a convenient one-stop interface for family-based tests of linkage disequilibrium.	http://www.sph.umich.edu/csg/abecasis/QTDT
QTL Cartographer	A suite of programs to map quantitative traits using a map of molecular markers.	http://statgen.ncsu.edu/qtlcart/cartographer.html
QUGENE: QUantitative GENEtics	A flexible platform for investigation of the characteristics of genetic models. The architecture of the software has two main levels: (1) the genotype-environment system engine and (2) the application modules. The engine is the platform on which the different systems for investigation are generated and the modules are used to conduct the simulation experiments.	http://pig.ag.uq.edu.au/qu-gene/
Relative	For relationship estimation, in particular between putative siblings when parents are untyped.	ftp://linkage.rockefeller.edu/software/relative/
RelCheck	For verifying relationships between all pairs of individuals in a linkage study. Allows for the presence of genotyping errors.	http://biosun01.biostat.jhsph.edu/~kbroman/software/
RHMAPPER: Radiation Hybrid MAPPER	An interactive program for radiation hybrid mapping. Uses a hidden Markov model for calculating maximum likelihood.	http://www.genome.wi.mit.edu/ftp/pub/software/rhmapper/

TABLE 25.1 (continued)

Name of Software	Features/Purpose	Web Site
SAGE: Statistical Analysis for Genetic Epidemiology	A software package containing more than twenty programs for use in genetic analysis of family and pedigree data.	http://darwin.cwru.edu/octance/sage/sage.php
Sib-Pair	For simple nonparametric genetic analysis of family data.	http://www2.qimr.edu.au/davidD/
Simibd	For performing nonparametric linkage analysis.	http://watson.hgen.pitt.edu/register/soft_doc.html
SimWalk2	For haplotype, parametric and nonparametric linkage, identity by descent, and mistyping analyses, using Markov chain, Monte Carlo, and simulated annealing algorithm.	http://watson.hgen.pitt.edu/register/soft_doc.html
SOLAR: Sequential Oligogenic Linkage Analysis Routines	A software package for genetic variance components analysis, including linkage analysis, quantitative genetic analysis, and covariate screening.	http://www.sfbr.org/sfbr/public/software/solar/index.html
SPERM	For the analysis of sperm typing data.	http://www.biomath.medsch.ucla.edu/faculty/klange/software.html
SPERMSEG	For analysis of segregation in single-sperm data.	http://galton.uchicago.edu/~mcpeek/software/spermseg/
SPLINK: Affected Sib Pairs LINKage Analysis	A program for sibling pair linkage analysis. Maximum likelihood subject to possible triangle restriction. Marker haplotypes based on several closely linked markers. Haplotype frequencies are estimated from the data.	http://www-gene.cimr.cam.ac.uk/clayton/software/
TDT-PC: Transmission Disequilibrium Test Power Calculator	To compute the statistical power of the Transmission/Disequilibrium Test (TDT), which is a powerful test for linkage in the presence of association.	http://biosun01.biostat.jhsph.edu/~wmchen/pc.html
TDT/S-TDT: Transmission Disequilibrium Test and Sibling-Transmission Disequilibrium Test	Provides separate results for TDT, S-TDT, and the combined (overall) test.	http://genomics.med.upenn.edu/spielman/TDT.htm

Transmit	For transmission disequilibrium testing. Marker haplotypes based on several closely linked markers. Parental genotype and/or haplotype phase may be missing.	http://www.gene.cimr.cam.ac.uk/clayton/software
2DMAP: 2-Dimensional Crossover-based MAP	For constructing two-dimensional crossover-based maps.	http://www.genlink.wustl.edu/software/
Typenext	To simulate marker data for untyped individuals to determine how much information each untyped individual would contribute if typed.	ftp://linkage.rockefeller.edu/software/typenext/
Vitesse	For likelihood calculation on pedigrees.	http://watson.hgen.pitt.edu/register/soft_doc.html
Web-Preplink	Prepares data files for Linkage using web interface.	http://linkage.rockefeller.edu/gui/webpreplink.html

Note: AFLP = Amplified fragment length polymorphism; LOD = Logarithm of odds; the odds are the likelihood that linkage exists relative to the likelihood that linkage does not exist; QTL = quantitative trait loci; RIL = recombinant inbred line; RFLP = restriction fragment length polymorphism; SNP = single nucleotide polymorphism; SSR = simple sequence repeat.

Chapter 26

Gregor

Todd Krone

Importance

An easy to use and valuable simulation tool to help plant breeders create population scenarios and subsequently observe the effects of selection via phenotype and/or genotype.

Originators

Dr. Nick Tinker and Dr. Diane E. Mather
Tinker, N.A. and Mather, D.E. (1993). GREGOR: Software for genetic simulation. *Journal of Heredity* 84:237.

Summary of One Scientist's Use of Gregor

This software program was found to be very useful when planning and executing backcrossing and breeding programs in maize. Its simplicity allows quick and efficient testing of various breeding questions. Many options and applications can be utilized in this program, but this brief summary does not allow complete coverage. I used Gregor for:

1. Testing the effects of varying assumptions on a backcrossing and breeding program. Many assumptions are made in a plant breeding program based on an understanding of the principles of plant breeding theory. Although theory gives an understanding of the effects that may result from varying assumptions, simulating a wide array of assumptions in Gregor provides a much deeper understanding of the effects. The many assumptions that can be tested using Gregor are

- number of genes affecting a trait,
- magnitude of gene effects,
- gene action,
- heritability,
- population size, and
- population type (e.g., F_2, doubled haploids, recombinant inbred lines [RILs], etc.).

Gregor gives several options for evaluation of the simulation, such as viewing graphical genotypes and summary statistics for any given trait. It is also very simple to export data to Microsoft Excel or other programs to evaluate the results more deeply. This process helps one understand the assumptions that are critical to the success of a program.

2. Testing the effect of varying assumptions on the effectiveness of gene mapping. The variables can be varied, as in the breeding program simulations (point number 1), along with marker number, distribution of markers, and marker dominance/additivity. Although Gregor can give summary statistics for markers of interest, it is found to be most useful when exporting data to Excel or Mapmaker for analysis. Theories can lead to good assumptions, but testing them in Gregor is a fast and inexpensive way to verify them.

3. Testing the effect of varying assumptions on selection for a trait via molecular markers. Once a breeding program is established, and molecular marker associations have been determined for a trait of interest, marker-assisted selection can be applied. This is a subheading of the breeding program as marker-assisted selection is simply another tool that can be used in a breeding program. In Gregor, molecular markers are listed as separate menu options from population and trait. The use of markers in breeding programs is not generally accepted as a common tool. Rather, it is seen as a developing tool. Thus, it is listed as a separate application. In many applications, marker-assisted selection may be appropriate. Prior to executing a selection scheme, it should be run through Gregor first to determine the appropriate marker number and distribution. This has been particularly useful in developing marker-assisted selection for backcross breeding.

Gregor is limited in that it is DOS based and restricts population size. However, other publicly available simulation software that can simulate breeding situations in such an effective and easy way are not known. Overall, the primary benefit of Gregor is that you can easily tailor simulations for any breeding scenario. In addition to having a sound knowledge of principles of genetics and breeding, the breeder should benefit from a virtual run of the breeding program. The software allows wise testing of practices prior to expending large sums of money and time executing a breeding program.

Software Available

Dr. Diane E. Mather, Department of Plant Science, McGill University, 21111 Lakeshore Road, Ste-Anne-de-Bellevue, Québec H9X 3V9, Canada. FAX: 514-398-7897. E-mail: <dianemather@mcgill.ca>. For a link to a Web site that describes Gregor, set your browser to <http://gnome. agrenv.mcgill.ca>.

Dr. Nicholas A. Tinker, Agriculture and Agri-Food, Canada, Biometrics and Bioproducts, ECORC, K. W. Neatby Building, Floor 2, Room 2056, 960 Carling Avenue, Ottawa, Ontario, K1A 0C6, CANADA.

Chapter 27

Analysis for an Experiment Designed As Augmented Lattice Square Design

Walter T. Federer

Importance

Augmented experiment designs are used internationally for screening large numbers of new genotypes used in early generations of plant-breeding programs. Any experiment design (complete block, incomplete block, row-column, or other) may be selected for the standard treatments replicated r times each. The blocks, incomplete blocks, or rows and columns are enlarged to accommodate the new treatments usually included only once in the experiment. The lattice square experiment design controls variation within each complete block in two directions (rows and columns). Augmented lattice square designs (ALSDs) are easily constructed as described by Federer, W. T. (2002) in "Construction and Analysis of an Augmented Lattice Square Design," *Biometrical Journal* 44(2):251-257. ALSDs can accommodate $c = 2k$ or $3k$ check or standard cultivars in $r = k$ complete blocks and $n = k^2(k-2)$ or $k^2(k-3)$ new genotypes. An ALSD with $k = 4 = r$, $c = 8$, and $n = 32$ illustrates a statistical analysis. A trend analysis using polynomial regression is used. The following data are presented for all sixty-four responses.

```
/*The infile for the data is auglsd8. The five columns of the data set
    below represent, respectively, Rep, Row, Col, Trt, and Yield. In
    the Trt column, checks are numbered 33-40 and new treatments are
    numbered 1-32*/

1   1   1   33   17
1   1   2    1    9
1   1   3    2    9
1   1   4   40   23
1   2   1   37   21
1   2   2   34   18
```

```
1  2  3   3   9
1  2  4   4   9
1  3  1   5   9
1  3  2  38  22
1  3  3  35  19
1  3  4   6   9
1  4  1   7   9
1  4  2   8  19
1  4  3  39  23
1  4  4  36  20
2  1  1  33  17
2  1  2   9   8
2  1  3  10   8
2  1  4  39  25
2  2  1  40  22
2  2  2  34  18
2  2  3  11  18
2  2  4  12  18
2  3  1  13  18
2  3  2  37  23
2  3  3  35  19
2  3  4  14  17
2  4  1  15  16
2  4  2  16  21
2  4  3  38  25
2  4  4  36  20
3  1  1  33  24
3  1  2  17  17
3  1  3  18  17
3  1  4  38  18
3  2  1  39  22
3  2  2  34  12
3  2  3  19  17
3  2  4  20  16
3  3  1  21  17
3  3  2  40  25
3  3  3  35  15
3  3  4  22  17
3  4  1  23  17
3  4  2  24  17
3  4  3  37  15
3  4  4  36  15
4  1  1  33  28
4  1  2  25  20
4  1  3  26  20
4  1  4  37  25
4  2  1  38  29
4  2  2  34  22
4  2  3  27  26
4  2  4  28  26
4  3  1  29  16
4  3  2  39  32
4  3  3  35  25
4  3  4  30  26
4  4  1  31  16
4  4  2  32  16
4  4  3  40  25
4  4  4  36  30
```

The SAS/GLM and SAS/MIXED codes for this data set are as follows:

```
options ls = 76;
proc iml;
   opn3= orpol(1:4,2); /* The 4 is the number of columns and 2 indi-
      cates that
      linear and quadratic polynomial regression coefficients are desired.
      */
   opn3[,1]= (1:4)`;
   op3 =opn3 ;      print op3; /* Print-out of coefficients. */
create opn3 from opn3[colname ={'COL' 'C1' 'C2'}]; append from opn3;
   close  opn3;run;
   opn4 =orpol(1:4,2); /* There are 4 rows and two regressions. */
   opn4[,1]=(1:4)` ;
   op4 =opn4;               print op4;
   create opn4 from opn4[colname ={'ROW' 'R1' 'R2'}]; append from opn4;
   close opn4; run;
data auglsd8;
   infile 'auglsd8.dat';
   input rep row col trt yield;
   if (trt>32) then new = 0; else new = 1;
/* This divides the 40 entries into 32 new treatments which are
considered as random effects and 8 checks which are fixed effects. */
   if (new) then trtn = 999; else trtn = trt;
data augbig;set auglsd8;
/* The regression coefficients are added to the data set. */
   idx = _n_;  run;
proc sort data = augbig;
   by COL; run;data augbig; merge augbig opn3; by COL; run;
proc sort data = augbig;
   by ROW; run; data augbig; merge augbig opn4; by ROW; run;
proc sort data = augbig; by idx; run;
proc glm   data = augbig;
   class  row col trt trtn rep;
   model yield = rep trt C1*rep R1*rep C1*R1*rep;
   lsmeans trt/out = lsmeans noprint; run;
proc sort data = lsmeans; by descending lsmean;
/* n is usually quite large and this statement arranges the fixed
effect means in descending order for viewing. */
proc print; run;
proc mixed data = augbig;
   class rep row col trt trtn;
   model yield = trtn/solution;
   random rep C1*rep R1*rep C1*R1*rep trt*new/solution;
/* These two statements obtain solutions for the various effects. */
   lsmeans trtn; make 'solutionr' out = sr noprint; run;
proc sort data = sr;
   by descending _est_;
/*The effect solutions are arranged from largest to smallest. */
proc print; run;
quit;
```

The output in the preceding example and program is presented in a modified version of the actual output. Following are the linear and quadratic coefficients:

```
                            OP3
                             1  -0.67082        0.5
                             2  -0.223607      -0.5
                             3   0.2236068     -0.5
                             4   0.6708204      0.5

                            OP4
                             1  -0.67082        0.5
                             2  -0.223607      -0.5
                             3   0.2236068     -0.5
                             4   0.6708204      0.5
```

                       Class Level Information
```
Class Levels    Values
ROW      4      1 2 3 4
COL      4      1 2 3 4
TRT     40      1 2 3 4 5 6 7 8 9 10 11 12 13 14 15 16 17 18 19 20 21
                22 23 24 25 26 27 28 29 30 31 32 33 34 35 36 37 38 39 40
TRTN     9      33 34 35 36 37 38 39 40 999 /* Checks numbered 33-40.*/
REP      4      1 2 3 4
               Number of observations in data set = 64
```

```
Dependent Variable: YIELD  Sum of          Mean
Source              DF     Squares         Square    F Value    Pr > F
Model               54     2022.63549      37.45621   5.73      0.0041
Error                9       58.84888       6.53876
Corrected Total     63     2081.48438
```

```
            R-Square          C.V.        Root MSE       YIELD Mean
            0.971727       13.62652        2.55710        18.7656
```

```
Dependent Variable: YIELD
Source              DF     Type I SS    Mean Square   F Value    Pr > F
REP                  3      634.92188    211.64063     32.37     0.0001
TRT                 39     1284.68750     32.94071      5.04     0.0070
C1*REP               4       52.76528     13.19132      2.02     0.1755
R1*REP               4        3.38948      0.84737      0.13     0.9677
C1*R1*REP            4       46.87135     11.71784      1.79     0.2145
Source              DF    Type III SS   Mean Square   F Value    Pr > F
REP                  3      240.67132     80.22377     12.27     0.0016
TRT                 39     1066.10112     27.33593      4.18     0.0137
C1*REP               4       37.86878      9.46720      1.45     0.2953
R1*REP               4       23.77297      5.94324      0.91     0.4985
C1*R1*REP            4       46.87135     11.71784      1.79     0.2145
```

```
Fixed effect means:OBS   _NAME_     TRT      LSMEAN     STDERR
                     1    YIELD       36     28.5625    3.98587
                     2    YIELD       22     27.5443    3.87268
                     3    YIELD       33     26.5625    3.98587
                     4    YIELD       30     24.3923    3.87268
                     5    YIELD       39     24.1171    1.67939
                     6    YIELD       20     23.9249    3.40407
                     7    YIELD       28     23.0745    3.40407
                     8    YIELD       40     23.0142    1.67939
                     9    YIELD       38     22.9711    1.67939
                    10    YIELD       19     22.3609    2.89414
```

```
11    YIELD    27    22.2280    2.89414
12    YIELD    18    22.0455    3.40407
13    YIELD    14    21.9964    3.87268
14    YIELD    17    21.7855    3.87268
15    YIELD    37    20.8976    1.67939
16    YIELD    35    20.5625    1.41148
17    YIELD    12    20.0757    3.40407
18    YIELD    11    19.6544    2.89414
19    YIELD    34    17.8125    1.41148
20    YIELD    25    17.2728    3.87268
21    YIELD    26    16.5147    3.40407
22    YIELD    24    15.8193    5.01784
23    YIELD     6    14.5671    3.87268
24    YIELD     1    14.3972    3.87268
25    YIELD     9    14.0444    3.87268
26    YIELD     8    13.7216    5.01784
27    YIELD    21    12.9400    5.01784
28    YIELD    16    12.6102    5.01784
29    YIELD     4    11.6749    3.40407
30    YIELD     2    11.5900    3.40407
31    YIELD     3    11.2568    2.89414
32    YIELD    10    10.5998    3.40407
33    YIELD    13    10.1342    5.01784
34    YIELD    23     8.6472    8.44815
35    YIELD    32     7.5988    5.01784
36    YIELD    29     7.0391    5.01784
37    YIELD     5     3.6367    5.01784
38    YIELD    31     3.5431    8.44815
39    YIELD    15    -0.5431    8.44815
40    YIELD     7    -3.1472    8.44815
```

```
                REML Estimation Iteration History
Iteration   Evaluations      Objective      Criterion
0           1             245.77696870
1           3             218.84012634     0.00762773
2           2             218.34805379     0.00012725
3           2             218.33482743     0.00000093
4           1             218.33472566     0.00000000
                 Convergence criteria met.
```

```
             Covariance Parameter Estimates (REML)

             Cov Parm          Estimate
             REP            11.99469629
             C1*REP          1.88018116
             R1*REP          3.98577174
             C1*R1*REP       4.69746155
             NEW*TRT         5.87662721
             Residual        8.41233535
```

```
                   Solution for Fixed Effects

Effect      TRTN      Estimate    Std Error    DF      t    Pr > |t|
INTERCEPT         15.84932467    1.85872722     3   8.53     0.0034

Effect      TRTN      Estimate    Std Error    DF      t    Pr > |t|
TRTN        33     6.02505345    1.80689098     9   3.33     0.0087
```

```
TRTN        34      1.78104271      1.61807612      9      1.10      0.2996
TRTN        35      3.51473327      1.61757976      9      2.17      0.0578
TRTN        36      4.97612515      1.80199562      9      2.76      0.0221
TRTN        37      5.08850236      1.67222656      9      3.04      0.0140
TRTN        38      7.28885283      1.67222656      9      4.36      0.0018
TRTN        39      9.63215629      1.67222656      9      5.76      0.0003
TRTN        40      8.35433917      1.67222656      9      5.00      0.0007
TRTN       999      0.00000000              .       .         .         .
```

Least Squares Means (*Check, fixed effect means.*)

Effect	TRTN	LSMEAN	Std Error	DF	t	Pr > |t|
TRTN	33	21.87437812	2.39139410	9	9.15	0.0001
TRTN	34	17.63036738	2.26946107	9	7.77	0.0001
TRTN	35	19.36405795	2.26934312	9	8.53	0.0001
TRTN	36	20.82544982	2.38836997	9	8.72	0.0001
TRTN	37	20.93782703	2.31267932	9	9.05	0.0001
TRTN	38	23.13817750	2.31267932	9	10.00	0.0001
TRTN	39	25.48148096	2.31267932	9	11.02	0.0001
TRTN	40	24.20366385	2.31267932	9	10.47	0.0001
TRTN	999	15.84932467	1.85872722	9	8.53	0.0001

(Solutions for random effects arranged in descending order. To obtain the means, add the intercept value to each of the effects below.)

OBS	_EFFECT_	REP	TRT	EST_	_SEPRED_	_DF_	T	_PT_
1	REP	4		4.85091339	1.86295604	9	2.60	0.0286
2	R1*REP	2		2.51920549	1.33546047	9	1.89	0.0919
3	NEW*TRT		8	2.16475642	1.94313336	9	1.11	0.2941
4	NEW*TRT		27	2.01517588	1.90869357	9	1.06	0.3186
5	NEW*TRT		28	1.85527717	1.93230928	9	0.96	0.3621
6	NEW*TRT		30	1.81072275	1.93311801	9	0.94	0.3734
7	C1*R1*REP	4		1.73101272	1.85245416	9	0.93	0.3745
8	NEW*TRT		16	1.54679253	1.94313336	9	0.80	0.4465
9	NEW*TRT		22	1.54645580	1.93311801	9	0.80	0.4443
10	C1*REP	4		1.25643193	1.09118038	9	1.15	0.2792
11	NEW*TRT		24	1.25209662	1.94313336	9	0.64	0.5354
12	NEW*TRT		11	1.23086038	1.90869357	9	0.64	0.5351
13	NEW*TRT		19	1.17618212	1.90869357	9	0.62	0.5530
14	NEW*TRT		12	1.11283909	1.93230928	9	0.58	0.5788
15	NEW*TRT		18	1.01457415	1.94260482	9	0.52	0.6141
16	R1*REP	1		1.01019332	1.33546047	9	0.76	0.4687
17	NEW*TRT		23	1.00479517	1.99102006	9	0.50	0.6259
18	NEW*TRT		20	0.93494169	1.93230928	9	0.48	0.6400
19	NEW*TRT		21	0.92045869	1.93273116	9	0.48	0.6452
20	NEW*TRT		17	0.88317994	1.94374142	9	0.45	0.6603
21	NEW*TRT		13	0.83974581	1.93273116	9	0.43	0.6742
22	NEW*TRT		14	0.40203254	1.93311801	9	0.21	0.8399
23	C1*REP	2		0.34480113	1.09118038	9	0.32	0.7592
24	C1*R1*REP	1		0.28936528	1.85245416	9	0.16	0.8793
25	NEW*TRT		33	0.00000000	2.42417557	9	0.00	1.0000
26	NEW*TRT		34	0.00000000	2.42417557	9	0.00	1.0000
27	NEW*TRT		35	0.00000000	2.42417557	9	0.00	1.0000
28	NEW*TRT		36	0.00000000	2.42417557	9	0.00	1.0000
29	NEW*TRT		37	0.00000000	2.42417557	9	0.00	1.0000
30	NEW*TRT		38	0.00000000	2.42417557	9	0.00	1.0000
31	NEW*TRT		39	0.00000000	2.42417557	9	0.00	1.0000
32	NEW*TRT		40	0.00000000	2.42417557	9	0.00	1.0000
33	C1*REP	1		-0.02662262	1.09118038	9	-0.02	0.9811
34	REP	2		-0.42232220	1.86295604	9	-0.23	0.8257
35	C1*R1*REP	3		-0.46971222	1.85245416	9	-0.25	0.8055
36	NEW*TRT		25	-0.53275905	1.94374142	9	-0.27	0.7902
37	NEW*TRT		26	-0.55027490	1.94260482	9	-0.28	0.7834

38	NEW*TRT		15	-0.60995214	1.99102006	9	-0.31	0.7663
39	R1*REP	3		-0.77362810	1.33546047	9	-0.58	0.5766
40	R1*REP	4		-0.91895723	1.33546047	9	-0.69	0.5087
41	NEW*TRT		31	-1.01254232	1.99102006	9	-0.51	0.6233
42	C1*REP	3		-1.02947994	1.09118038	9	-0.94	0.3701
43	C1*R1*REP	2		-1.32767888	1.85245416	9	-0.72	0.4917
44	NEW*TRT		2	-1.38564933	1.94260482	9	-0.71	0.4937
45	NEW*TRT		29	-1.39513647	1.93273116	9	-0.72	0.4887
46	NEW*TRT		1	-1.42624812	1.94374142	9	-0.73	0.4818
47	NEW*TRT		32	-1.45720674	1.94313336	9	-0.75	0.4724
48	NEW*TRT		4	-1.56655327	1.93230928	9	-0.81	0.4384
49	NEW*TRT		3	-1.58335059	1.90869357	9	-0.83	0.4282
50	REP	3		-1.67547560	1.86295604	9	-0.90	0.3919
51	NEW*TRT		5	-1.76704354	1.93273116	9	-0.91	0.3844
52	NEW*TRT		6	-1.78805600	1.93311801	9	-0.92	0.3791
53	NEW*TRT		7	-1.91714185	1.99102006	9	-0.96	0.3608
54	NEW*TRT		9	-2.24587207	1.94374142	9	-1.16	0.2777
55	NEW*TRT		10	-2.47310037	1.94260482	9	-1.27	0.2349
56	REP	1		-2.75311559	1.86295604	9	-1.48	0.1736

Chapter 28

Augmented Row-Column Design and Trend Analyses

Walter T. Federer
Russell D. Wolfinger

Purpose

To obtain estimates of augmented treatments under a mixed model.

Data

The fifteen-row by twelve-column designed data set, as outlined in Federer (1998), is used in this chapter. There are two checks repeated $r = 30$ times each and 120 new or augmented treatments each included once. Since the row-column design was not connected, in the sense that not all row, column, and treatment effects have solutions, it was necessary to use functions of row and column effects. Orthogonal polynomial regressions up to tenth degree for columns and up to twelfth degree for rows were computed. Those regressions with F-values lower than the 25 percent level were omitted from the model. Since row-column orientation may not be in the same direction as gradients in the experiment, interactions of row and column regressions were employed to account for the variation.

The treatments are divided into fixed effects (checks) and random effects (augmented treatments). An ordering of treatment effects from highest to lowest is useful since large numbers of augmented treatments are usually encountered in this type of screening experiment. The following SAS program constructs orthogonal polynomial coefficients.

References

Federer, W.T. (1998). Recovery of interblock, intergradient, and intervariety information in incomplete block and lattice rectangle designed experiments. *Biometrics* 54(2):471-481.

Wolfinger, R.D., Federer, W.T., and Cordero-Brana, O. (1997). Recovering information in augmented designs, using SAS PROC GLM and PROC MIXED. *Agronomy Journal* 89:856-859.

SAS Code

```
/* ---Create orthogonal polynomial regression coefficients.---*/
proc iml;
   opn12=orpol(1:12,10); /*---12 rows and up to tenth degree coeffi-
      cients---*/
   opn12[,1] = (1:12)`;
   op12=opn12;
   create opn12 from opn12[colname={'COL' 'C1' 'C2' 'C3' 'C4' 'C5'
'C6' 'C7' 'C8' 'C9' 'C10'}]; append from opn12;
   close opn12; run;
   opn15 = orpol(1:15,12); /*---15 columns , up to 12ᵗʰ degree coeffi-
      cients---*/
   opn15[,1]=(1:15)`;
   op15 = opn15;
   create opn15 from opn15[colname={'ROW' 'R1' 'R2' 'R3' 'R4' 'R5'
'R6' 'R7' 'R8' 'R9' 'R10' 'R11' 'R12'}];  append from opn15;
close opn15;
run;
/* The data set augmerc1.dat contained responses for grain weight and
      eight other characteristics of the 122 wheat genotypes (treat-
      ments) and comes from site number 1. */
data augsite1;
   infile 'augmerc1.dat';
   input site col row trt grainwt ca cb cc cd ra rb rc rd;
/* The following statements partition the 122 treatments into two
      sets, checks (fixed) and new (treated as random). */
   if (trt>120) then new = 0; else new = 1;  if (new) then trtn= 999;
      else trtn=trt;
/* The following steps create the data set augbig for analyses. */
data augbig; set augsite1;
   idx = _n_;
run;
proc sort data = augbig;
   by col; run;
data augbig;
   merge augbig opn12;
   by col;      run;
proc sort data = augbig;
   by row;  run;
data augbig;
   merge augbig opn15;
   by row;  run;
proc sort data = augbig;
```

```
   by idx;
run;
/* Exploratory model selection resulted in the following model for
     this data set. Residuals may also be obtained. */
proc glm data = augbig;
   class row col trt trtn;
   model grainwt =C1 C2 C3 C4 C6 C8 R1 R2 R4 R8 R10 R1*C1
   R1*C2 R1*C3 trt;
   output out = subres R = resid; proc print; /*---Printed in augbig--
   -*/
run;
/* "info nobound" may be included in the following statement if this
     type of solution for variance components is desired.  Also, if
     ANOVA solutions for the variance components are desired, the
     PARMS procedure statement may be used after the random statement
     in PROC MIXED.  REML solutions may not be appropriate for mean
     squares with few degrees of freedom.  The trt*new in the random
     statement is used when augmented treatments are treated as random
     effects. */
proc mixed data = augbig;
   class row col trt trtn;
   model grainwt = trtn/solution;
   random R1 R2 R4 R8 R10 C1 C2 C3 C4 C6 C8 R1*C1 R1*C2 R1*C3
   trt*new / solution;
   lsmeans trtn;
   make 'solutionr' out = sr noprint;
run;
/* The following statements arrange the solutions in descending order.
     */
proc sort data = sr;
   by descending _EST_ ;
proc print;
run;
```

Using the data and program described above, an abbreviated output is
    given below:

Class	Levels	Values
ROW	15	1 2 3 4 5 6 7 8 9 10 11 12 13 14 15
COL	12	1 2 3 4 5 6 7 8 9 10 11 12
TRT	122	1 2 3 4 5 6 7 8 9 10 11 12 13 14 15 16 17 18 19 20 21 22
		23 24 25 26 27 28 29 30 31 32 33 34 35 36 37 38 39 40 41
		42 43 44 45 46 47 48 49 50 51 52 53 54 55 56 57 58 59 60
		61 62 63 64 65 66 67 68 69 70 71 72 73 74 75 76 77 78 79
		80 81 82 83 84 85 86 87 88 89 90 91 92 93 94 95 96 97 98
		99 100 101 102 103 104 105 106 107 108 109 110 111 112
		113 114 115 116 117 118 119 120 121 122

TRTN   3   121 122 999   /* Checks are number 121 and 122. */

Dependent Variable: GRAINWT

Source	DF	Sum of Squares	Mean Square	F Value	Pr > F
Model	135	1685564.291	12485.661	3.62	0.0001
Error	44	151761.820	3449.132		
Corrected Total	179	1837326.111			

R-Square	C.V.	Root MSE	GRAINWT Mean
0.917401	6.664109	58.72931	881.2778

Dependent Variable: GRAINWT

Source	DF	Type I SS	Mean Square	F Value	Pr > F
C1	1	7.053	7.053	0.00	0.9641
C2	1	78620.049	78620.049	22.79	0.0001
C3	1	31357.514	31357.514	9.09	0.0043
C4	1	35185.066	35185.066	10.20	0.0026
C6	1	15954.687	15954.687	4.63	0.0370
C8	1	88778.180	88778.180	25.74	0.0001
R1	1	130227.001	130227.001	37.76	0.0001
R2	1	3182.964	3182.964	0.92	0.3420
R4	1	34117.771	34117.771	9.89	0.0030
R8	1	20274.909	20274.909	5.88	0.0195
R10	1	16821.594	16821.594	4.88	0.0325
C1*R1	1	138479.979	138479.979	40.15	0.0001
C2*R1	1	61605.531	61605.531	17.86	0.0001
C3*R1	1	13248.961	13248.961	3.84	0.0564
TRT	121	1017703.032	8410.769	2.44	0.0005

Source	DF	Type III SS	Mean Square	F Value	Pr > F
C1	1	12952.986	12952.986	3.76	0.0591
C2	1	48712.489	48712.489	14.12	0.0005
C3	1	42867.475	42867.475	12.43	0.0010
C4	1	22613.228	22613.228	6.56	0.0140
C6	1	31220.232	31220.232	9.05	0.0043
C8	1	77300.177	77300.177	22.41	0.0001
R1	1	28677.708	28677.708	8.31	0.0061
R2	1	12832.205	12832.205	3:72	0.0602
R4	1	4992.843	4992.843	1.45	0.2354
R8	1	20170.221	20170.221	5.85	0.0198
R10	1	15068.496	15068.496	4.37	0.0424
C1*R1	1	52885.122	52885.122	15.33	0.0003
C2*R1	1	24976.581	24976.581	7.24	0.0100
C3*R1	1	7998.357	7998.357	2.32	0.1350
TRT	121	1017703.032	8410.769	2.44	0.0005

The MIXED Procedure
Covariance Parameter Estimates (REML)

Cov Parm	Estimate
R1	9685.7206435
R2	147.76385914
R4	2382.8694456
R8	1165.7087131
R10	1055.8582605
C1	0.00000000
C2	5739.9404029
C3	2401.8203763
C4	1702.2868779
C6	1375.7002135
C8	6678.6086902
R1*C1	125150.99858
R1*C2	62327.029930
R1*C3	7191.3342088
NEW*TRT	2880.2792533
Residual	4385.0838420

Solution for Fixed Effects

Effect	TRTN	Estimate	Std Error	DF	t	Pr > |t|
INTERCEPT		887.85653363	7.78342790	44	114.07	0.0001
TRTN	121	22.56422549	14.52418333	44	1.55	0.1275
TRTN	122	-62.03676062	14.42385015	44	-4.30	0.0001

```
                          Least Squares Means
       Effect   TRTN      LSMEAN      Std Error       DF      t        Pr > |t|
       TRTN     121    910.42075912  12.23674167     44     74.40      0.0001
       TRTN     122    825.81977301  12.15736600     44     67.93      0.0001
       TRTN     999    887.85653363   7.78342790     44    114.07      0.0001
```

15 highest new treatment effects

```
OBS   _EFFECT_    TRT      _EST_           _SEPRED_       _DF_     T        _PT_
  1   R1*C1             345.50045585    76.02914755       44     4.54     0.0001
  2   R1                 95.98535922    21.73762881       44     4.42     0.0001
  3   NEW*TRT      60     86.34108807    42.34708520       44     2.04     0.0475
  4   C8                 79.38460062    19.40852139       44     4.09     0.0002
  5   NEW*TRT      21     62.77643482    43.26254487       44     1.45     0.1539
  6   R1*C3              62.42608658    57.39674225       44     1.09     0.2827
  7   NEW*TRT      11     58.69112792    43.23892480       44     1.36     0.1816
  8   NEW*TRT      99     56.51372478    42.26198657       44     1.34     0.1880
  9   NEW*TRT       2     54.08461760    42.75771851       44     1.26     0.2126
 10   NEW*TRT      35     49.26910001    42.25202699       44     1.17     0.2499
 11   NEW*TRT     118     49.05376892    42.18955585       44     1.16     0.2512
 12   NEW*TRT      58     48.88460726    42.12266951       44     1.16     0.2521
 13   NEW*TRT     111     46.12302563    42.56352985       44     1.08     0.2844
 14   C3                 45.39408465    18.47150180       44     2.46     0.0180
 15   R4                 44.40038633    20.28481411       44     2.19     0.0340
 16   NEW*TRT      46     44.15378862    42.18922373       44     1.05     0.3010
 17   NEW*TRT     120     44.13816913    42.42913604       44     1.04     0.3039
 18   NEW*TRT      61     42.88822110    42.53422618       44     1.01     0.3188
 19   NEW*TRT      38     39.16602034    42.24545264       44     0.93     0.3589
 20   NEW*TRT      82     38.75002538    42.56505326       44     0.91     0.3676
 21   NEW*TRT      90     37.87365601    42.33458245       44     0.89     0.3759
```

...................*Random effects 21 to 119 deleted.*

15 lowest new treatment effects.

```
120   NEW*TRT       5    -36.65859703    42.24564023       44    -0.87     0.3902
121   NEW*TRT      23    -36.94625121    43.22631392       44    -0.85     0.3973
122   NEW*TRT     107    -40.26108079    42.27021281       44    -0.95     0.3461
123   NEW*TRT      55    -40.71903829    42.17831063       44    -0.97     0.3396
124   NEW*TRT      42    -40.77369290    42.69072074       44    -0.96     0.3447
125   NEW*TRT      56    -41.41021039    42.21331886       44    -0.98     0.3320
126   NEW*TRT      28    -41.70085605    42.31372687       44    -0.99     0.3298
127   NEW*TRT      17    -49.15984900    42.44192942       44    -1.16     0.2530
128   NEW*TRT       6    -52.33527614    42.15283284       44    -1.24     0.2210
129   NEW*TRT      51    -57.85952344    42.47160123       44    -1.36     0.1800
130   C2                -73.26618742    19.28735943       44    -3.80     0.0004
131   NEW*TRT      52    -75.09659823    42.62419762       44    -1.76     0.0850
132   NEW*TRT      43    -80.06306299    42.56340110       44    -1.88     0.0666
133   NEW*TRT      44    -80.23408721    42.42893008       44    -1.89     0.0652
134   NEW*TRT      81    -85.99507895    42.43715136       44    -2.03     0.0488
135   NEW*TRT      50    -91.58752547    42.56624934       44    -2.15     0.0370
136   R1*C2            -238.5012759     73.78434826       44    -3.23     0.0023
```

Chapter 29

PROC GLM and PROC MIXED Codes for Trend Analyses for Row-Column-Designed Experiments

Walter T. Federer
Russell D. Wolfinger

Purpose

The program outlined in this chapter may be used for a variety of response models in row-column experiments. The example used to illustrate the steps in this program is a randomized complete block design (RCBD) that was laid out as an eight-row by seven-column field experiment. The experiment with data is described in Federer and Schlottfeldt (1954). The data include twenty plant heights in centimeters for seven different treatments. Since the experiment was planted as an eight-row by seven-column arrangement, an RCBD analysis may not be appropriate. The SAS code is written to compare five different response models that account for the spatial variation present. Variation orientation was different than the row-column layout.

SAS PROC GLM and PROC MIXED codes are presented for standard textbook analyses of variance for RCBD and for row-column design. These are followed by codes for trend analyses using standardized orthogonal polynomial regressions for rows and columns and for interaction of row and column regressions. A trend model using row, column, and interactions of row and column regressions appears to control the variation for this experiment. A PROC GLM analysis of variance and residuals is useful in exploratory model selection that takes account of spatial variation in the experiment. A PROC MIXED analysis is then used to recover information from the random effects (Federer, 1998; Federer and Wolfinger, 1998).

References

Federer, W.T. (1998). Recovery of interblock, intergradient, and intervariety information in incomplete block and lattice rectangle designed experiments. *Biometrics* 54(2):471-481.

Federer, W.T. and Schlottfeldt, C.S. (1954). The use of covariance to control gradients in experiments. *Biometrics* 10:282-290.

Federer, W.T. and Wolfinger, R.D. (1998). SAS PROC GLM and PROC MIXED code for recovering inter-effect information. *Agronomy Journal* 90:545-551.

SAS Code

```
/*---input the data---*/
data colrow;
   input height row col trt;
   /*---rescale data for stability---*/
   y = height/1000;
   datalines;
1299.2   1   1   6
 875.9   1   2   7
 960.7   1   3   4
1004.0   1   4   3
1173.2   1   5   1
1031.9   1   6   2
1421.1   1   7   5
1369.2   2   1   2
 844.2   2   2   5
 968.7   2   3   6
 975.5   2   4   7
1322.4   2   5   3
1172.6   2   6   1
1418.9   2   7   4
1169.5   3   1   1
 975.8   3   2   5
 873.4   3   3   3
 797.8   3   4   7
1069.7   3   5   2
1093.3   3   6   6
1169.6   3   7   4
1219.1   4   1   6
 971.7   4   2   1
 607.6   4   3   7
1000.0   4   4   4
1343.3   4   5   2
 999.4   4   6   5
1181.3   4   7   3
1120.0   5   1   6
 827.0   5   2   7
 671.9   5   3   4
 972.2   5   4   3
1083.7   5   5   1
1146.9   5   6   2
```

```
  993.8    5   7   5
 1031.5    6   1   7
  846.5    6   2   2
  667.8    6   3   4
  853.6    6   4   3
 1087.1    6   5   1
  990.2    6   6   5
 1021.9    6   7   6
 1076.4    7   1   2
  917.9    7   2   1
  627.6    7   3   5
  776.4    7   4   6
  960.4    7   5   3
  852.4    7   6   7
 1006.2    7   7   4
 1099.6    8   1   4
  947.4    8   2   5
  787.1    8   3   2
  898.3    8   4   1
 1174.9    8   5   3
 1003.3    8   6   6
  947.6    8   7   7
run;

/*---code to construct orthogonal polynomials---*/
proc iml;
   /*---7 columns and up to 6th degree polynomials---*/
   opn4=orpol(1:7,6);
   opn4[,1] = (1:7)`;
   op4= opn4;
   create opn4 from opn4[colname={'col' 'c1' 'c2' 'c3' 'c4' 'c5'
      'c6'}];
   append from opn4;
   close opn4;
   /*---8 rows and up to 7th degree polynomials---*/
   opn3=orpol(1:8,7);
   opn3[,1] = (1:8)`;
   op3 = opn3;
   create opn3 from opn3[colname={'row' 'r1' 'r2' 'r3' 'r4' 'r5'
      'r6' 'r7'}];
   append from opn3;
   close opn3;
run;
/*---merge in polynomial coefficients---*/
data rcbig;
   set colrow;
   idx = _n_;
proc sort data=rcbig;
   by col;
data rcbig;
   merge rcbig opn4;
   by col;
proc sort data=rcbig;
   by row;
data rcbig;
   merge rcbig opn3;
   by row;
```

```
proc sort data = rcbig;
   by idx;
run;
/*---3d plot of data, one can also substitue row and column variables
     as well as residuals for
     y to see how they model the trend---*/
proc g3d data=rcbig;
   plot row*col=y / rotate=20;
run;
/*---standard rcbd analysis with rows as blocks; treatments are not
     significantly different---*/
/*---fixed-effects row model for RCBD---*/
proc glm data=rcbig;
   class row col trt;
   model y = row trt;
   output out=subres r=resid;
run;
/*---standard row-column analysis fits much better than RBCD, and now
     treatment 7 is significantly different---*/
/*---fixed-effects row-column model---*/
proc glm data=rcbig;
   class row col trt;
   model y = row col trt;
   output out=subres r=resid;
run;
/*---model for random differential gradients within rows; does not fit
     as well as row-column model, but results are similar---*/
/*---fixed-effects model for gradients within rows ---*/
proc glm data=rcbig;
   class row col trt;
   model y = trt row c2*row c3*row c4*row;
   output out=subres r=resid;
run;
/*---Fixed-effects polynomial model; it may be that a trend and analy-
     sis is desired in that only certain polynomial regressions are
     needed to explain the row and column variation. Also, since spa-
     tial variation may not be in the row-column orientation of the
     experiment, interactions of regressions may be needed to account
     for this type of spatial variation. Of the 13 polynomial regres-
     sions for rows and columns and the 16 interactions ci*rj, for i,
     j = 1, 2, 3, and 4, those that had F-values greater than F at the
     25% level were retained in the response model.---*/
proc glm data=rcbig;
   class row col trt;
   model y = trt c1 c2 c3 c5 r1 r2 r3 r5 r6 r7 c1*r1 c2*r1 c2*r3 c3*r2
      c4*r1 c4*r2;
   output out=subres r=resid;
run;
/*---random polynomial coefficient  model---*/
proc mixed data=rcbig;
   class row col trt;
   model y = trt / ddfm=res;
   random c1 c2 c3 c5 r1 r2 r3 r5 r6 r7 c1*r1 c2*r1 c2*r3 c3*r2 c4*r1
      c4*r2;
   lsmeans trt / diff adjust=tukey;
run;
```

```
/*---Since the row and column variations were quite un-patterned,
     i.e., only c4, c6, and r4 were not in the model, the following
     analysis may be more appropriate for this data set.---*/
proc glm data=rcbig;
   class row col trt;
   model y = row col trt c1*r1 c2*r1 c2*r3 c3*r2 c4*r1 c4*r2;
run;
/*---combination  model---*/
proc mixed data=rcbig;
   class row col trt;
   model y = trt / ddfm=res;
   random row col  c1*r1 c2*r1 c2*r3 c3*r2 c4*r1 c4*r2
   repeated / type=sp(exp)(row col) subject=intercept;
   lsmeans trt / diff adjust=tukey;
run;
```

An abbreviated output from this code is presented below:

RCBD ANOVA

Dependent Variable: Y

Source	DF	Sum of Squares	Mean Square	F Value	Pr > F
Model	13	0.66219035	0.05093772	1.69	0.1004
Error	42	1.26958627	0.03022824		
Corrected Total	55	1.93177662			

R-Square	C.V.	Root MSE	Y Mean
0.342788	17.17205	0.173863	1.012475

Dependent Variable: Y

Source	DF	Type I SS	Mean Square	F Value	Pr > F
ROW	7	0.38831490	0.05547356	1.84	0.1056
TRT	6	0.27387545	0.04564591	1.51	0.1985
Source	DF	Type III SS	Mean Square	F Value	Pr > F
ROW	7	0.38831490	0.05547356	1.84	0.1056
TRT	6	0.27387545	0.04564591	1.51	0.1985

Row-column ANOVA

Dependent Variable: Y

Source	DF	Sum of Squares	Mean Square	F Value	Pr > F
Model	19	1.66711058	0.08774266	11.93	0.0001
Error	36	0.26466604	0.00735183		
Corrected Total	55	1.93177662			

R-Square	C.V.	Root MSE	Y Mean
0.862993	8.468638	0.085743	1.012475

Source	DF	Type I SS	Mean Square	F Value	Pr > F
ROW	7	0.38831490	0.05547356	7.55	0.0001
COL	6	1.15907213	0.19317869	26.28	0.0001
TRT	6	0.11972355	0.01995392	2.71	0.0281
Source	DF	Type III SS	Mean Square	F Value	Pr > F
ROW	7	0.38831490	0.05547356	7.55	0.0001
COL	6	1.00492023	0.16748671	22.78	0.0001
TRT	6	0.11972355	0.01995392	2.71	0.0281

Gradients within rows ANOVA

Dependent Variable: Y

Source	F	Sum of Squares	Mean Square	F Value	Pr > F
Model	37	1.72819875	0.04670807	4.13	0.0011
Error	18	0.20357788	0.01130988		
Corrected Total	55	1.93177662			

	R-Square	C.V.	Root MSE	Y Mean
	0.894616	10.50376	0.106348	1.012475

Dependent Variable: Y

Source	DF	Type I SS	Mean Square	F Value	Pr > F
TRT	6	0.27387545	0.04564591	4.04	0.0098
ROW	7	0.38831490	0.05547356	4.90	0.0030
C2*ROW	8	0.60283912	0.07535489	6.66	0.0004
C3*ROW	8	0.32440799	0.04055100	3.59	0.0116
C4*ROW	8	0.13876129	0.01734516	1.53	0.2142
Source	DF	Type III SS	Mean Square	F Value	Pr > F
TRT	6	0.25638292	0.04273049	3.78	0.0130
ROW	7	0.38831490	0.05547356	4.90	0.0030
C2*ROW	8	0.59754712	0.07469339	6.60	0.0004
C3*ROW	8	0.32649657	0.04081207	3.61	0.0113
C4*ROW	8	0.13876129	0.01734516	1.53	0.2142

Trend ANOVA
Dependent Variable: Y

Source	DF	Sum of Squares	Mean Square	F Value	Pr > F
Model	22	1.79302842	0.08150129	19.38	0.0001
Error	33	0.13874820	0.00420449		
Corrected Total	55	1.93177662			

	R-Square	C.V.	Root MSE	Y Mean
	0.928176	6.404311	0.064842	1.012475

Dependent Variable: Y

Source	DF	Type I SS	Mean Square	F Value	Pr > F
TRT	6	0.27387545	0.04564591	10.86	0.0001
C1	1	0.09681321	0.09681321	23.03	0.0001
C2	1	0.53598746	0.53598746	127.48	0.0001
C3	1	0.22278336	0.22278336	52.99	0.0001
C5	1	0.13314475	0.13314475	31.67	0.0001
R1	1	0.27808763	0.27808763	66.14	0.0001
R2	1	0.02147675	0.02147675	5.11	0.0305
R3	1	0.04373966	0.04373966	10.40	0.0028
R5	1	0.02033078	0.02033078	4.84	0.0350
R6	1	0.01185195	0.01185195	2.82	0.1026
R7	1	0.01086024	0.01086024	2.58	0.1175
C1*R1	1	0.00973558	0.00973558	2.32	0.1376
C2*R3	1	0.01107563	0.01107563	2.63	0.1141
C3*R2	1	0.04705541	0.04705541	11.19	0.0021
R1*C4	1	0.04578624	0.04578624	10.89	0.0023
R2*C4	1	0.00916801	0.00916801	2.18	0.1492
C2*R1	1	0.02125631	0.02125631	5.06	0.0313

Source	DF	Type III SS	Mean Square	F Value	Pr > F
TRT	6	0.16044158	0.02674026	6.36	0.0002
C1	1	0.06777963	0.06777963	16.12	0.0003
C2	1	0.44309828	0.44309828	105.39	0.0001
C3	1	0.24999420	0.24999420	59.46	0.0001
C5	1	0.13222351	0.13222351	31.45	0.0001
R1	1	0.27808763	0.27808763	66.14	0.0001
R2	1	0.02147675	0.02147675	5.11	0.0305
R3	1	0.04373966	0.04373966	10.40	0.0028
R5	1	0.02033078	0.02033078	4.84	0.0350
R6	1	0.01185195	0.01185195	2.82	0.1026
R7	1	0.01086024	0.01086024	2.58	0.1175
C1*R1	1	0.00914040	0.00914040	2.17	0.1498
C2*R3	1	0.01580043	0.01580043	3.76	0.0611
C3*R2	1	0.04870965	0.04870965	11.59	0.0018
R1*C4	1	0.04431490	0.04431490	10.54	0.0027
R2*C4	1	0.01028565	0.01028565	2.45	0.1273
C2*R1	1	0.02125631	0.02125631	5.06	0.0313

```
            Covariance Parameter Estimates (REML)
            Cov Parm        Estimate
            C1           0.00843481
            C2           0.06534973
            C3           0.03944736
            C5           0.01928089
            R1           0.03912510
            R2           0.00246616
            R3           0.00564660
            R5           0.00230245
            R6           0.00109118
            R7           0.00094951
            C1*R1        0.00559139
            C2*R1        0.01769383
            C2*R3        0.01540992
            C3*R2        0.04762647
            R1*C4        0.04172363
            R2*C4        0.00559275
            Residual     0.00421378
```

Least Squares Means

Effect	TRT	LSMEAN	Std Error	DF	t	Pr > |t|
TRT	1	1.03145832	0.02506657	33	41.15	0.0001
TRT	2	1.03632328	0.02409811	33	43.00	0.0001
TRT	3	1.08344910	0.02517848	33	43.03	0.0001
TRT	4	1.06286153	0.02574839	33	41.28	0.0001
TRT	5	0.95488139	0.02447435	33	39.02	0.0001
TRT	6	1.01891389	0.02524623	33	40.36	0.0001
TRT	7	0.89943749	0.02437852	33	36.89	0.0001

Row-column and interaction of regressions or combination ANOVA
Dependent Variable: Y

Source	DF	Sum of Squares	Mean Square	F Value	Pr > F
Model	25	1.79923177	0.07196927	16.29	0.0001
Error	30	0.13254485	0.00441816		
Corrected Total	55	1.93177662			

R-Square	C.V.	Root MSE	Y Mean
0.931387	6.565027	0.066469	1.012475

Dependent Variable: Y

Source	DF	Type I SS	Mean Square	F Value	Pr > F
ROW	7	0.38831490	0.05547356	12.56	0.0001
COL	6	1.15907213	0.19317869	43.72	0.0001
TRT	6	0.11972355	0.01995392	4.52	0.0023
C1*R1	1	0.00957865	0.00957865	2.17	0.1513
R1*C2	1	0.01825578	0.01825578	4.13	0.0510
C2*R3	1	0.00785874	0.00785874	1.78	0.1923
C3*R2	1	0.04166095	0.04166095	9.43	0.0045
R1*C4	1	0.04499265	0.04499265	10.18	0.0033
R2*C4	1	0.00977442	0.00977442	2.21	0.1473

Source	DF	Type III SS	Mean Square	F Value	Pr > F
ROW	7	0.38831490	0.05547356	12.56	0.0001
COL	6	1.01906239	0.16984373	38.44	0.0001
TRT	6	0.11791625	0.01965271	4.45	0.0025
C1*R1	1	0.00939749	0.00939749	2.13	0.1551
R1*C2	1	0.02030565	0.02030565	4.60	0.0403

```
C2*R3                       1       0.01290053      0.01290053       2.92     0.0978
C3*R2                       1       0.04269878      0.04269878       9.66     0.0041
R1*C4                       1       0.04417127      0.04417127      10.00     0.0036
R2*C4                       1       0.00977442      0.00977442       2.21     0.1473
```

Covariance Parameter Estimates (REML)

```
                Cov Parm     Subject              Estimate
                ROW                             0.00729090
                COL                             0.02179930
                C1*R1                           0.00584283
                R1*C2                           0.01598859
                C2*R3                           0.01084891
                C3*R2                           0.04046662
                R1*C4                           0.04157133
                R2*C4                           0.00474734
                SP(EXP)INTERCEPT               -0.00000000
                Residual                        0.00443729
```

Least Squares Means

```
    Effect   TRT       LSMEAN       Std Error    DF         t    Pr > |t|
    TRT       1     1.03279947      0.06849923    49     15.08     0.0001
    TRT       2     1.04085965      0.06827608    49     15.24     0.0001
    TRT       3     1.07188050      0.06898348    49     15.54     0.0001
    TRT       4     1.05156492      0.06912807    49     15.21     0.0001
    TRT       5     0.96546152      0.06879786    49     14.03     0.0001
    TRT       6     1.02168355      0.06853511    49     14.91     0.0001
    TRT       7     0.90307540      0.06835031    49     13.21     0.0001
```

Differences of Least Squares Means

```
    Effect  TRT  _TRT    Difference     Std Error   DF         t   Pr > |t|
    TRT      1    2     -0.00806018     0.03504670   49     -0.23     0.8191
    TRT      1    3     -0.03908103     0.03618394   49     -1.08     0.2854
    TRT      1    4     -0.01876545     0.03958021   49     -0.47     0.6375
    TRT      1    5      0.06733794     0.03740107   49      1.80     0.0780
    TRT      1    6      0.01111592     0.03811781   49      0.29     0.7718
    TRT      1    7      0.12972407     0.03761943   49      3.45     0.0012
    TRT      2    3     -0.03102085     0.03841115   49     -0.81     0.4232
    TRT      2    4     -0.01070527     0.03929203   49     -0.27     0.7864
    TRT      2    5      0.07539812     0.03608589   49      2.09     0.0419
    TRT      2    6      0.01917610     0.03569990   49      0.54     0.5936
    TRT      2    7      0.13778425     0.03651435   49      3.77     0.0004
    TRT      3    4      0.02031558     0.03754102   49      0.54     0.5909
    TRT      3    5      0.10641897     0.04097063   49      2.60     0.0124
    TRT      3    6      0.05019695     0.03892509   49      1.29     0.2033
    TRT      3    7      0.16880510     0.03807134   49      4.43     0.0001
    TRT      4    5      0.08610340     0.03927030   49      2.19     0.0331
    TRT      4    6      0.02988137     0.03847642   49      0.78     0.4411
    TRT      4    7      0.14848952     0.03787633   49      3.92     0.0003
    TRT      5    6     -0.05622202     0.03756983   49     -1.50     0.1409
    TRT      5    7      0.06238613     0.03639929   49      1.71     0.0929
    TRT      6    7      0.11860815     0.03565169   49      3.33     0.0017
```

Differences of Least Squares Means
```
            Adjustment        Adj P
            Tukey-Kramer     1.0000
            Tukey-Kramer     0.9310
            Tukey-Kramer     0.9991
```

```
Tukey-Kramer   0.5541
Tukey-Kramer   0.9999
Tukey-Kramer   0.0187
Tukey-Kramer   0.9831
Tukey-Kramer   1.0000
Tukey-Kramer   0.3749
Tukey-Kramer   0.9981
Tukey-Kramer   0.0074
Tukey-Kramer   0.9980
Tukey-Kramer   0.1492
Tukey-Kramer   0.8534
Tukey-Kramer   0.0010
Tukey-Kramer   0.3182
Tukey-Kramer   0.9862
Tukey-Kramer   0.0048
Tukey-Kramer   0.7455
Tukey-Kramer   0.6103
Tukey-Kramer   0.0260
```

Chapter 30

SAS/GLM and SAS/MIXED
for Trend Analyses Using Fourier
and Polynomial Regression for Centered
and Noncentered Variates

Walter T. Federer
Murari Singh
Russell D. Wolfinger

Purpose

Spatial variations that are cyclic in nature should have statistical procedures to account for their occurrence. Since Fourier polynomial regression
is a procedure that fits cyclic variations, a code is given in this chapter for
such analyses. The data set used to illustrate the code's application is an
eight-row, seven-column experiment on tobacco plant heights by Federer
and Schlottfeldt (1954). The experiment was designed as a randomized
complete block design, but was laid out as an eight-row by seven-column
design instead. The laying out of an RCBD in a row-column arrangement
is appropriate. However, the analysis needs to take the layout and any other
type of variation into account. The spatial variation in the experimental
area was noncyclical and not entirely row-column oriented. The Fourier
regression model would not be expected to perform well with this data set
because the variation was not cyclic. If it is desired to use noncentered
polynomial regression, a code for this is also given in this chapter. Note
that which regressions to retain in the model will need to be determined
from a Type I rather than a Type III or IV analysis.

Codes for trend analyses are presented in the following order: Fourier
regression trend analysis (FRTA), noncentered variate polynomial regression trend analysis (NPRTA), randomized complete block design (RCBD),

row-column design (RCD), orthogonal polynomial (centered variates) regression trend analysis (PRTA), and a mixture of row-column and orthogonal polynomial regression trend analyses. The last is considered to be the appropriate model for this data set. Since only three orthogonal polynomials—degrees 4 and 6 in columns and degree 4 in rows, c4, c6, and r4—were omitted in the next to last analysis, it was decided to use the last analysis listed as an appropriate model to explain the spatial variation. For this model, the blocking effect parameters were taken to be random for the SAS/MIXED procedure and treatment estimates and means were obtained. The code is useful for exploratory model selection in patterning spatial variation.

References

Federer, W.T. (1998). Recovery of interblock, intergradient, and intervariety information in incomplete block and lattice rectangle designed experiments. *Biometrics* 54(2):471-481.
Federer, W.T. and Schlottfeldt, C.S. (1954). The use of covariance to control gradients in experiments. *Biometrics* 10:282-290 [Errata, *Biometrics* 11:251, 1955].

SAS Codes

```
/*--The SAS codes for obtaining standard textbook RCBD and RCD analy-
     sis, FRTA, NPRTA, and PRTA analyses  are given below:-- */
data colrow;
 infile 'colrow.dat';
 input Yield row col Trt;

/*--code for Fourier polynomials, FRTA--*/
           NTrt = 7; Nrow = 8; Ncol = 7;
Frc1 = Sin(2*3.14159*col/Ncol)    ;
Frc2 = Cos(2*3.14159*col/Ncol)    ;
Frr1 = Sin(2*3.14159*row/Nrow)    ;
Frr2 = Cos(2*3.14159*row/Nrow)    ;
/*--code for non-centered polynomials, NPRTA--*/
pc1= col; pc2=col**2;pc3=col**3;pc4=col**4;pc5=col**5;pc6=col**6;
pr1= row; pr2=row**2;pr3=row**3;pr4=row**4;pr5=row**5;
pr6=row**6; pr7 = row**7 ;
 run;
/*--code for ANOVA using Fourier series, FRTA--*/
proc glm data = colrow ;
 class Trt row col ;
 model Yield = Trt Frc1 Frc2 Frr1 Frr2 Frc1*Frr1 Frc1*Frr2
               Frc2*Frr1 Frc2*Frr2;
run;
/*--code for ANOVA using non-centered polynomials, NPRTA--*/

proc glm data = colrow;
```

```
 class Trt row col ;
 model  Yield = Trt pc1 pc2 pc3 pc4 pc5 pc6 pr1 pr2 pr3 pr4 pr5 pr6
pr7 pc1*pr1 pc2*pr1 pc2*pr3 pc3*pr2 pc4*pr1 pc4*pr2;  run;

/*--code to construct orthogonal polynomials--*/
Proc iml;
/*--7 columns and up to 6th degree polynomials--*/
opn4=orpol(1:7,6);
opn4[,1]=(1:7)`;
 op4=opn4;
 create opn4 from opn4[colname={'col' 'c1' 'c2' 'c3' 'c4' 'c5' 'c6'}];
 append from opn4;
 close opn4;
/*--8 rows and up to 7th degree polynomials--*/
 opn3=orpol(1:8,7);
 opn3[,1]=(1:8)`;
 op3=opn3;
 create opn3 from opn3[colname={'row' 'r1' 'r2' 'r3' 'r4' 'r5'
 'r6' 'r7'}] ;
 append from opn3;
 close opn3;  run;
/*--merge in polynomial coefficients--*/
 data rcbig;
 set colrow;
 idx = _n_;
 proc sort data = rcbig;
  by col ;
 data rcbig ;
  merge rcbig opn4;
  by col ;
 proc sort data = rcbig;
  by row ;
 data rcbig ;
  merge rcbig opn3;
  by row ;
 proc sort data = rcbig ;
  by idx ;
run;
/*--ANOVA for randomized complete blocks(rows), RCBD--*/
Proc Glm data = rcbig ;
 Class row Trt ;
 Model Yield = row  Trt ;
 run ;
/*--ANOVA for row-column design, RCD--*/
Proc Glm data = rcbig ;
 Class row  col  Trt ;
 Model Yield = row  col  Trt ;
run;

/*--ANOVA using orthogonal polynomials after omitting regressors
which had an F-value less than the 25% level, PRTA--*/
 Proc Glm data = rcbig ;
  Class Trt row  col  ;
  Model Yield = Trt c1 c2 c3 c5 r1 r2 r3 r5 r6 r7 c1*r1 c2*r1
  c2*r3 c3*r2 c4*r1 c4*r2;
  run ;
```

```
/*--ANOVA for mixture of row-column and orthogonal polynomial
regression trend analysis--this is the preferred analysis--*/
Proc Glm data = rcbig ;;
  Class row  col  Trt ;
  Model Yield = row  col  Trt c1*r1 c2*r1 c2*r3 c3*r2 c4*r1 c4*r2 ;
 Run ;
/*--random blocking effects and fixed Trt effects--*/
Proc Mixed data = rcbig ;
  Class row  col  Trt ;
  Model Yield = Trt/solution ;
  Random row  col  c1*r1 c2*r1 c2*r3 c3*r2 c4*r1 c4*r2 ;
  Lsmeans Trt ;
run ;
```

SAS Program Output (Abbreviated)

General Linear Models Procedure

Dependent Variable: YIELD

Source	DF	Sum of Squares	Mean Square	F Value	Pr > F
Model	14	1363645.120	97403.223	7.03	0.0001
Error	41	568131.505	13856.866		
Corrected Total	55	1931776.625			

R-Square	C.V.	Root MSE	YIELD Mean
0.705902	11.62648	117.7152	1012.475

Dependent Variable: YIELD

Source	DF	Type I SS	Mean Square	F Value	Pr > F
TRT	6	273875.4500	45645.9083	3.29	0.0097
FRC1	1	48018.4941	48018.4941	3.47	0.0698
FRC2	1	702583.1192	702583.1192	50.70	0.0001
FRR1	1	301163.4604	301163.4604	21.73	0.0001
FRR2	1	7263.2834	7263.2834	0.52	0.4732
FRC1*FRR1	1	2486.5375	2486.5375	0.18	0.6741
FRC1*FRR2	1	26380.3457	26380.3457	1.90	0.1751
FRC2*FRR1	1	107.9593	107.9593	0.01	0.9301
FRC2*FRR2	1	1766.4703	1766.4703	0.13	0.7229

Source	DF	Type III SS	Mean Square	F Value	Pr > F
TRT	6	233771.3341	38961.8890	2.81	0.0220
FRC1	1	17663.3134	17663.3134	1.27	0.2655
FRC2	1	718308.7485	718308.7485	51.84	0.0001
FRR1	1	301163.4356	301163.4356	21.73	0.0001
FRR2	1	7263.2583	7263.2583	0.52	0.4732
FRC1*FRR1	1	1924.3838	1924.3838	0.14	0.7113
FRC1*FRR2	1	26805.6766	26805.6766	1.93	0.1718
FRC2*FRR1	1	62.7531	62.7531	0.00	0.9467
FRC2*FRR2	1	1766.4703	1766.4703	0.13	0.7229

Dependent Variable: YIELD

Source	DF	Sum of Squares	Mean Square	F Value	Pr > F
Model	25	1727731.702	69109.268	10.16	0.0001
Error	30	204044.923	6801.497		
Corrected Total	55	1931776.625			

R-Square	C.V.	Root MSE	YIELD Mean
0.894374	8.145504	82.47119	1012.475

Dependent Variable: YIELD

Source	DF	Type I SS	Mean Square	F Value	Pr > F
TRT	6	273875.4500	45645.9083	6.71	0.0001
PC1	1	96813.2065	96813.2065	14.23	0.0007
PC2	1	535987.4627	535987.4627	78.80	0.0001
PC3	1	222783.3577	222783.3577	32.76	0.0001
PC4	1	17076.3332	17076.3332	2.51	0.1236
PC5	1	130081.1699	130081.1699	19.13	0.0001
PC6	1	2178.7015	2178.7015	0.32	0.5756
PR1	1	278087.6327	278087.6327	40.89	0.0001
PR2	1	21476.7478	21476.7478	3.16	0.0857
PR3	1	43739.6582	43739.6582	6.43	0.0167
PR4	1	1967.8963	1967.8963	0.29	0.5946
PR5	1	20330.7758	20330.7758	2.99	0.0941
PR6	1	11851.9481	11851.9481	1.74	0.1968
PR7	1	10860.2508	10860.2508	1.60	0.2161
PC1*PR1	1	9578.6523	9578.6523	1.41	0.2446
PC2*PR1	1	18255.7824	18255.7824	2.68	0.1118
PC2*PR3	1	518.4962	518.4962	0.08	0.7844
PC3*PR2	1	64.3080	64.3080	0.01	0.9232
PC4*PR1	1	27989.2845	27989.2845	4.12	0.0515
PC4*PR2	1	4214.5876	4214.5876	0.62	0.4374

Source	DF	Type III SS	Mean Square	F Value	Pr > F
TRT	6	99655.69428	16609.28238	2.44	0.0483
PC1	1	98.68785	98.68785	0.01	0.9049
PC2	1	23.75368	23.75368	0.00	0.9533
PC3	1	67.88482	67.88482	0.01	0.9211
PC4	1	561.62765	561.62765	0.08	0.7758
PC5	1	1513.16254	1513.16254	0.22	0.6406
PC6	1	2820.59147	2820.59147	0.41	0.5245
PR1	1	17782.73420	17782.73420	2.61	0.1164
PR2	1	16210.23452	16210.23452	2.38	0.1331
PR3	1	14741.72628	14741.72628	2.17	0.1514
PR4	1	13480.96650	13480.96650	1.98	0.1695
PR5	1	12426.54810	12426.54810	1.83	0.1866
PR6	1	11559.77509	11559.77509	1.70	0.2023
PR7	1	10860.25084	10860.25084	1.60	0.2161
PC1*PR1	1	3426.85586	3426.85586	0.50	0.4833
PC2*PR1	1	6236.00944	6236.00944	0.92	0.3460
PC2*PR3	1	1843.47705	1843.47705	0.27	0.6065
PC3*PR2	1	633.86885	633.86885	0.09	0.7623
PC4*PR1	1	20811.22981	20811.22981	3.06	0.0905

```
PC4*PR2              1        4214.58759      4214.58759      0.62      0.4374
Dependent Variable: YIELD
                              Sum of            Mean
Source              DF        Squares          Square    F Value    Pr > F
Model               13      662190.3521      50937.7194     1.69     0.1004
Error               42     1269586.2729      30228.2446
Corrected Total     55     1931776.6250
```

	R-Square	C.V.	Root MSE	YIELD Mean
	0.342788	17.17205	173.8627	1012.475

General Linear Models Procedure

Dependent Variable: YIELD

Source	DF	Type I SS	Mean Square	F Value	Pr > F
ROW	7	388314.9021	55473.5574	1.84	0.1056
TRT	6	273875.4500	45645.9083	1.51	0.1985

Source	DF	Type III SS	Mean Square	F Value	Pr > F
ROW	7	388314.9021	55473.5574	1.84	0.1056
TRT	6	273875.4500	45645.9083	1.51	0.1985

General Linear Models Procedure

Dependent Variable: YIELD

```
                              Sum of            Mean
Source              DF        Squares          Square    F Value    Pr > F
Model               19     1667110.584       87742.662    11.93     0.0001
Error               36      264666.041        7351.834
Corrected Total     55     1931776.625
```

	R-Square	C.V.	Root MSE	YIELD Mean
	0.862993	8.468638	85.74284	1012.475

General Linear Models Procedure

Dependent Variable: YIELD

Source	DF	Type I SS	Mean Square	F Value	Pr > F
ROW	7	388314.902	55473.557	7.55	0.0001
COL	6	1159072.132	193178.689	26.28	0.0001
TRT	6	119723.549	19953.925	2.71	0.0281

Source	DF	Type III SS	Mean Square	F Value	Pr > F
ROW	7	388314.902	55473.557	7.55	0.0001
COL	6	1004920.232	167486.705	22.78	0.0001
TRT	6	119723.549	19953.925	2.71	0.0281

General Linear Models Procedure

Dependent Variable: YIELD

```
                              Sum of            Mean
Source              DF        Squares          Square    F Value    Pr > F
Model               22     1793028.425       81501.292    19.38     0.0001
Error               33      138748.200        4204.491
Corrected Total     55     1931776.625
```

	R-Square	C.V.	Root MSE	YIELD Mean

```
        0.928176    6.404311   64.84205      1012.475
              General Linear Models Procedure
```

Dependent Variable: YIELD

Source	DF	Type I SS	Mean Square	F Value	Pr > F
TRT	6	273875.4500	45645.9083	10.86	0.0001
C1	1	96813.2065	96813.2065	23.03	0.0001
C2	1	535987.4627	535987.4627	127.48	0.0001
C3	1	222783.3577	222783.3577	52.99	0.0001
C5	1	133144.7539	133144.7539	31.67	0.0001
R1	1	278087.6327	278087.6327	66.14	0.0001
R2	1	21476.7478	21476.7478	5.11	0.0305
R3	1	43739.6582	43739.6582	10.40	0.0028
R5	1	20330.7758	20330.7758	4.84	0.0350
R6	1	11851.9481	11851.9481	2.82	0.1026
R7	1	10860.2434	10860.2434	2.58	0.1175
C1*R1	1	9735.5843	9735.5843	2.32	0.1376
C2*R1	1	20003.6652	20003.6652	4.76	0.0364
C2*R3	1	11087.8978	11087.8978	2.64	0.1139
C3*R2	1	47865.7583	47865.7583	11.38	0.0019
R1*C4	1	45098.6300	45098.6300	10.73	0.0025
R2*C4	1	10285.6523	10285.6523	2.45	0.1273

Source	DF	Type III SS	Mean Square	F Value	Pr > F
TRT	6	160441.5837	26740.2639	6.36	0.0002
C1	1	67779.6280	67779.6280	16.12	0.0003
C2	1	443098.2755	443098.2755	105.39	0.0001
C3	1	249994.1996	249994.1996	59.46	0.0001
C5	1	132223.5073	132223.5073	31.45	0.0001
R1	1	278087.6327	278087.6327	66.14	0.0001
R2	1	21476.7478	21476.7478	5.11	0.0305
R3	1	43739.6582	43739.6582	10.40	0.0028
R5	1	20330.7758	20330.7758	4.84	0.0350
R6	1	11851.9481	11851.9481	2.82	0.1026
R7	1	10860.2434	10860.2434	2.58	0.1175
C1*R1	1	9140.3988	9140.3988	2.17	0.1498
C2*R1	1	21256.3073	21256.3073	5.06	0.0313
C2*R3	1	15800.4345	15800.4345	3.76	0.0611
C3*R2	1	48709.6471	48709.6471	11.59	0.0018
R1*	1	44314.8960	44314.8960	10.54	0.0027
R2*C4	1	10285.6523	10285.6523	2.45	0.1273

```
              General Linear Models Procedure
```

Dependent Variable: YIELD

Source	DF	Sum of Squares	Mean Square	F Value	Pr > F
Model	25	1799231.771	71969.271	16.29	0.0001
Error	30	132544.854	4418.162		
Corrected Total	55	1931776.625			

R-Square	C.V.	Root MSE	YIELD Mean
0.931387	6.565027	66.46925	1012.475

```
              General Linear Models Procedure
```

Dependent Variable: YIELD

Source	DF	Type I SS	Mean Square	F Value	Pr > F
ROW	7	388314.902	55473.557	12.56	0.0001
COL	6	1159072.132	193178.689	43.72	0.0001

```
TRT             6       119723.549      19953.925       4.52    0.0023
C1*R1           1         9578.652       9578.652       2.17    0.1513
R1*C2           1        18255.782      18255.782       4.13    0.0510
C2*R3           1         7858.737       7858.737       1.78    0.1923
C3*R2           1        41660.946      41660.946       9.43    0.0045
R1*C4           1        44992.650      44992.650      10.18    0.0033
R2*C4           1         9774.420       9774.420       2.21    0.1473
```

Source	DF	Type III SS	Mean Square	F Value	Pr > F
ROW	7	388314.902	55473.557	12.56	0.0001
COL	6	1019062.385	169843.731	38.44	0.0001
TRT	6	117916.250	19652.708	4.45	0.0025
C1*R1	1	9397.488	9397.488	2.13	0.1551
R1*C2	1	20305.653	20305.653	4.60	0.0403
C2*R3	1	12900.535	12900.535	2.92	0.0978
C3*R2	1	42698.784	42698.784	9.66	0.0041
R1*C4	1	44171.272	44171.272	10.00	0.0036
R2*C4	1	9774.420	9774.420	2.21	0.1473

```
          Covariance Parameter Estimates (REML)
          Cov Parm         Estimate
          ROW            7290.8987058
          COL           21799.310745
          C1*R1          5842.8474878
          R1*C2         15988.580408
          C2*R3         10848.115411
          C3*R2         40466.589248
          R1*C4         41571.338509
          R2*C4          4747.3414225
          Residual       4437.2974998
```

Solution for Fixed Effects

Effect	TRT	Estimate	Std Error	DF	t	Pr > |t|
INTERCEPT		903.07551781	68.35032779	6	13.21	0.0001
TRT	1	129.72384541	37.61944012	30	3.45	0.0017
TRT	2	137.78424774	36.51437318	30	3.77	0.0007
TRT	3	168.80483626	38.07134440	30	4.43	0.0001
TRT	4	148.48914899	37.87631974	30	3.92	0.0005
TRT	5	62.38611199	36.39931336	30	1.71	0.0969
TRT	6	118.60818492	35.65171322	30	3.33	0.0023
TRT	7	0.00000000	.	.	.	.

Tests of Fixed Effects

Source	NDF	DDF	Type III F	Pr > F
TRT	6	30	4.64	0.0019

Least Squares Means

Effect	TRT	LSMEAN	Std Error	DF	t	Pr > |t|
TRT	1	1032.7993632	68.49924871	30	15.08	0.0001
TRT	2	1040.8597656	68.27610042	30	15.24	0.0001
TRT	3	1071.8803541	68.98349917	30	15.54	0.0001
TRT	4	1051.5646668	69.12808466	30	15.21	0.0001
TRT	5	965.46162980	68.79788141	30	14.03	0.0001
TRT	6	1021.6837027	68.53512010	30	14.91	0.0001
TRT	7	903.07551781	68.35032779	30	13.21	0.0001

Chapter 31

PROC GLM and PROC MIXED for Trend Analysis of Incomplete Block- and Lattice Rectangle-Designed Experiments

Walter T. Federer
Russell D. Wolfinger

Purpose

For resolvable row-column or lattice rectangle designs, a variety of analysis options are given in this chapter. These programs are for randomized complete block designs, incomplete block designs with rows (columns) as blocks, standard textbook analysis, differential gradients within rows (columns), and trend analysis using orthogonal polynomial regression functions of the rows and columns and their interactions. The example used in this chapter pulls data from Table 12.5 of W. G. Cochran and G. M. Cox's 1957 book *Experimental Designs.*

There are sixteen insecticide treatments arranged in four rows and four columns within each of the five complete blocks (replicates) to form a balanced lattice square. The data are means of three counts of plants infected with boll weevil. The trend analysis is the most appropriate analysis for these data. The code can also be used for incomplete block design by either deleting the row or the column category.

SAS Code

```
options ls = 76;
proc iml;
    opn4=orpol(1:4,3); /* 4 columns and 3 regressions. */
    opn4[,1] = (1:4)`;
    op4= opn4;    print op4;
    create opn4 from opn4[colname={'COL' 'C1' 'C2' 'C3'}];
    append from opn4;
```

```
        close opn4;
run;
    opn3=orpol(1:4,3); /* 4 rows and 3 regressions. */   opn3[,1] =
        (1:4)`;
    op3 = opn3;    print op3;
    create opn3 from opn3[colname={'ROW' 'R1' 'R2' 'R3'}];
    append from opn3;
    close opn3;
run;
data lsgr;
    infile 'lsgr1645.dat'; /* Name of data file. */
    input count rep ROW COL treat;
data lsbig; /* Name of lsgr after adding 6 polynomial regressions. */
    set lsgr;
    idx = _n_;   run;
proc sort data= lsbig;
    by     COL ;  run;
data lsbig;
    merge lsbig opn4;
    by     COL;  run;
proc sort data = lsbig;
    by     ROW;  run;
data lsbig;
    merge lsbig opn3;
    by     ROW;   run;
proc sort data = lsbig; by idx; run;
proc print; run;

/*In the codes below, a fixed-effects model is given first using PROC
    GLM; this is followed by a code for a random-effects mode using
    PROC MIXED. */
/* Randomized complete block design analysis. */
proc glm data = lsbig;   class rep row col treat;
    model count = rep treat;
run;
proc mixed data = lsbig;
    class rep row col treat;
    model count = treat;
    random rep;
    lsmeans treat;
run;

/* Incomplete block (row) analysis. */
proc glm data = lsbig;   class rep row col treat;
    model yield = rep row(rep) treat;
run;
proc mixed data = lsbig;
    class rep row col treat;
    model count = treat;
    ramdom rep row(rep);
    lsmeans treat;
run;

/* Standard textbook lattice square analysis. */
proc glm data = lsbig;   class rep row col treat;
    model count = rep treat row(rep) col(rep);
run;
```

```
proc mixed data - lsbig;
   class rep row col treat;
   model count = treat;
   random rep row(rep) col(rep);
   lsmeans treat;
run;
```

```
/* Differential linear gradients within rows analysis.  Quadratic and
     cubic gradients did not appear to be present for these data.
     This analysis is deemed appropriate for the data in Table 12.3 of
     W. G/ Cochran and G. M. Cox's 1957 book entitled Experimental De-
     signs, but not for this example. */
proc glm data = lsbig;    class rep row col treat;
   model count = rep treat row(rep) C1*row(rep);
run;
proc mixed data = lsbig;
   class rep row col treat;
   model count = treat/solution;
   random rep row(rep) C1*row(rep);
   lsmeans treat;run;/* Trend analysis using polynomial regressions
     and their interactions. */
proc glm data = lsbig;    class rep row col treat;
   model count = rep treat r1*rep  r2*rep  c1*rep  c1*r1*rep
     c2*r1*rep c2*r2*rep c3*r2*rep;run;
proc mixed  data = lsbig;    class rep row col treat;
   model count = treat/solution;
   random rep r1*rep r2*rep c1*rep c1*r1*rep c2*r1*rep c2*r2*rep
     c3*r2*rep;
   lsmeans treat;
run;
```

An abbreviated part of the output of the previous code follows.

```
/* Linear, quadratic, and cubic polynomial regression coefficients. */
                         OP3
                          1 -0.67082          0.5 -0.223607
                          2 -0.223607        -0.5  0.6708204
                          3  0.2236068       -0.5 -0.67082
                          4  0.6708204        0.5  0.2236068
/* Randomized complete block analysis. */
Dependent Variable: COUNT
                             Sum of        Mean
Source               DF     Squares        Square      F Value    Pr > F
Model                19    1275.76500      67.14553       1.73    0.0564
Error                60    2332.77300      38.87955
Corrected Total      79    3608.53800

Source               DF    Type I SS    Mean Square   F Value    Pr > F
REP                   4      31.56300       7.89075      0.20    0.9358
TREAT                15    1244.20200      82.94680      2.13    0.0200

Source               DF    Type III SS   Mean Square   F Value    Pr > F
REP                   4      31.56300       7.89075      0.20    0.9358
TREAT                15    1244.20200      82.94680      2.13    0.0200

/*Standard textbook analysis. */
Dependent Variable: COUNT
```

Source	DF	Sum of Squares	Mean Square	F Value	Pr > F
Model	49	2928.37008	59.76265	2.64	0.0029
Error	30	680.16792	22.67226		
Corrected Total	79	3608.53800			

Source	DF	Type I SS	Mean Square	F Value	Pr > F
REP	4	31.56300	7.89075	0.35	0.8433
TREAT	15	1244.20200	82.94680	3.66	0.0012
ROW(REP)	15	1093.01550	72.86770	3.21	0.0032
COL(REP)	15	559.58958	37.30597	1.65	0.1197

Source	DF	Type III SS	Mean Square	F Value	Pr > F
REP	4	31.56300	7.89075	0.35	0.8433
TREAT	15	319.45208	21.29681	0.94	0.5350
ROW(REP)	15	1026.75583	68.45039	3.02	0.0049
COL(REP)	15	559.58958	37.30597	1.65	0.1197

/ Differential linear gradients within rows and replicates. This is
    an appropriate analysis for the data in Table 12.3 of Cochran and
    Cox (1957). Experimental Designs but not for this data set. */*
Dependent Variable: COUNT

Source	DF	Sum of Squares	Mean Square	F Value	Pr > F
Model	54	3134.27638	58.04216	3.06	0.0016
Error	25	474.26162	18.97046		
Corrected Total	79	3608.53800			

R-Square	C.V.	Root MSE	COUNT Mean
0.868572	39.94048	4.35551	10.9050

Source	DF	Type I SS	Mean Square	F Value	Pr > F
REP	4	31.56300	7.89075	0.42	0.7955
TREAT	15	1244.20200	82.94680	4.37	0.0006
ROW(REP)	15	1093.01550	72.86770	3.84	0.0015
C1*ROW(REP)	20	765.49588	38.27479	2.02	0.0488

Source	DF	Type III SS	Mean Square	F Value	Pr > F
REP	4	31.563000	7.890750	0.42	0.7955
TREAT	15	347.188383	23.145892	1.22	0.3202
ROW(REP)	15	884.112744	58.940850	3.11	0.0060
C1*ROW(REP)	20	765.495883	38.274794	2.02	0.0488

/ Polynomial regression trend analysis considered appropriate for
    this example.*/*
Dependent Variable: COUNT

Source	DF	Sum of Squares	Mean Square	F Value	Pr > F
Model	54	3384.66411	62.67896	7.00	0.0001
Error	25	223.87389	8.95496		
Corrected Total	79	3608.53800			

R-Square	C.V.	Root MSE	COUNT Mean
0.937960	27.44139	2.99248	10.9050

Source	DF	Type I SS	Mean Square	F Value	Pr > F
REP	4	31.56300	7.89075	0.88	0.4893
TREAT	15	1244.20200	82.94680	9.26	0.0001
R1*REP	5	845.22838	169.04568	18.88	0.0001
R2*REP	5	246.02137	49.20427	5.49	0.0015
C1*REP	5	434.80006	86.96001	9.71	0.0001
R1*C1*REP	5	118.25808	23.65162	2.64	0.0475
R1*C2*REP	5	174.90489	34.98098	3.91	0.0094

```
R2*C2*REP              5      156.38244     31.27649     3.49    0.0157
R2*C3*REP              5      133.30389     26.66078     2.98    0.0305
```

Source	DF	Type III SS	Mean Square	F Value	Pr > F
REP	4	31.563000	7.890750	0.88	0.4893
TREAT	15	421.496008	28.099734	3.14	0.0056
R1*REP	5	809.864820	161.972964	18.09	0.0001
R2*REP	5	177.016456	35.403291	3.95	0.0089
C1*REP	5	352.057631	70.411526	7.86	0.0001
R1*C1*REP	5	96.281349	19.256270	2.15	0.0923
R1*C2*REP	5	204.368728	40.873746	4.56	0.0043
R2*C2*REP	5	138.464513	27.692903	3.09	0.0262
R2*C3*REP	5	133.303894	26.660779	2.98	0.0305

Cov Parm	Estimate
REP	0.00000000
R1*REP	53.23803826
R2*REP	9.80562147
C1*REP	23.76781000
R1*C1*REP	9.98969710
R1*C2*REP	45.69858595
R2*C2*REP	31.54099937
R2*C3*REP	21.49617387
Residual	9.00079394

Tests of Fixed Effects

Source	NDF	DDF	Type III F	Pr > F
TREAT	15	25	3.41	0.0033

Least Squares Means

Effect	TREAT	LSMEAN	Std Error	DF	t	Pr > |t|
TREAT	1	5.10314265	1.63982935	25	3.11	0.0046
TREAT	2	13.43151284	1.75595605	25	7.65	0.0001
TREAT	3	9.78194530	1.71248681	25	5.71	0.0001
TREAT	4	11.59100160	1.70310420	25	6.81	0.0001
TREAT	5	12.04012050	1.77463190	25	6.78	0.0001
TREAT	6	6.46087629	1.71679001	25	3.76	0.0009
TREAT	7	4.87293305	1.65381037	25	2.95	0.0069
TREAT	8	11.43810953	1.88342899	25	6.07	0.0001
TREAT	9	9.89142127	1.65449886	25	5.98	0.0001
TREAT	10	15.19391731	1.92490202	25	7.89	0.0001
TREAT	11	15.29420036	1.80297734	25	8.48	0.0001
TREAT	12	11.46771203	1.69521361	25	6.76	0.0001
TREAT	13	10.39896987	1.63737478	25	6.35	0.0001
TREAT	14	15.23542928	1.71894217	25	8.86	0.0001
TREAT	15	8.54488157	1.66229114	25	5.14	0.0001
TREAT	16	13.73382654	1.67968857	25	8.18	0.0001

Chapter 32

Partitioning Crop Yield into Genetic Components

Vasilia A. Fasoula
Dionysia A. Fasoula

Importance

To increase efficiency in plant breeding by selecting for high yield and stability from the early generations of selection. Partitioning crop yield into genetic components increases efficiency and offers the following advantages: (1) yield and stability genes are selected early in the program, rather than late-generation testing where most genes are irretrievably lost; (2) breeders can identify and cross, in early generations, complementary lines for genes that control the three components of crop yield and combine them in one line; and (3) breeders can develop density-independent cultivars especially favored by farmers.

Genetic Components of Crop Yield

1. *Genes controlling yield potential per plant.* These genes contribute to the production of density-independent cultivars by expanding the lower limit of the optimal productivity density range.
2. *Genes conferring tolerance to biotic and abiotic stresses.* These genes enhance the production of density-independent cultivars by expanding the upper limit of the optimal productivity density range.
3. *Genes controlling responsiveness to inputs.* These genes enable cultivars to exploit optimal growing conditions.

Parameters That Determine the Genetic Components

1. *The progeny mean yield per plant* $(\bar{x})$ evaluates and selects genes contributing to higher yield.
2. *The progeny standardized mean* $(\bar{x}/s)$ evaluates and selects genes contributing to stability of performance.
3. *The progeny standardized selection differential* $\left[(\bar{x}_{sel} - \bar{x})/s\right]$ evaluates and selects genes that exploit nonstress environments, where $\bar{x}$ is the progeny mean, s is the progeny phenotypic standard deviation, and $\bar{x}_{sel}$ is the mean yield of the selected plants at a predetermined selection pressure.

Conditions of Selection

To partition crop yield into genetic components, it is essential to perform selection under the following conditions:

1. *Absence of competition.* This condition increases response to selection by reducing the masking effect of competition on single-plant heritability and by optimizing the range of phenotypic expression.
2. *Enhanced gene fixation.* This condition is essential for (1) reducing the masking effect of heterozygosity on single-plant heritability, (2) exploiting additive alleles, and (3) increasing genetic advance through selection.
3. *Multiple environment evaluation.* This condition exposes progenies to the environmental diversity encountered over the target area of adaptation and improves heritability by allowing selection for reduced genotype-by-environment interaction and increased responsiveness to inputs.
4. *Utilization of the honeycomb selection designs.* Comparable evaluation of progenies across the target area of adaptation requires designs that fulfill four conditions: (1) effective sampling of environmental diversity, (2) concurrent selection among and within progenies, (3) joint selection for broad as well as specific adaptation, and (4) application of high selection pressures.
5. *Nonstop selection.* This condition refers to the constant improvement of the crop yield and quality of released and adapted cultivars. Continuous selection after the release of cultivars is imposed by the con-

stant need to eliminate undesirable mutations while exploiting desirable ones.

Originators

Fasoula, V.A. and Fasoula, D.A. (2000). Honeycomb breeding: Principles and applications. *Plant Breeding Reviews* 18:177-250.

Software Available

Batzios, D.P. and Roupakias, D.G. (1997). HONEY: A microcomputer program for plant selection and analysis of the honeycomb designs. *Crop Science* 37:744-747 (program free of charge).
For software distribution, contact Dimitrios P. Batzios, Variety Research Institute, 57400 Sindos, Greece. Tel. + 302310796-264. FAX + 302310796-343. E-mail: <Varinst@spark.net.gr>. The software refers to honeycomb breeding as it appears in Fasoula and Fasoula (1995) and Fasoula and Fasoula (2000).

Contact

Vasilia A. Fasoula, Center for Applied Genetic Technologies, 111 Riverbend Road, University of Georgia, Athens, GA 30602-6810, USA. E-mail: <vfasoula@uga.edu>.

EXAMPLE

The following example (Table 32.1) demonstrates the partitioning of crop yield into genetic components and shows the evaluation of twenty F_4 cotton lines plus one check tested in a replicated-honeycomb trial (Fasoulas and Fasoula, 1995) across two locations with a total of 100 replications (fifty replications per location). This example is included in the available software and utilizes data from Batzios (1997).

Lines were evaluated in the absence of competition in two locations for the three genetic components of crop yield: (1) yield per plant, (2) tolerance to stresses, and (3) responsiveness to inputs. The check (line 21) represents the cotton cultivar Sindos 80 developed in Greece. The three ge-

TABLE 32.1. Evaluation of 20 F$_4$ cotton lines selected for high yield per plant

Progeny lines	Mean yield per plant (g)		Tolerance to stresses		Responsiveness to inputs	
	$(\bar{x})$	(%)*	$(\bar{x}/s)$	(%)*	$(\bar{x}_{sel} - \bar{x})/s$	(%)*
15	315.0	100	2.42	100	1.86	91
11	305.8	97	2.08	86	1.67	82
3	298.1	95	2.22	92	1.71	83
5	272.2	86	1.59	66	1.71	84
12	259.9	82	1.72	71	1.69	82
17	250.1	79	1.90	79	1.70	83
18	245.5	78	2.36	97	1.71	85
7	236.8	75	2.07	86	1.75	85
1	235.5	75	1.62	67	1.76	86
13	235.4	75	1.54	64	1.69	83
6	216.6	69	2.03	84	1.62	79
14	215.5	68	2.07	86	1.67	81
19	208.3	66	1.71	71	1.44	70
2	199.7	63	1.20	50	1.58	77
8	198.0	63	1.64	68	1.83	89
21(ck)	184.5	58	1.61	67	1.93	94
16	179.4	57	1.81	75	1.74	85
20	169.6	54	1.18	49	1.58	77
9	167.9	53	1.23	51	2.05	100
4	159.3	51	1.25	52	1.82	89
10	130.4	41	1.29	53	1.78	87

*Percent of highest value

netic components of crop yield were calculated from 100 values of single plants representative of each progeny line grown in the honeycomb experiments.

Conclusions

An analysis of Table 32.1 data leads to several conclusions that are relevant to the benefits of partitioning crop yield into genetic components:

1. Percent range of expression: 41 to 100 for the first component, 49 to 100 for the second component, and 70 to 100 for the third component. Thus, in this genetic material, genes controlling yield per plant and tolerance to stresses showed the largest variation.
2. The check cultivar (21) has the following relative values compared to the best line: 58 percent for the first component, 67 percent for the second component, and 94 percent for the third. Evidently, for the check cultivar, the genetic components of crop yield ranks in relative importance as: responsiveness to inputs → tolerance to stresses → yield per plant.
3. The best F_4 line on the basis of the component evaluation is line 15. Three best plants were selected from line 15 and seven other plants were selected from the most superior lines of Table 32.1, and their progenies were tested as $F_{4:6}$ lines in evaluation trials. These 10 best $F_{4:6}$ lines derived by the honeycomb methodology, along with the 10 best $F_{4:6}$ lines derived by the conventional methodology, were evaluated in randomized complete block (RCB) trials. The lines derived from honeycomb breeding outperformed the lines derived from conventional breeding (Batzios, 1997; Batzios et al., 2001). More specifically, in the RCB trials, the three best $F_{4:6}$ lines (15-1, 15-2, and 15-3), derived from line 15, produced the following yield superiority in percent of the check cultivar Sindos 80.

The best $F_{4:6}$	Yield (%)
15-1	154
15-2	141
15-3	128
Sindos 80	100

Lines 15-1 and 15-2 ranked first and outyielded the other eighteen lines. This indicates that honeycomb selection for superior component and quality performance across the target area of adaptation can be a safe and efficient way to exploit desirable genes in every generation to substantially increase efficiency.
4. Selection for the three crop yield components and quality across the target area of adaptation at all stages of the breeding program makes regional testing unnecessary and halves the time required to release a cultivar.

5. Given that selection is based on genetic components of crop yield, the developed cultivars are density independent, an advantage favored greatly by farmers since density-independent cultivars perform well at a greater range of plant densities.
6. If during evaluation and selection no lines show satisfactory relative superiority, promising lines that complement each other for the three components of crop yield and quality are crossed to obtain desirable recombinant lines.

A Noteworthy Relation

When evaluation is performed in the absence of competition, an important relation is revealed between the genetic components of crop yield and the parameters of the general response equation. Thus, starting from the general response equation

$$R = i \cdot h^2 \cdot \sigma_p \qquad \text{(Falconer, 1989)}$$

and substituting heritability by its equivalent σ_g^2 / σ_p^2, the equation becomes

$$R = i \cdot \frac{\sigma_g}{\sigma_p} \cdot \sigma_g \qquad \text{(Falconer, 1989)}$$

where i is the intensity of selection or the standardized selection differential, h^2 is the heritability, σ_p is the phenotypic standard deviation, and σ_g is the genotypic standard deviation.

Iliadis (1998) estimated the correlation coefficient between σ_g and $\bar{x}$ (progeny mean yield per plant) in chickpea grown in the absence of competition. This correlation coefficient ($r = 0.95$) was high. This suggests that when evaluation is practiced in the absence of competition that maximizes both $\bar{x}$ and σ_g, $\bar{x}$ may replace σ_g and the equation becomes

$$R = i \cdot \frac{\bar{x}}{\sigma_p} \cdot \bar{x}.$$

The new formula is a product of (1) the progeny standardized selection differential, (2) the progeny standardized mean, and (3) the progeny mean, i.e., the product of the three genetic components of crop yield (Fasoula and Fasoula, 2000), described in detail previously.

REFERENCES

Batzios, D.P. (1997). Effectiveness of selection methods in cotton (*Gossypium hirsutum* L.) breeding. Doctoral thesis, Department of Genetics and Plant Breeding, Aristotelian University, Thessaloniki, Greece.

Batzios, D.P. and Roupakias, D.G. (1997). HONEY: A microcomputer program for plant selection and analysis of the honeycomb designs. *Crop Science* 37:744-747.

Batzios, D.P., Roupakias, D.G., Kechagia, U., and Galanopoulou-Sendouca, S. (2001). Comparative efficiency of honeycomb and conventional pedigree methods of selection for yield and fiber quality in cotton (*Gossypium* spp.). *Euphytica* 122:203-221.

Falconer, D.S. (1989). *Introduction to quantitative genetics.* John Wiley & Sons, New York.

Fasoula, V.A. and Fasoula, D.A. (2000). Honeycomb breeding: Principles and applications. *Plant Breeding Reviews* 18:177-250.

Fasoulas, A.C. and Fasoula, V.A. (1995). Honeycomb selection designs. *Plant Breeding Reviews* 13:87-139.

Iliadis, C.G. (1998). Evaluation of a breeding methodology for developing germplasm in chick-pea (*Cicer arietinum* L.). Doctoral thesis, Department of Genetics and Plant Breeding, Aristotelian University, Thessaloniki, Greece.

Short Note 1

Inbreeding Coefficient in Mass Selection in Maize

Fidel Márquez-Sánchez

Importance

To know how much inbreeding is being generated through the course of mass selection.

Definitions

Inbreeding coefficient at generation *t:* the amount of inbreeding that has been accumulated up to cycle *t* of selection.

$$F(MS)_t = (1/2nm)\left[1 + 2m(n-1)F_{t-1} + 2(m-1)F_{t-2} + F_{t-3}\right]$$

where $F(MS)_t$ = inbreeding coefficient at cycle *t* of mass selection; n = number of open-pollinated ears that make the seed balanced composite from where the population under selection originates; m = number of seeds per open-pollinated ear.

Originator

Márquez-Sánchez, F. (1998). Expected inbreeding with recurrent selection in maize: I. Mass selection and modified ear-to-row selection. *Crop Science* 38(6):1432-1436.

Contact

Dr. Fidel Márquez-Sánchez. E-mail: <fidelmqz@hotmail.com>.

Observations

n and m must be adjusted by the variance effective number,

$$N_{e(v)} = N[4s/(1+2s)],$$

where $N = nm$, and s is the selection pressure (Crossa and Venkovsky, 1997). The adjustment is made as follows:

$$n' = [N_{e(v)}Q]^{\frac{1}{2}}$$
$$m' = [N_{e(v)}/Q]^{\frac{1}{2}}, \text{ where } Q = n/m.$$

In the case of modified ear-to-row selection the actual number of plants (N) in the selection plot must first be adjusted by the inbreeding effective number,

$$N_{e(f)} = 4N_{fr}N_{mr}/(N_{fr}+N_{mr}),$$

where N_{fr} and N_{mr} are the numbers of female and male rows, respectively, of the detasseling-selection plot (Falconer, 1961).

REFERENCES

Crossa, J. and Venkovsky, R. (1997). Variance effective population size for two-stage sampling of monoecious species. *Crop Science* 37:14-26.
Falconer, D.S. (1961). *Introduction to Quantitative Genetics*. The Ronald Press Company, New York.

Short Note 2

Regression of Forage Yield Against a Growth Index As a Tool for Interpretation of Multiple Harvest Data

Jeffery F. Pedersen

Purpose

Concise representation of multiple harvest forage data and as a graphical aid in assessing the value of a forage variety across an entire growing season.

Originator

Pedersen, J.F., Moore, K.J., and van Santen, E. (1991). Interpretive analyses for forage yield trial data. *Agronomy Journal* 83:774-776.

Contact

Dr. J. F. Pedersen, USDA-ARS, 344 Keim Hall, University of Nebraska, Lincoln, NE 68583-0937, USA. E-mail: <jfp@unlserve.unl.edu>.

EXAMPLE

```
DATA ONE;
INPUT REP YEAR MONTH $ LINE $ KGHA;
CARDS;
1                    87          Apr       AUVigor        2487
1                    87          May       AUVigor        1228
1                    87          Jun       AUVigor         407
1                    87          Dec       AUVigor          78
```

```
1                         87        Apr      Johnston          793
1                         87        May      Johnston          985
1                         87        Jun      Johnston          514
1                         87        Dec      Johnston            0
2                         87        Apr      AUVigor          2491
2                         87        May      AUVigor          1232
2                         87        Jun      AUVigor           411
2                         87        Dec      AUVigor            82
2                         87        Apr      Johnston          797
2                         87        May      Johnston          989
2                         87        Jun      Johnston          518
2                         87        Dec      Johnston            0;

PROC SORT;
BY MONTH YEAR LINE;
RUN;

* The following PROC GLM does not contribute directly to the stability
    analysis.  It tests for differences due to MONTH*YEAR (environ-
    ment) LINE and the LINE*MONTH*YEAR interaction *;

PROC GLM;
CLASS MONTH YEAR REP LINE;
MODEL KGHA=MONTH*YEAR REP(MONTH*YEAR) LINE LINE*MONTH*YEAR;
MEANS LINE MONTH*YEAR LINE*MONTH*YEAR;
RUN;

* The following PROC MEANS outputs a data set named TWO with KGHA
    means across reps *;

PROC MEANS NOPRINT;
BY MONTH YEAR LINE;
VARIABLES KGHA;
OUTPUT OUT=TWO MEAN=KGHA;
RUN;

PROC PRINT DATA=TWO;
RUN;

* The following PROC GLM  is used to put MONTH*YEAR(environmental)
    means onto the data set.  MONTH*YEAR mean=COLM.  The new data
    set is named THREE.  The ANOVA generated is not otherwise used
    for data interpretation*;

PROC GLM;
CLASS MONTH YEAR;
MODEL KGHA=MONTH*YEAR;
OUTPUT OUT=THREE PREDICTED=COLM;
RUN;

* The following PROC GLM puts the grand mean (YBAR) and the value for
    the environmental mean minus the grand mean(COLEF) on a data set
    named FOUR.  COLEF is the environmental index for the stability
    analysis *;

PROC GLM;
MODEL COLM=;
```

```
OUTPUT OUT=FOUR PREDICTED=YBAR RESIDUAL=COLEF;
RUN;

PROC PRINT;
RUN;

* The following PROC GLM calculates the regression coefficient for
    KGHA on COLEF (the environmental index) for each line. Estimates
    of XBAR and b are given for each line in the ANOVA. Predicted
    values=YHAT and residuals=RY.  Outputed data set= FIVE.  *;

* To test H0: b=1,  t= (estimate - 1) / SE
 df= df shown for COLEF*LINE in ANOVA *;

PROC GLM;
CLASSES LINE;
MODEL KGHA=LINE COLEF*LINE / P NOINT SOLUTION;
OUTPUT OUT=FIVE PREDICTED=YHAT RESIDUAL=RY;
RUN;

PROC PLOT;
PLOT YHAT*COLEF=LINE;
RUN;

PROC SORT;
BY LINE;
RUN;

PROC PLOT;
BY LINE;
PLOT KGHA*COLEF='*'  YHAT*COLEF=LINE/OVERLAY;
RUN;

*  The following PROC MEANS generates the raw sum of squares (USS) for
    deviations of the means from regression on environment index.
    Variance of deviation from regression can be calculated as USS /
    n-2  *;

*  To test H0: F=1

                r (USS/#obs-2) (#lines/#lines-1)
                -------------------------------
                   Error MS (#lines/#lines-1)

where r = #replications
#lines/#lines-1 = correction factor
Error MS = error ms from last ANOVA
df= df #obs-2, pooled error df (from 1st PROC GLM)  *;

PROC MEANS USS;
BY LINE;
VARIABLES RY;
RUN;
```

Short Note 3

Tolerance Index

Lajos Bona

Purpose

To identify the tolerance level of tested cereal (or other plant) cultivars/entries. The simple formula outlined in this chapter was applied for evaluation of small-grain cereal entries for acid soil tolerance, but it can serve as a useful tool for other traits as well. Among and within species, ranking and numerical evaluation of a range of entries will be reliable based on tolerance index *(Ti)*.

Definitions

Tolerance index refers to the characteristic production (grain yield, biomass, root or shoot length, etc.) of a genotype in a given stress environment (e.g., acid soil) relative to a nonstress environment (e.g., improved or limed acid soil).

$$Ti_{GY} = ALRL_{(-L)} / ALRL_{(+L)}$$

where, Ti = tolerance index (for grain yield) for a certain genotype, $ALRL_{(-L)}$ = calculated mean longest root length of a genotype in unlimed (–L) acid soil (production in stress environment), $ALRL_{(+L)}$ = calculated mean longest root length of a genotype in limed (+L) acid soil (production in nonstress environment).

or

$$Ti_{GY} = AGY_{(-L)} / AGY_{(+L)}$$

where, Ti = tolerance index (for grain yield) for a certain genotype, $AGY_{(-L)}$ = observed grain yield of a genotype in unlimed (–L) acid soil (production in stressed environment), $AGY_{(+L)}$ = observed grain yield of a genotype in limed (+L) acid soil (production in nonstressed environment).

Originator

Bona, L., Wright, R.J., and Baligar, V.C. (1991). A rapid method for screening cereals for acid soil tolerance. *Cereal Research Communications* 19:465-468.

Short Note 4

Computer Program to Calculate Population Size

Leví M. Mansur

Purpose

To calculate population size necessary to recover any number of individuals exhibiting a trait.

Definitions

Sedcole (1977) provided four methods to calculate the total number of plants needed to obtain one or more segregants with desired genes for a given probability of success. The following formula gives an accurate result:

$$n = \left\{ \left[2(r - 0.5) + z^2 (1 - q) \right] + z \left[z^2 \left(1 - q \right)^2 \right) + 4(1 - q)(r - 0.5) \right]^{1/2} \right\} / 2q$$

where n = total number of plants needed, r = required number of plants with desired genes, q = frequency of plants with desired genes, p = value that is function of (p).

Originator

Sedcole, J.R. (1977). Number of plants necessary to recover a trait. *Crop Science* 17:667-668.

Software Available

Mansur, L.M., Hadder, K., and Suárez, J.C. (1990). Computer program to calculate population size necessary to recover any number of individuals exhibiting a trait. *Journal of Heredity* 81:407-440 (software free of charge). E-mail: Leví Mansur at <levi@entelchile.net>.

Example

Data to be analyzed $r = 10$, $P = 0.95$, $q = 0.25$, germination rate $= 0.8$. The number of progenies that must be grown is $N = 75$.

Index

2DMAP program, 277

Abiotic stress, 321
Acid soil tolerance, 335
ACT (analysis of complex traits) program, 266
Additive genetic correlation, 109, 110, 113
Additive genetic covariance(s), 110, 111, 113
Additive genetic effects, 77
Additive genetic variance, 97, 98, 110
Additive-by-additive genetic model, 51
Additive-by-environment effect, 68
Additive-dominance model, 39, 43, 44, 45, 55, 118
Additive-dominance-epistasis model, 51, 57
Additive-effect QTL, 228
AFLP (amplified fragment-length polymorphism), 267
ALLASS (allele association) program, 266
Allele frequency estimation, 273
ALP (automated linkage preprocessor) program, 266
AMMI (Additive Main Effects and Multiplicative Interaction), 145
Analysis of variance, 110
Analyze program, 266
Animal genetic model, 67
Animal model, 70
APM (affected pedigree-member method) program, 266
Arlequin program, 266
ASPEX (affected sibling pairs exclusion map) program, 266
Association analysis, 273

Augmented experiment designs, 283
Augmented lattice square design, 283
Augmented row-column design, 291
Augmented treatments, 291
AUP (adjusted unbiased prediction), 26, 42
Automatic sequencers, 271
Autoregression, 137
Autoscan program, 267
Autosomal genes
    additive effect, 68
    dominance effect, 68
Average number of alleles, 263

Backcross breeding, 280
Balanced data, endosperm and seed models, 21
Balanced lattice square, 137, 146, 315
Balanced lattice square experiment, 146
Beta program, 267
Bin mapping, 272
Biotechnology, 153
Biotic stress, 321
Biplot, 193, 195, 201
BLOCK program, 267
BLUP (best linear unbiased prediction), 181
Bootstrap
    for genetic diversity parameter calculation, 263
    for mean, 261, 263
    for standard error, 261
Borel program, 267, 274
Broad adaptation, 123
Broad inference(s), 181, 184

Canonical analysis, 156, 169
Carter-Falconer mapping function, 255, 256
CarthaGene program, 267
CASPAR program, 267
Categorical variables, 153
Centered variates, 307
Ceph2Map program, 267
Chickpea, 326
Chi-square method, 231
Chromosome drawing, 270
Classifying observations into homogeneous groups, 153
Clump program, 267
Coancestry coefficient(s), 98, 99, 100, 108, 186
Coefficient of coancestry, 98, 99, 108, 186
Coincidence, 255, 260
Combin program, 267
COMDS program, 267
Competition, 322
Complete block design, 283
Composite interval mapping, 272, 273
Conditional distribution, 115
Conditional genetic effects, 116
Conditional genetic model, 213
Conditional genetic variance components, 173
Conditional linkage analysis, 267
Conditional mapping of QTL, 213
Conditional phenotypic data, 215
Conditional QTL-by-environment effect, 216
Conditional random variables, 115
Conditional variance analysis, 119
Conditional variance components, 116
Conditional variance-covariance matrix, 116
Contingency table, 267
Continuous variables, 153, 156
CoPE program, 268
Correlated response, 109
Correlation coefficient estimation, 171, 177
Coupling phase linkage, 232, 233

Covariance, 29, 55, 109, 179
Covariance analysis, 34, 73, 98, 110
Covariance analysis output, 47, 61, 75
Covariance between seed and plant traits, 27, 28
Covariance component estimation, 31, 45, 57, 171, 177
Covariance components, 21, 39, 43, 51, 53
Covariance estimation, 109
Covariate screening, 276
Covariates, 124
CRI-MAP program, 268
CRI-MAP-PVM program, 268
Crop yield components, 321
Cyrillic program, 268
Cytoplasm effect, 23
Cytoplasm general heritability, 25
Cytoplasm general response, 26
Cytoplasm interaction heritability, 25
Cytoplasm interaction response, 26
Cytoplasm-by-environment effect, 23

Data preparation for Linkage program, 277
Degrees of freedom simulation, 137
Density-independent cultivars, 321, 326
Developmental analysis, 115
Developmental genetic model analysis, 119, 173, 177
Developmental traits, 171, 174, 213
dGene program, 268
Diallel analysis
    additive-dominance model, 39
    additive-dominance-epistasis model, 51, 67
    animal model, 67
    endosperm model, 21
    genotype-by-environment effect model, 67
    maternal effect model, 67
    sexlinked model, 67
Diallel crosses, 171-173
Diallel defined, 2
Diallel mating design, 1

Diallel method 1, 1-3, 7
Diallel method 2, 1, 2, 15
Diallel method 3, 1, 2, 16
Diallel method 4, 1, 2, 18
Diallel model analysis, 174
DIALLEL-SAS, *xiii*, 1-3
Dinucleotide markers, 270
Diploid seed models, 22, 173
Direct additive effect, 23
Direct additive-by-environment effect, 23
Direct dominance effect, 23
Direct dominance-by-environment effect, 23
Direct effects (path analysis), 89
Direct general heritability, 25
Direct interaction heritability, 25
Direct interaction response, 26
DMap program, 268
DNA sequences, 266
Dominance genetic effects, 77
Dominance genetic variance, 97, 98
Dominance heterosis, 42, 54
Dominance-by-environment effect, 68
Double coancestry coefficient, 98-100, 108
Doubled haploid (DH) population, 213, 219, 222, 271, 280

Early selection, 321
Ear-to-row selection, 330
EBLUP (empirical best linear unbiased prediction), 183
ECLIPSE program, 268, 274
Eclipse2, 268
Eclipse3, 268
Ecovalence, 123
Effective number of alleles, 263
Effective population size calculation, 273
EH program, 268
Endosperm, 119
Environmental variance, 130
Epistasis, 51, 57, 77
Epistatic effects, 77, 213, 219
Epistatic QTL, 220, 228

Epistatic-by-environment interaction effect, 220, 221
ERPA program, 268
ETDT program, 269
Expectation maximization (EM), 154
Experimental error variance, 131

Family-based association test, 269
FASTLINK program, 269
FAST-MAP program, 269
FBAT program, 269
Firstord program, 269
Fisher program, 269
Fixed-effects model, 181
Forage quality, 111
Forage yield, 331
Fourier regression and trend analysis, 307
Frequency analysis, 162
Full-sib families, 98-100, 108, 271

Gamble notation, 77
GAP program, 269
GAS program, 269
GASP program, 269
GASSOC program, 270
Gaussian model, 153
GCA (general combining ability), 1, 2
GENDIPLD for seed model, 29, 30
Gene fixation, 322
Gene linkage relationships, 231
Gene mapping, 280
Gene segregation, 231, 232
Genealogical relationships, 267
Genealogy, 274
GeneCheck program, 270
Gene-environment interaction, 270
Gene-gene interaction, 270
Genehunter program, 270
Genehunter-Imprinting program, 270
General heritability, 25, 42, 53, 70
General heterosis, 26, 27, 42, 54
General linear model, 1
General response, 26, 42, 53, 70

Generation means analysis, 77, 79, 81
Genetic advance, 322
Genetic analysis of pedigree data, 276
Genetic components of crop yield, 321
Genetic correlations, 109, 111
Genetic counseling, 273
Genetic data management, 275
Genetic diversity, 261-263
Genetic effect prediction, 55, 171, 176
Genetic effects, 55, 77
Genetic likelihood computing, 268
Genetic linkage analysis, 273
Genetic linkage map, 271, 272
Genetic linkage software on the Internet, 265
Genetic map distance definition, 255
Genetic mapping using backcrosses/RIL, 272
Genetic marker data analysis, 268
Genetic model, 21, 39, 51, 67, 174
Genetic partitioning for diploid seed model, 22
Genetic partitioning for triploid endosperm model, 22
Genetic resource conservation, 153
Genetic risk, 268, 271
Genetic variance components, 69, 97, 99, 108
Genetic variance partitioning, 273
Genetic variation, 115
Genetics of complex diseases, 267
GenMath, 256-258
GenoDB program, 270
Genome-wide scan, 266, 267
Genomic data, 262, 269
Genotype-by-environment effect, 39, 40, 51, 68, 98
Genotype-by-environment interaction, *xiii,* 129, 182, 193, 275, 322
  and AMMI, 145
  in diallel analysis, 21
  in heterosis, 26
  and stability, 123, 124
Genotype-by-environment interaction variance, 131

Genotypic standard deviation, 326
Genotyping checks, 266
Genotyping errors, 275
Genotyping microsatellite markers, 269
GENTRIPL program for endosperm traits, 30
GGE (Genotype + Genotype-by-environment interaction), 193, 201
GGE biplot, 201
GGT program, 270
Gibb's sampling, 267
Goodness-of-fit testing, 231
Graphical representation of relationship, 270
GREGOR limitations, 281
GREGOR program, 279
Griffing's diallel methods, 1-3
Growth index, 331
GRR program, 270
GSCAN program, 270

Haldane's mapping function, 255, 256, 258, 259
Half-sib families, 98, 109-112
Haplo program, 270
Haplotype aligning, 268
Haplotype frequency, 268, 276
Haplotype reconstruction, 270
HAPPY program, 270
Hardy program, 271, 274
Hardy Weinberg equilibrium, 271
Hayman methodology, 77, 79
Heritability, 21, 67, 322, 326
  components, 25, 39, 41, 53, 70
  and shrinkage, 182
Heterogametic progeny, 68
Heterogeneity, 124
Heterosis, 29, 45
Heterosis analysis, 35
Heterosis analysis output, 48, 62
Heterosis based on population mean, 43
Heterosis components, 26, 42, 54, 55, 70
Heterosis prediction, 31, 57, 175
Hidden Markov model, 275

Honeycomb breeding, 325
Honeycomb experiment, 324
Honeycomb selection, 322
Honeycomb trial, 323

Identical by descent, 266, 276
Imprinted genes, 270
Inbreeding coefficient, 329
Inbreeding effective number, 330
Incomplete block design, 283, 315
Indirect effects, 89
Input response, 325
Insect resistance, 77
InSegT program, 274
Interaction effects, 40
Interaction heritability, 25, 42, 53, 70
Interaction heterosis, 26, 27, 42, 54
Interaction response, 26, 42, 53, 70
Interference, 271

Jackknife, 57, 119, 175, 228, 261
    for genetic diversity parameter
        calculation, 263
    for mean/standard error, 262
Jackknifing, 44, 73, 171, 173, 174
JoinMap program, 271

KINDRED program, 271
Kosambi mapping function, 255, 256
Kriging, 137

Lattice rectangle design, 315
LDB program, 271
Least significant difference (LSD), 186
Likelihood calculation, 277
Likelihood ratio, 221
Linkage, 253, 271-273, 275-277
    coupling phase, 231, 232
    genetic correlation, 109
    repulsion phase, 231

Linkage and mapping, *xiii,* 265
Linkage disequilibrium, 266, 270, 275
Linkage mapping of markers, 273
Linkage maps, 267
Linkage program, 269, 271
Linkage test, 272
Linkbase program, 271
Locating polymorphic markers, 271
Loci ordering, 269
LOD score, 269-272, 274, 277
Logistic regression analysis, 269
Loki program, 271, 274
LUP (linear unbiased prediction), 26, 42

Magnesium concentration, 111
Mahalanobis distance, 168
Maize
    days to anthesis, 156, 279
    ear height, 156
    grain weight, 156
    plant height, 156
    silking, 156
Maize breeding, 279
Map drawing, 272
Map Manager Classic program, 272
Map Manager QT program, 272
Map Manager QTX program, 272
Map/Map+/Map+H program, 271
MAPL program, 272
Mapmaker, 272, 280
Mapmaker/Exp program, 272
Mapmaker/HOMOZ program, 272
Mapmaker/QTL program, 272
Mapping disease genes, 269
Mapping functions, 255, 256, 258, 260, 272
Mapping QTL, 272, 273
    epistatic effects, 219
    QTL-by-environment effects, 219
Mapping quantitative traits, 272, 275
Mapping software on the Internet, 265
MapPop program, 272
Mapqtl program, 273
Marker grouping, 272
Marker haplotypes, 277

Marker-assisted selection, 280
Marker-by-environment interaction effect, 220
Marker-by-marker interaction effect, 220
Marker-by-marker-by-environment effect, 220
Marker-typing incompatibility, 274
Markov chain Monte Carlo (MCMC), 267
Mass selection in maize, 329
Maternal additive effect, 23
Maternal additive-by-environment effect, 23
Maternal dominance effect, 23
Maternal dominance-by-environment effect, 23
Maternal effect, 2, 24, 68
Maternal general heritability, 25
Maternal general response, 26
Maternal interaction heritability, 25
Maternal interaction response, 26
Maternal-by-environment effect, 68
Mating design, 1, 21, 39, 51, 67
Matrix programming language of SAS, *xiii*
MCLEEPS program, 273, 274
Mean cross product, 110
MEGA2 program, 273
Megaenvironment identification, 193
Mendel program, 273
MFLINK program, 273
Microsatellite DNA fragments, 266
MIM program, 273
MINQUE, 24, 206
    equations, 41
    unbiased variance estimation, 53, 69, 116
Mixed linear model, 23, 40, 52, 68, 220
    augmented treatments, 291
    time-dependent traits, 115
Mixed model approaches, 97, 116, 171
    BLUPs, 181
    unbalanced data, 206
Mixed-model-based composite interval mapping (MCIM), 215, 221
MLD program, 273
MLM (modified location model), 153
Model-free linkage analysis, 273

Molecular marker map, 275
Molecular markers, *xiii*, 280
Monte Carlo estimation of P, 271
Monte Carlo methods, 267, 269, 273, 276
Monte Carlo multipoint linkage analysis, 271
MORGAN program, 273, 274
Morgan unit definition, 255
MSPE (minimum mean squared prediction error), 181
Multienvironment experiments, 1, 39, 51, 322
Multilocus linkage maps, 268
MultiMap program, 273
Multiple harvest data, 331
Multiple marker alleles, 270
Multiple pairwise linkage analysis, 271
Multiple trait analysis, 206, 211
Multipoint exclusion mapping, 266
Multipoint linkage analysis, 270
Multipoint LOD score calculation, 272
Multipoint mapping, 255
Multipoint maximum likelihood distance estimation, 267

Narrow-sense heritability, 97, 186
Nei's expected heterozygosity, 263
Neutral detergent fiber, 111
Nonadditivity, 124
Noncentered polynomial regression, 307
Noncentered variates, 307
Nonmaternal effect, 2
Nonparametric linkage analysis, 266-268, 273, 276
Nonstop selection, 322
NOPAR program, 273
Normal deviates, 137
Normal mixed model, 182
North Carolina design I, 111
North Carolina design II, 111

Orthogonal polynomial regression, 291, 297, 308, 315

PANGAEA program, 274
PAP program, 274
Parametric analysis, 270
Pascoe-Morton mapping function, 255, 256, 259
Path coefficient analysis, 89
PATHSAS, 89, 91, 95
PED program, 274
PedCheck program, 274
Pedfiddler program, 274
PedHunter program, 274
Pedigree data analysis, 269
Pedigree drawing, 268, 269, 271, 274, 275
Pedigree graphs, 274
Pedigree plotting, 275
Pedigree/Draw program, 274
PedJava program, 274
Pedpack program, 274
PedPlot program, 275
PEDRAW/WPEDRAW program, 275
PEDSYS program, 275
Phenotypic standard deviation, 326
Physical maps, 271
Pleiotropy, 109
Pointer program, 275
Polygenic inheritance, 269
Polymorphic loci, 262
Polynomial regression, 137, 307
Population genetics software, 266
Population size, 280, 281, 337
Prediction error variances, 183
Prediction methods, 57, 119
Preproc, 268
Principal component analysis, 137
Principal components, 145, 146
Probability density function (PDF), 153
PROC INBREED, 99
Progeny program, 275
Program listing
    diallel method 1, 3
    diallel method 2, 15
    diallel method 3, 16
    diallel method 4, 18

QTDT program, 275
QTL (quantitative trait loci), 213, 265
QTL analysis, 272
QTL Cartographer program, 275
QTL effects, 228
QTL mapping, 219
QTL testing by linear regression, 270
QTL-by-environment interaction, 213, 215, 221
QTL-epistasis model, 220
Quantitative genetic analysis, 276
Quantitative genetic models, 171
Quantitative trait loci (QTL), 213, 265
Quantitative traits, 89, 115
QUGENE program, 275

Radiation hybrid mapping, 271, 275
Random effects, 182, 291
Random factors, 28
Randomization plan, 137
Randomized complete block design, 112, 297, 307, 315, 325
Reciprocal crosses, 1
Reciprocal effect, 2
Recombination frequencies, 231, 255, 259
Recombination rate, 271
Regional test models, 207
Regional trial analysis, 171, 177, 205
Relationship estimation, 275
Relative program, 275
RelCheck program, 275
REML (restricted maximum likelihood), 53, 97, 129
REML program, 103, 107
Replications-within-environment variance, 131
Resampling techniques, 261
Response to inputs, 321, 322
Restriction fragment length polymorphism (RFLP), 156
RHMAPPER program, 275
RIL (recombinant inbred lines), 213, 219, 271, 280
Row-column design, 283, 297, 308, 315

SAGE program, 276
SASGENE program, 232, 233, 235, 236
SAS_STABLE program, 129
SCA (specific combining ability), 1
Scaling constants, 193
Seed model, 21, 22, 27, 29, 30
Seed model analysis, 172, 173, 176
Seed quality, 27
Seed traits, 171
Segregation analysis, 267, 269, 271, 273, 275
Segregation ratio computation, 272
Selection conditions, 322
Selection differential, 322, 326, 327
Selection effect, 279
Selection for yield and stability, 321
Selection indices, 109
Selection intensity, 326
Selection response, 21, 39, 51, 67, 97
Selection response equation, 326
Selective mapping, 272
Semiparametric linkage analysis, 270
Sex linkage, 119
Sexlinked additive-by-environment effect, 68
Shrinkage estimators, 181
Shrinkage factors, 182
Sibling transmission disequilibrium test, 276
Siblingship inference, 267
Sib-Pair program, 276
Simibd program, 276
Simple sequence repeat (SSR), 267
Simulating degrees of freedom, 137
Simulating marker data, 277
Simulation of genetic data, 269
Simulation tool for breeding, 279
SimWalk2 program, 276
Single nucleotide polymorphism (SNP), 266
Single trait analysis, 209
Single trait test, 31
Single-gene goodness-of-fit, 253
Single-sperm data, 276
Singular values, 193
Smoothing, 137

SNP (single nucleotide polymorphism), 266
SOLAR program, 276
Spatial variation, 145, 307, 308
SPERM program, 276
Sperm typing, 276
SPERMSEG program, 276
SPLINK program, 276
SREG (sites regression), 201
SSR (simple sequence repeat), 267
Stability estimation, 178
Stability for multiple traits, 179
Stability variance, *xiii,* 123, 124, 129
Statistical property of sum method, 110
Sum method for additive genetic correlation, 109, 111
Sum method for estimating covariance, 110

TDT-PC program, 276
TDT/S-TDT program, 276
Testing assumptions, 279, 280
Testing breeding questions, 279
Time-dependent traits, 115
Tobacco, 307
Tolerance index, 335, 336
Tolerance to stress, 325
Transmission disequilibrium test, 269, 270, 274, 276, 277
Transmit program, 277
Trend analysis, 308
    AMMI, 145
    augmented design, 291
    Fourier and polynomial regression, 307
    incomplete block and lattice-rectangle experiments, 315
    row-column-designed experiments, 297
Triploid endosperm model, 22
Triploid seed models, 173
Two-dimensional crossover-based maps, 277
Two-point linkage analysis, 267
Two-point LOD score, 269
Typenext program, 277

Unbalanced data
   endosperm model, 21
   regional trials, 205
   seed model, 21
Unbalanced data analysis, 206
Unbalanced data handling, 171
Unbiased estimation of covariance, 41
Unconditional data, 118

Variance analysis, 45, 58, 73, 110
Variance component, 53, 55, 175
Variance component estimation, 45, 57,
     171, 176, 179

Variance components, 21, 39, 43, 51, 67
Variance estimation, 70
Vitesse program, 277

Web-Preplink program, 277

Yield and stability genes, 321
Yield potential, 321

Z-score computation, 270